Methods in Molecular Biology

Series Editor
John M. Walker
School of Life and Medical Sciences
University of Hertfordshire
Hatfield, Hertfordshire, UK

For further volumes:
http://www.springer.com/series/7651

For over 35 years, biological scientists have come to rely on the research protocols and methodologies in the critically acclaimed Methods in Molecular Biology series. The series was the first to introduce the step-by-step protocols approach that has become the standard in all biomedical protocol publishing. Each protocol is provided in readily-reproducible step-by step fashion, opening with an introductory overview, a list of the materials and reagents needed to complete the experiment, and followed by a detailed procedure that is supported with a helpful notes section offering tips and tricks of the trade as well as troubleshooting advice. These hallmark features were introduced by series editor Dr. John Walker and constitute the key ingredient in each and every volume of the Methods in Molecular Biology series. Tested and trusted, comprehensive and reliable, all protocols from the series are indexed in PubMed.

Riboregulator Design and Analysis

Edited by

James Chappell

Biosciences Department, Rice University, Houston, TX, USA

Melissa K. Takahashi

Department of Biology, California State University, Northridge, Northridge, CA, USA

Humana Press

Editors
James Chappell
Biosciences Department
Rice University
Houston, TX, USA

Melissa K. Takahashi
Department of Biology
California State University, Northridge
Northridge, CA, USA

ISSN 1064-3745 ISSN 1940-6029 (electronic)
Methods in Molecular Biology
ISBN 978-1-0716-2420-3 ISBN 978-1-0716-2421-0 (eBook)
https://doi.org/10.1007/978-1-0716-2421-0

This Humana imprint is published by the registered company Springer Science+Business Media, LLC, part of Springer Nature.
The registered company address is: 1 New York Plaza, New York, NY 10004, U.S.A.

Preface

Since the last decade, RNA has become a powerful biomolecule for implementing gene control and engineering living systems for discovery and application. A versatile toolbox of naturally inspired and entirely synthetic RNA regulatory mechanisms now exists, able to enact regulation through controlling transcription, translation, and nucleic acid modification. These riboregulators include RNA-only systems such as small RNAs and riboswitches, and ribonucleoprotein systems such as those based upon CRISPR-Cas. Coupled with these advances has been an expansion of synthetic biology applications that seek to use RNA technologies to create diagnostics, construct synthetic gene circuits, create cellular recording devices, and control metabolic flux. Along with this, advances in experimental and computational structure characterization techniques are revolutionizing our fundamental understanding of how natural RNA has evolved and empowering our ability to rationally design synthetic RNA *de novo*. Taken together, RNA is an increasingly powerful biomolecule for programming cellular functions.

The present book, *Methods in Molecular Biology*, describes the latest advances in riboregulators and techniques to study, evolve, and design them. We start with a series of chapters focused on the design and application of small RNA (sRNA) regulator systems that control transcription and translation. These riboregulators are providing a versatile and increasingly diverse set of technologies for expression control and advanced applications that include diagnostics and creation of synthetic gene circuitry. Next, we describe techniques to create switchable riboregulator systems called riboswitches that allow for conditional control gene expression in response to the binding of user-specified molecular and physical stimuli. The third part of the book highlights recent technologies that leverage RNA-guided CRISPR-Cas systems to edit the epigenome, control gene expression, and create diagnostics. Finally, we end on a selection of chapters describing computational and experimental techniques to investigate the sequence-structure-function relationship of RNA systems that can both advance fundamental understanding and rational design of riboregulators.

This book is aimed to assist both scientists and engineers interested in the design and application of riboregulators. We would like to thank all authors for contributing to this book and hope that it will help drive further innovation in riboregulator research.

Houston, TX, USA — *James Chappell*
Northridge, CA, USA — *Melissa K. Takahashi*

Contents

Contributors

CHASE L. BEISEL • *Helmholtz Institute for RNA-based Infection Research (HIRI), Helmholtz-Centre for Infection Research (HZI), Würzburg, Germany; Medical Faculty, University of Würzburg, Würzburg, Germany*
JAMES CHAPPELL • *Department of BioSciences, Rice University, Houston, TX, USA; Department of Bioengineering, Rice University, Houston, TX, USA*
SOMA CHAUDHARY • *Biodesign Center for Molecular Design and Biomimetics, The Biodesign Institute, and the School of Molecular Sciences, Arizona State University, Tempe, AZ, USA*
LYDIA M. CONTRERAS • *McKetta Department of Chemical Engineering, University of Texas at Austin, Austin, TX, USA*
ANDREW D. ELLINGTON • *Department of Molecular Biosciences, University of Texas at Austin, Austin, TX, USA*
ANNA ENDER • *Institute for Biochemistry, Leipzig University, Leipzig, Germany*
SVEN FINDEIß • *Bioinformatics Group, Department of Computer Science and Interdisciplinary Center for Bioinformatics, Leipzig University, Leipzig, Germany*
KRISHAN KUMAR GOLA • *Department of Electrical Engineering, IIT Delhi, New Delhi, India*
THOMAS E. GOROCHOWSKI • *School of Biological Sciences, University of Bristol, Bristol, UK*
F. VERONICA GRECO • *School of Biological Sciences, University of Bristol, Bristol, UK*
ALEXANDER A. GREEN • *Molecular Biology, Cell Biology and Biochemistry Graduate Program, Graduate School of Arts and Sciences, Boston University, Boston, MA, USA; Biodesign Center for Molecular Design and Biomimetics, The Biodesign Institute, and the School of Molecular Sciences, Arizona State University, Tempe, AZ, USA; Department of Biomedical Engineering, Boston University, Boston, MA, USA*
CLAIRE S. GRIERSON • *School of Biological Sciences, University of Bristol, Bristol, UK*
ROSA S. GUERRA-RESENDEZ • *Systems, Synthetic, and Physical Biology PhD Program, Rice University, Houston, TX, USA*
CARMEN HERNANDEZ-PEREZ • *Department of Biology, California State University Northridge, Northridge, CA, USA*
ISAAC B. HILTON • *Systems, Synthetic, and Physical Biology PhD Program, Rice University, Houston, TX, USA; Department of Bioengineering, Rice University, Houston, TX, USA; Department of BioSciences, Rice University, Houston, TX, USA*
SEONGHO HONG • *Department of Life Sciences, Pohang University of Science and Technology, Pohang, Gyeongbuk, South Korea*
BOYANG HUA • *Department of Molecular Biology and Genetics, Johns Hopkins Medical School, Baltimore, MD, USA*
THEA IRVINE • *School of Biological Sciences, University of Bristol, Bristol, UK*
SUNGHO JANG • *Department of Bioengineering and Nano-Bioengineering, Incheon National University, Incheon, Korea; Division of Bioengineering, College of Life Sciences and Bioengineering, Incheon National University, Incheon, Korea; Research Center for Bio Materials & Process Development, Incheon National University, Incheon, Korea*
CHUNLEI JIAO • *Helmholtz Institute for RNA-based Infection Research (HIRI), Helmholtz-Centre for Infection Research (HZI), Würzburg, Germany*

Gyoo Yeol Jung • *School of Interdisciplinary Bioscience and Bioengineering, Pohang University of Science and Technology, Pohang, Gyeongbuk, Korea; Department of Chemical Engineering, Pohang University of Science and Technology, Pohang, Gyeongbuk, Korea*

Maria Claudia Villegas Kcam • *Department of BioSciences, Rice University, Houston, TX, USA*

Jongmin Kim • *Department of Life Sciences, Pohang University of Science and Technology, Pohang, Gyeongbuk, South Korea*

Minsun Kim • *School of Interdisciplinary Bioscience and Bioengineering, Pohang University of Science and Technology, Pohang, Gyeongbuk, Korea*

Janice Kramat • *Department of Biology, Technical University Darmstadt, Darmstadt, Germany*

Baiyang Liu • *Graduate Program in Systems, Synthetic, and Physical Biology, Rice University, Houston, TX, USA*

Ewelina M. Małecka • *T.C. Jenkins Department of Biophysics, Johns Hopkins University, Baltimore, MD, USA*

Griffin McCutcheon • *Biodesign Center for Molecular Design and Biomimetics, The Biodesign Institute, and the School of Molecular Sciences, Arizona State University, Tempe, AZ, USA; Department of Biomedical Engineering, Boston University, Boston, MA, USA*

Tae Seok Moon • *Department of Energy, Environmental and Chemical Engineering, Washington University in St. Louis, St. Louis, MO, USA*

Mario Mörl • *Institute for Biochemistry, Leipzig University, Leipzig, Germany*

Dongwon Park • *Department of Life Sciences, Pohang University of Science and Technology, Pohang, Gyeongbuk, South Korea*

Abhilash Patel • *Department of Bioengineering, Imperial College, London, UK*

Shaunak Sen • *Department of Electrical Engineering, IIT Delhi, New Delhi, India*

Trevor R. Simmons • *McKetta Department of Chemical Engineering, University of Texas at Austin, Austin, TX, USA*

Peter F. Stadler • *Bioinformatics Group, Department of Computer Science and Interdisciplinary Center for Bioinformatics, Leipzig University, Leipzig, Germany; Max Planck Institute for Mathematics in the Science, Leipzig, Germany; Institute for Theoretical Chemistry, University of Vienna, Vienna, Austria; Santa Fe Institute, Santa Fe, NM, USA*

Eric J. Strobel • *Department of Biological Sciences, The University at Buffalo, Buffalo, NY, USA*

Beatrix Suess • *Department of Biology, Technical University Darmstadt, Darmstadt, Germany; Centre of Synthetic Biology, Technical University Darmstadt, Darmstadt, Germany*

Courtney E. Szyjka • *Department of Biological Sciences, The University at Buffalo, Buffalo, NY, USA*

Melissa K. Takahashi • *Department of Biology, California State University Northridge, Northridge, CA, USA*

Sarah A. Woodson • *T.C. Jenkins Department of Biophysics, Johns Hopkins University, Baltimore, MD, USA*

Kaiyue Wu • *Molecular Biology, Cell Biology and Biochemistry Graduate Program, Graduate School of Arts and Sciences, Boston University, Boston, MA, USA; Department of Biomedical Engineering, Boston University, Boston, MA, USA*

Zhaoqing Yan • *Molecular Biology, Cell Biology and Biochemistry Graduate Program, Graduate School of Arts and Sciences, Boston University, Boston, MA, USA; Department of Biomedical Engineering, Boston University, Boston, MA, USA*
Raphael Angelo I. Zambrano • *Department of Biology, California State University Northridge, Northridge, CA, USA*

Chapter 1

RNP-Based Control Systems for Genetic Circuits in Synthetic Biology Beyond CRISPR

Trevor R. Simmons, Andrew D. Ellington, and Lydia M. Contreras

Abstract

Ribonucleoproteins (RNPs) are RNA-protein complexes utilized natively in both prokaryotes and eukaryotes to regulate essential processes within the cell. Over the past few years, many of these native systems have been adapted to provide control over custom genetic targets. Engineered RNP-based control systems allow for fine-tune regulation of desired targets, by providing customizable nucleotide-nucleotide interactions. However, as there have been several engineered RNP systems developed recently, identifying an optimal system for various bioprocesses is challenging. Here, we review the most successful engineered RNP systems and their applications to survey the current state of the field. Additionally, we provide selection criteria to provide users a streamlined method for identifying an RNP control system most useful to their own work. Lastly, we discuss future applications of RNP control systems and how they can be utilized to address the current grand challenges of the synthetic biology community.

Key words Ribonucleoproteins, Genetic circuits, Posttranscriptional regulation, CRISPR-Cas, Engineered RNP control

1 Introduction

In recent years, synthetic biology has seen the developments of standardized genetic tools that allow regulation across a diverse set of systems, from *Escherichia coli* or *Saccharomyces cerevisiae* to mouse or human cell lines. These tools typically rely on transcriptional, translational, and other common signaling pathways and have been utilized to control various bioprocesses including engineering microbial metabolism to produce value-added chemicals [1–3], editing genomes directly [4], enabling cells to track and respond to environmental signals or stimuli [5], and allowing organisms to dynamically regulate their own metabolisms in response to stress [6].

In particular, ribonucleoprotein (RNP)-based control systems are proving increasingly valuable for the design and construction of orthogonal regulatory systems in organisms because they can be

James Chappell and Melissa K. Takahashi (eds.), *Riboregulator Design and Analysis*,
Methods in Molecular Biology, vol. 2518, https://doi.org/10.1007/978-1-0716-2421-0_1,

tuned via nucleotide-level interactions to direct metabolic pathways and cellular responses. As their name implies, RNP-based systems consist of an RNA-binding protein (RBP) complexed with an RNA transcript. Such complexes are frequently used by nature to control key aspects of gene regulation [7], providing a useful starting point for engineering the regulation of cellular responses. This methods chapter reviews key RNP-based control systems in both prokaryotic and eukaryotic systems that have been adapted as genetic circuitry for orthogonal control. We detail useful selection criteria for choosing RNP-based control systems for bioprocessing applications and point out future directions for expanding their use and versatility.

2 Current Systems

There are a wide variety of native RNPs [8, 9], but only a few that have been adapted to genetic circuitry and orthogonal regulation. Because of differences between the basic molecular biology of prokaryotes and eukaryotes, we consider RNP-based circuitry that involves these two domains separately, although the field is beginning to move toward more general and universal methods for engineering RNP control.

2.1 Engineered RNPs in Prokaryotes

There are a variety of native RNPs that have key regulatory functions in prokaryotes, such as the SRP [10, 11], RNase P [12], RNA-binding proteins (RBPs) complexed with RNAs and PNPase [13]; some of these have been adapted for use within engineered circuits. In addition, more general control by widely used CRISPR RNPs has been developed (*see* Fig. 1).

2.1.1 RNA-Binding Protein (RBP)-RNA Regulated Systems

Some of the earliest engineered RNP-controlled systems come from reprogramming bacterial RNase P ribozymes, particularly the RNase P from *E. coli*. Natively, RNase P cleaves tRNA precursors to generate the 5′ termini of mature tRNA molecules [14]. The RNP utilizes a guide sequence, which hybridizes to its target RNA, and directs cleavage of the target. Li et al. [15] first demonstrated that by engineering specific "external guide sequences" (EGS), one could specifically target RNase P cleavage of desired mRNA transcripts. The authors utilized RNase P from *E. coli* and engineered EGS to cleave mRNA sequences of Β-galactosidase from *E. coli* and nuclease A in vitro. Additionally, they demonstrated almost 90% reduction in Β-galactosidase activity using the engineered EGS-RNase P complex. After this breakthrough, work began to expand the capability of RNase P targeting of mRNA targets. Guerrier-Takada et al. [16] engineered external guide sequences with RNase P from *E. coli* to target chloramphenicol resistance genes in cultures of *E. coli*. The authors eliminated nearly all growth of chloramphenicol-resistant cultures that

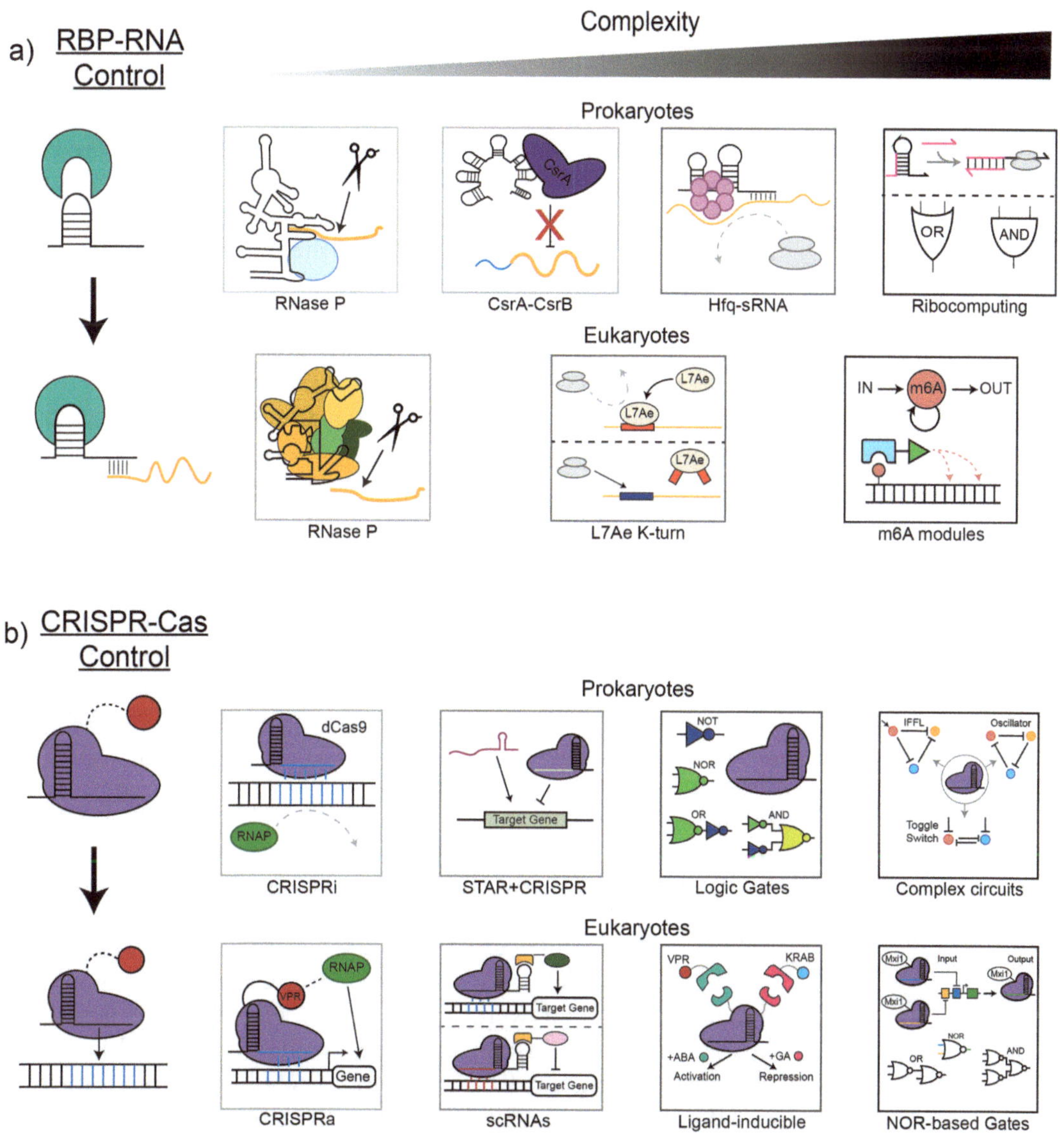

Fig. 1 Ribonucleoprotein-derived genetic control schemes developed in prokaryotic and eukaryotic systems. (**a**) Examples of RNP control schemes that utilize RNA-binding proteins (RBPs) and a guide RNA to achieve desired regulation arranged with increasing genetic circuit complexity. (**b**) Examples of RNP control schemes that utilize a CRISPR-Cas scaffold as a basis for developing engineered RNP-based control arranged with increasing genetic circuit complexity. Abbreviations are as follows: *m6A* N6-methyladenosine, *RNAP* RNA polymerase, *STARs* small transcription activating RNAs, *IFFL* incoherent feedforward loop, *VPR* VP64-p65-Rta domain, *KRAB* Kruppel-associated box, *ABA* abscisic acid, and *GA* gibberellin

contained plasmid-based chloramphenicol acetyltransferase (CAT) by expressing two CAT external guide sequences to drive cleavage via RNase P. The two EGS were put under control of an IPTG-inducible promoter, thus creating one of the earliest "NOT" gates, which targeted a drug-resistant phenotype *in E. coli*. Specifically, in this simple single input logic gate, if IPTG is present as an input, chloramphenicol resistance of the *E. coli* cultures is effectively eliminated (i.e., in the presence of the inducer, the output was "negated"). Additionally, this genetic control scheme was adapted for human health applications. Cobaleda et al. [17] engineered two EGS to target oncogene products expressed in leukemia cells that inhibit apoptosis. By expressing the EGS and RNase P from *E. coli* in a mammalian cell model containing the two oncogene products, they demonstrated at least a 96% reduction in cell viability for cells expressing the two oncogenes, while minimizing off-target effects to only a 5% reduction in cell viability. This study was the first to engineer an EGS-RNase P complex that directly targeted and reduced oncogene expression in a mammalian cancer cell line. Subsequently, Trang et al. [18] used the RNase P RNP complex from *E. coli* to develop an engineered external guide sequence (EGS) to target the ICP4 mRNA transcript, a key protein involved in human herpes simplex virus 1 (HSV-1). Using the engineered EGS-RNase P complex, they achieved an 87% reduction in ICP4 mRNA and an 80% reduction in ICP4 protein in $^{\varphi}$CRE fibroblast cells. Additionally, cells expressing the engineered EGS-RNase P complex demonstrated a 1000-fold reduction in viral growth 18 h after exposure, while an inactive version of the complex demonstrated no reduction in viral growth. Recently, Yang et al. [19] developed a mutant RNase P ribozyme from *E. coli* by screening a library of rationally designed mutants. The identified mutant demonstrated 50-fold improved cleavage of viral mRNA from human cytomegalovirus (HCMV) relative to the wild-type version. In vivo, the mutant RNase P ribozyme achieved a 98% reduction in HCMV mRNA, which was 13% greater than the wild-type RNase P. This engineered RNase P was then introduced to the human cell line U373MG that were infected with HCMV. Five days after introducing the RNase P ribozymes to the infected cell lines, the cells with the mutant RNase P demonstrated 3500-fold reduction in HCMV mRNA, while cells containing the WT RNase P only showed a 100-fold reduction in HCMV mRNA.

Since the initial development of RNase P-regulated genetic control, development of novel prokaryotic RNP control systems has expanded to global RNA-binding proteins (RBPs) to posttranscriptionally regulate translation of mRNA targets. Two prevalent examples include the complexing of the Hfq or CsrA bacterial proteins and a small noncoding RNA (sRNA) [20]; we refer to these as RBP-sRNA regulated systems. The most common example of this RNP control scheme is Hfq-sRNA regulation. In this system, the Hfq protein binds to an sRNA and facilitates binding to a specific mRNA target, allowing for the sRNAs to hybridize and directly regulate the posttranscriptional fate of the mRNA itself [21]. This mechanism has been adapted to create many

Hfq-sRNA control systems in bacteria. One key example is from Na et al. in which the authors employed the Hfq-sRNA-mRNA interaction to downregulate expression of desired genes within *E. coli* [22]. The authors achieved this by engineering the MicC sRNA (an sRNA involved in regulation of the EnvZ-OmpR two-component signaling pathway [23]) with a custom seeding region to trans-actively downregulate a desired mRNA target sequence. They demonstrated the utility of the Hfq-MicC control system in a bioprocessing application by improving de novo production of tyrosine and cadaverine in *E. coli*. The authors improved overall product yield by downregulating several genes involved in competing metabolic pathways for tyrosine and cadaverine production. Noh et al. expanded upon this system by developing a MicC sRNA with tunable transcription levels [24]. This provided a system with a greater dynamic range of MicC sRNA transcriptions, which allowed for tunable repression of target genes. They utilized these modified constructs to improve overall titers of proline and putrescine in *E. coli* by knocking down genes involved in competing metabolic pathways of the two compounds. Importantly, systems like these, which have been previously reviewed [1], allow the user to fine-tune control over expression of desired genes by providing a customizable nucleotide seed sequence in the modified MicC sRNA. Moreover, users do not have to spend time engineering the genome to knockout or knockdown these genetic targets. An additional example of utilizing Hfq-sRNA posttranscriptional regulation comes from Lahiry et al. [25], in which the authors rationally redesigned the DsrA sRNA of *E. coli* to target mRNA transcripts involved in the n-butanol synthesis pathway of *Clostridium acetobutylicum*. The authors utilized the regulatory mechanism of the Hfq-sRNA-mRNA RNP complex to downregulate two key mRNA sequences shown to reduce n-butanol production. The authors iteratively redesigned the native stem loops of the DsrA sRNA to ultimately achieve over fourfold reduction in expression of both target genes. This work provides another framework for exploiting nucleotide-level interactions to drive RNP-based control within the Hfq-sRNA system.

In addition to Hfq, the bacterial global regulator protein CsrA has been utilized in an engineered RNP-based control system for posttranscriptional regulation of mRNA targets. CsrA is best known to prevent translation of mRNA targets via binding in the 5′ untranslated region (UTR) and occlude ribosome binding, although native, alternative mechanisms of activation and binding to other mRNA regions have also been reported for CsrA [26]. Omics studies of the Csr system such as Sowa et al. [27] and Leistra et al. [28] have identified multiple candidate genes that could be posttranscriptionally upregulated by CsrA. Moreover, recent work from Renda et al. [29] confirmed first the first mRNA target posttranscriptionally activated due to binding of CsrA in the 5′ UTR. Thus, CsrA sits as a potential control node for simultaneous activation and repression of genetic targets. The RBP is primarily regulated by two small RNAs, CsrB and CsrC, which can each sequester multiple copies of CsrA and block formation of the RNP complex between CsrA and its target mRNAs [26]. In Leistra et al., the authors rationally engineered mutants of the CsrB sRNA of *E. coli* to have a gradient of stronger or weaker affinity for CsrA [30]. The authors created 20 total mutants, 10 with weaker affinity for CsrA than WT CsrB and 10 with stronger affinity. The strongest CsrB mutant demonstrated almost 3.5-fold stronger affinity with CsrA. Lastly, they demonstrated that the CsrB mutants could be utilized to control CsrA-regulated metabolic processes, such as glutamate production. Overexpressing the strongest CsrB mutant led to a 1.75-fold increase in glutamate production relative to overexpression of WT CsrB. This work is one of the first systems to engineer the Csr system and offer an approach to control the formation of CsrA-based RNP complexes by engineering competing RNA-CsrA interactions. This system offers a potential future opportunity to engineer tunable RNP-based control systems utilizing CsrA.

While not an RNA-binding protein, the ribosome can also be considered a RNP at the system level and, as such, affords broad opportunities for regulatory control. Genetic circuits have been developed to control translation by blocking or promoting formation of the ribosomal RNP complex. One key example comes from Green et al. [31], in which the authors developed "ribocomputing" devices utilizing sRNAs to unwind toehold switch hairpins that contained the ribosome binding site for a downstream gene of interest. In this scheme, when the toehold switch was unwound, the ribosome could bind and begin translation. The authors developed several devices of increasing complexity. They began with an individual "OR," as well as an "AND" logic gate. The "OR" gate consisted of two consecutive hairpins such that when provided with an sRNA complementary to *either* hairpin, the RBS was freed and allowed the ribosome to bind. The "AND" logic gate utilized a single hairpin that required two sRNAs to hybridize and rearrange the hairpin; only upon binding of both sRNAs, the RBS was liberated for ribosome binding. Next, they constructed multi-input gates such as a 4-sRNA "AND" gate, as well as a 6-sRNA "OR" gate. Ultimately, they constructed a 12-sRNA input circuit that utilized five consecutive "AND" gates. Each of these constructed gates achieved a range between 25-fold and 41-fold activations of exogenous GFP. The type of flexibility of this ribocomputing system is an enormous asset given the potential distinct advantages of "OR" and "AND" logic gates for gene regulation in different context. For example, work from Amalfitano et al. [32] utilized these ribocomputing gates to control a glucose reporter that can be measured by store bought glucose meters. These circuits were utilized to specifically detect the presence of bacterial RNA from *Salmonella typhi*, the bacterium responsible for typhoid fever, at clinically relevant levels. Moreover, these circuits were then modified to specifically detect viral RNA from SARS-CoV-2 within an hour, while not being activated by any other control viral RNAs present. These ribocomputing circuits may provide the opportunity for portable diagnostics for use in clinical and affordable human health applications.

2.1.2 CRISPR-Cas-Regulated Systems

The second and potentially most well-known category of RNP-based control in bacteria are engineered CRISPR-Cas systems. Natively, CRISPR-Cas systems utilize a Cas protein complexed with a crRNA (CRISPR RNA, complementary to the DNA target) and tracrRNA (transactivating crRNA, a scaffold for Cas9 protein binding) for targeted genome editing [4]. While native CRISPR-Cas systems were known for at least two decades [33, 34], their engineering utility was first presented by Jinek et al., where they demonstrated targeted genome editing in *E. coli* by utilizing a Cas9 nuclease protein from *S. pyogenes* and a single guide RNA (gRNA) chimera adapted from a fusion of the tracrRNA and crRNA [35]. Since then, many iterations of CRISPR-Cas systems have been developed for targeted gene editing as well as precise control of gene expression. One of the biggest breakthroughs to CRISPR-Cas systems was the introduction of a catalytically inactive Cas9 protein (dCas9) from Qi et al. [36]; here, the authors introduced two point mutations to remove Cas9 endonuclease activity, effectively creating a customizable DNA-binding protein (DBP). Using this construct, the authors developed the CRISPR interference (CRISPRi) system. The dCas9 protein and gRNA formed an RNP that operated as a DNA-binding complex that could repress various targeted genes based on the seeding sequence of the gRNA. They were able to demonstrate almost 1000-fold repression of specific target genes in *E. coli* and demonstrated the ability to target multiple genes simultaneously by using multiple gRNAs [36]. The development of dCas9 has allowed for novel CRISPR-Cas9 regulatory tools. As an example, Nielsen et al. expanded the versatility of CRISPR-Cas regulation in genetic control schemes [37] by utilizing dCas9 to construct multiple genetic circuits within *E. coli*. The authors first constructed an arabinose-responsive "NOT" gate, such that in the presence of arabinose, the gRNA sequence was expressed and complexed with dCas9 to transcriptionally repress a desired target gene by binding within the promoter sequence of the target gene. The authors then built more complex gates such as two consecutive "NOT" gates, a "NOR" gate (where the lack of two imposed conditions leads to a desired output), an "OR" gate, as well as an "AND" gate. The two consecutive "NOT" gates utilized anhydrotetracycline (aTc), and 2,4-diacetylphloroglucinol (DAPG)-inducible promoters to achieve a double inverter circuit. In the presence of aTc, the gRNA sequence was transcribed to direct dCas9 repression of the downstream target. Upon addition of DAPG, a gRNA sequence was expressed that blocked the aTc promoter sequence via dCas9, alleviating repression of the downstream target. The "NOR" gate utilized a DAPG and arabinose-responsive promoter to control expression of a gRNA to target a downstream gene with dCas9. In the presence of either or both compounds, the downstream target was repressed. The "OR" gate combined the "NOR" and "NOT" gates consecutively to activate the downstream target in the presence of DAPG or arabinose. Lastly, the "AND" gate combined the arabinose-responsive and DAPG-responsive "NOT" gates into the "NOR" gate to generate an "AND" gate, such that only in the presence of both arabinose and DAPG would the downstream target be activated. These engineered circuits measured at least a 100-fold change in target gene expression for each gate, using plasmid-based red fluorescent protein (RFP) as a reporter. Finally, the authors applied the "OR" gate to control the *malT* gene and demonstrated control of the lambda phage susceptibility phenotype under the control of Ara and/or DAPG inducers. Along with interference, methods to activate genes using CRISPR-Cas components (CRISPRa) in bacteria were developed. Bikard et al. first demonstrated activation of endogenous target genes by fusing dCas9 to the omega subunit of RNAP in *E. coli* [38]. While novel, activation using CRISPR proved to be more difficult, as the authors could only achieve a maximum of 23-fold activation of endogenously expressed GFP, which was comparable to other transcriptional activation systems at the current time [39, 40]. Importantly, Dong et al. [41] expanded upon the CRISPR activation regulatory scheme originally developed by Bikard et al. To achieve enhanced CRISPR-based activation in bacterial systems, the authors utilized modified gRNAs fused to an RNA hairpin that binds an RNA-binding protein (RBP), which itself is fused to a transcriptional activator protein domain. A transcriptional activator protein can be defined as a protein or protein domain, which natively recruits RNA polymerase machinery [42]. After screening several bacterial transcriptional activators, they found the SoxS activator fusion could achieve almost a three-fold activation in expression of targeted endogenous genes relative to the original CRISPR. The complex was modified to improve CRISPR activation activity, while minimizing native SoxS DNA binding affinity. Lastly, the authors demonstrated both aTc and arabinose-inducible genetic regulation of multiple genetic targets using the CRISPRa constructs and the CRISPRi system from Qi et al. in tandem. This work was one of the first demonstrations of simultaneous activation and repression of endogenous genes in bacterial systems. These developments provided novel opportunities to engineer complex genetic circuits from CRISPR-based regulation.

Recently, Santos-Moreno et al. [43] greatly expanded upon the complexity of CRISPRi-regulated genetic circuits. The authors developed the classic toggle switch, feedforward loop, and oscillator in *E. coli.* The toggle switch expressed two gRNA sequences that responded to either arabinose or AHL. In the absence of either inducer, two gRNAs were expressed to repress each other, which was measured by low fluorescence of super-folder green fluorescent protein (sfGFP). This was achieved by constitutively expressing two gRNA sequences. The first gRNA sequence (sgRNA-1) repressed the second gRNA sequence (sgRNA-4) and the *sfGFP* gene, while the second gRNA (sgRNA-4) repressed sgRNA-1. Thus, the two cancelled each other out and kept sfGFP expression low. Concurrently, a second copy of sgRNA-1 was controlled by an arabinose-inducible promoter, and a second copy of sgRNA-4 was under an AHL-inducible promoter. In the presence of AHL, the switch was toggled to high activity by induction of the second copy of sgRNA-4; this was indicated by high sfGFP fluorescence. Conversely, in the presence of arabinose, the second copy of sgRNA-1 was expressed to maintain repression of sfGFP, measured by low fluorescence. The authors developed an arabinose-inducible feedforward loop, in which a single gRNA repressed two targets: an sfGFP sequence and a second sgRNA. The second sgRNA also repressed the sfGFP sequence. Thus, across an arabinose induction gradient, there is a small concentration that allows for a spike in sfGFP fluorescence, while all other arabinose concentrations lead to basal levels of sfGFP fluorescence. Lastly, the authors constructed the "Repressilator" circuit, termed the "CRISPRlator," in which three sgRNAs consecutively repressed each other. This work was the first demonstration of using CRISPRi to construct the toggle switch, incoherent feedforward loop, and oscillator, three classic genetic circuits originally published by Gardner et al. [44] (toggle switch), Magnan et al. [45] (incoherent feedforward loop), and Elowitz et al. [46] (oscillator), respectively.

Along with developing CRISPR-based genetic control schemes in model organisms, such as *E. coli*, there has been development of regulatory systems in more industrially relevant prokaryotes. One of the best examples comes from Gordon et al. [47]. The authors developed a CRISPRi toolbox for the cyanobacteria *Synechococcus sp.* Strain PCC 7002 to repress endogenous gene targets by adapting the CRISPRi system previously developed by Qi et al. [36]. They placed the sgRNA and dCas9 constructs under an anhydrotetracycline (aTc)-inducible promoter, effectively creating an aTc-responsive "NOT" gate. The authors first demonstrated sevenfold repression in endogenously expressed EYFP using the "NOT" gate. Next, they applied the regulatory system to redirect carbon flux for improved lactate production in cyanobacteria. By repressing glutamine synthesis I (*glnA*), they were able to achieve over 8 mM in lactate synthesis, which was the highest titer of lactate photosynthetically produced. Additionally, the rate of lactate production was twofold greater than that of the nonengineered, wild-type strain. This work demonstrates the versatility of CRISPR-based regulatory circuits and the ability to utilize RNP control in industrially relevant microbes. Concurrently, Cleto et al. [48] developed a CRISPRi system for the industrial microbe *Corynebacterium glutamicum*, which is heavily used for large-scale batch production of many amino acids. The authors developed an IPTG-responsive "NOT" gate to control repression of endogenous target genes. Next, they utilized the CRISPRi system to knock-down genes in glycolysis to redirect carbon flux to improve production of L-lysine and L-glutamine. By using the CRISPRi system to downregulate the *pgi* or *pyk* genes, the authors achieve a 1.31-fold and 3.04-fold increase of L-lysine and L-glutamine, respectively, relative to a wild-type strain. The utilization of the CRISPRi-based gate demonstrated the versatility and ease for gene regulation without genomic editing in a key industrial microbe, which could help speed up time required for industrial strain engineering.

Additionally, CRISPR-based control schemes have been utilized in tandem with additional methods of genetic control to produce novel circuities. One paramount system that exemplifies this comes from works by Chappell et al. 2015 [49] and Chappell et al. 2017 [50], in which they developed a scheme to control transcription termination of RNA polymerase (RNAP) on its target DNA sequences. The authors developed and improved upon the small transcription activating RNAs (STARs) control system, respectively. The authors developed sRNA sequences to disrupt a transcription terminator that was placed upstream of a gene of interest. Addition of the STAR allowed for 94-fold activation of the downstream gene. The RNAP complex was not disrupted by the termination sequence, allowing for transcription elongation of the downstream target genes. These STARs were utilized to create RNA-based logic gates for transcription and regulate multiple genes in metabolic pathways, as well as used in tandem with CRISPRi to create a feedforward control circuit to drive accelerated CRISPRi response.

2.2 Engineered RNPs in Eukaryotes

Eukaryotic systems contain both homologous RNPs found in prokaryotes, such as RNase P, as well as additional native eukaryote RNPs including the spliceosome [51], telomerases [52], and snRNPs [53]. Some of these systems have been adapted for targeted genetic control in eukaryotes. Additionally, as was the case with prokaryotes, engineered RNP schemes based on CRISPR-Cas control are now abundant.

2.2.1 Engineered RNA-Binding Protein-RNA-Regulated Systems

Just as in prokaryotic systems, some of the first engineered RNP-based control schemes in eukaryotes involved the RNase P complex. Primarily, RNase P from humans utilized in complex with engineered external guide sequences (EGS) to cleave specific mRNA targets. Moreover, with the foundation of utilizing RNase P from *E. coli* to target viral mRNA transcripts, most systems adapted the human RNase P complex to achieve similar results. Jiang et al. [54] were one of the first to utilize engineered external guide sequences (EGS) to direct human RNase P to target and cleave the protease mRNA sequence from human cytomegalovirus (HCMV). The engineered target sequence was 35-fold more active directing RNase P cleavage of the target mRNA sequence in vitro than when using a wild-type-derived EGS. Additionally, the engineered EGS-RNase P complex achieved a 95% reduction in expression of the protease and a 4000-fold reduction in viral growth in U373MG cells, compared to the wild-type-derived EGS, which only achieved an 85% reduction of HCMV protease and 150-fold reduction in viral growth. This work opened the door for utilizing RNase P RNP complexes as a means to treat human health issues, such as HCMV infections.

Concurrently, implementing native RNA-binding proteins (RBPs) for custom genetic regulation began emerging as a popular field of interest for developing RNP-based control schemes. Most often, specific RBPs are fused to another protein editor that is typically utilized to modify desired RNA transcripts. The review by Shotwell et al. [55] provides an in-depth review of the selection and design of RBPs in these regulatory systems. One of the first examples of this approach comes from Saito et al. [56]. The authors developed a modular synthetic translational repressor and activator to regulate translation of exogenous mRNA transcripts, utilizing the archaea-derived L7Ae-box C/D kink-turn motif RNP to control translation. This RNP consists of the archaeal RNA-binding protein, L7Aa, which binds a short guide RNA sequence that contains two conserved motifs: box C (RUGAUGA) and box D (CUGA). The constructed repressor design consists of the C/D guide RNA motif sequence fused directly upstream of the start codon of a desired transcript. When expressed, the ribosomal L7Ae protein binds the guide RNA C/D box motif in mRNA fusion transcript and blocks translation of the downstream gene. The authors demonstrated a 90% reduction in translation of a box C/D-enhanced green fluorescent protein (EGFP) target fusion, measured by a 90% reduction in EGFP fluorescence. To design the activator, the authors replaced the original C/D box motif with an anti-C/D box motif, the complementary sequence, directly upstream of the EGFP target gene. Concurrently, the original C/D box sequence was constitutively expressed, such that the two sequences could hybridize and the C/D turn motif would block translation through RNA interference. However, when the L7Ae protein is expressed, the protein binds the C/D RNA sequence and alleviates the RNA interference and translational repression of the transcript. Using this scheme, they achieved over 70% activation of the anti-C/D motif EGFP transcript, measured by EGFP fluorescence, relative to fluorescence of only expressing the unmodified EGFP transcript. Additionally, they demonstrated this system could simultaneously repress and activate two different targets, showing nearly 100% repression of a model DsRed transcript and almost 100% activation of another model EGFP transcript. Lastly, L7Ae was put under an anhydrotetracycline (aTc)-inducible promoter to create a tetracycline-responsive "NOT" gate using the repressor construct in HeLa cells, showing titratable repression of an EGFP transcript. This study was one of the first to utilize a specific RNA-binding protein and an RNA sequence in eukaryotic cells to regulate expression of a desired mRNA transcript. Utilizing RBPs in the aforementioned manner to directly engineer epitranscriptome control in eukaryotes has been of particular interest over the last few years. One of the most exciting examples comes from Liu et al. (2019) [57], in which the authors focus on an engineered N^6-methyladenosine (m6A) writer that allows for directed m6A edits on desired mRNA transcripts [45]. The authors developed the novel m6A writer by fusing the human METTL3/METTL14 heterodimer, a native m6A writer, to the dCas9 enzyme and were able to selectively target mRNA sequences with gRNAs. Additionally, the authors developed a novel m6A eraser by fusing either human ALKBH5 or FTO to dCas9 to exploit the native eraser capabilities of these two proteins to selectively remove m6A modifications. Using this system, they demonstrated selective addition and removal of m6A modifications within the human *Malat1* mRNA sequences, which drove structural remodeling of the sequence and downstream interactions with the human heterogeneous nuclear ribonucleoprotein C (HNRNPC). Removal of an m6A modification within a hairpin of the *Malat1* transcript restructures the hairpin, such that HNRNPC cannot bind. Removal of the m6A modification by the engineered m6A eraser reduced binding by 50%. These circuits could be used in the future to potentially control downstream regulatory effects of transcripts containing m6A modifications in humans. In addition to utilizing human m6A writer/eraser dCas9 fusions, others have imported components from prokaryotes to achieve similar novel RBP fusion regulatory systems, such as in Park et al. (2019) [58]. The authors developed a synthetic "read-write module" to detect or add m6A modifications to mRNA transcripts utilizing multiple fusion proteins. The module first contained an "initiator" complex, which consisted of DNA adenine methyltransferase (Dam) *from E. coli* fused to an engineered zinc finger protein domain that binds specific DNA sequence in the AAV1 locus in the human genome. The zinc finger coordinated the Dam protein adjacent to the AAV1 locus, such that Dam protein could add an m6A modification to a GATC motif in the locus. Next, they built the "reader" portion by fusing the DNA binding domain of Dpn1 from *S. pneumoniae* to DNA effector domains (VP64 for activation, KRAB for repression). Dpn1 would bind to the specific m6A modifications, while the DNA effector domain would "read" out transcription of AAV1 locus, which contained a copy of EGFP integrated into the genome. To create the "write" portion, the authors fused Dam to DpnI, such that the write module could "write" additional m6A modifications to the AAV1 locus. The full "read-write" module demonstrated inducible transcriptional activation of EGFP in the AAV1 locus, measured by EGFP fluorescence. Upon induction, the "initiator" added a single m6A modification, the "writer" bound to the single modification, and added additional m6A modifications to downstream GATC motifs. The multiple m6A modifications enhanced recruitment of the "reader" to activation expression of EGFP from the AAV1 locus. The full module demonstrated a 30-fold activation or repression of EGFP depending if the VP64 or KRAB domain was utilized in the "reader." The authors lastly utilized this three-component system to confer "epigenetic memory" to the genome via m6A modifications that were conserved across daughter cells, meaning new cells contained the parent cell m6A modifications, which activated EGFP expression from the AAV1 locus. The cells containing the three-component system retained at least 50% of the population continued to upregulate EGFP transcription for ~14.3 days, relative to cells utilizing inducible plasmid-based transcription factor activator proteins, which was only 2.8 days.

2.2.2 CRISPR-Cas-Regulated Systems

The most well-studied category of engineered RNP genetic circuits in eukaryotes involves CRISPR-Cas machinery. After Jinek et al. [35] demonstrated the use of CRISPR-Cas9 for gene editing in bacteria, work from Cong et al. [59] and Mali et al. [60] quickly followed to establish functional CRISPR-Cas9 in eukaryotic cells for precise gene editing. In Cong et al., the authors demonstrated Cas9 from *S. pyogenes* can be programmed to target multiple genomic locations in human 293FT cells, using variable gRNA sequences. Additionally, they demonstrated this system could utilize multiple gRNA sequences simultaneously for multiplexed genome modifications. Work from Mali et al. [60] demonstrated that a similar CRISPR-Cas9 system could achieve homologous recombination in 293T, K562, and induced pluripotent stem cells using Cas9 from *S. pyogenes* and engineered gRNAs. Additionally, they constructed a library of 190,000 predicted gRNA sequences which covers 40.5% of the human genome.

After demonstrating functional CRISPR-based systems in eukaryotes, the first major development in CRISPR-regulated genetic circuits in eukaryotes came from work by Qi et al. [36], which was detailed in the previous section. Along with developing a CRISPRi system in bacteria, they demonstrated the CRISPRi system could also work in HEK293 cells with at least twofold repression of an endogenous EGFP-coding region. With Qi et al. demonstrating that dCas9 could regulate desired exogenous and endogenous transcripts in eukaryotes, the first CRISPR-driven genetic control schemes appeared. Kiani et al. [61] developed some of the first CRISPR-based repressors in HEK293T cells. They achieved over 100-fold repression of exogenous EYFP by using the engineered gRNA-dCas9 RNP to target the promoter of the EYFP gene. They next layered two repressor constructs consecutively, creating one of the first double inverter circuits in eukaryotes. The inner layer expressed "gRNA-a" to complex with dCas9 and repress EYFP transcription, while the outer layer expressed "gRNA-b" to repress transcription of the inner gRNA. This double inverter achieved a 27-fold change in activity of EYFP. Lastly, the authors created a doxycycline-dependent "NOT" gate by regulating gRNA transcription via a doxycycline-responsive promoter. The doxycycline-dependent "NOT" gate demonstrated 2–30-fold repression of plasmid-based EYFP-based doxycycline induction concentration in HEK293T cells.

To improve repression strength by dCas9 in eukaryotes, work was done to fuse dCas9 to various effector domains. Gilbert et al. [62] demonstrated that fusing the chromatin modifier Kruppel-associated box (KRAB) domain could achieve at least 15-fold repression of endogenous EGFP in HEK293 cells; the rationale for the dCas9-KRAB fusion confers localization of the KRAB domain via dCas9 to allow for chromatin remodeling and silence transcription through histone deacetylation, meaning the target sequence would become inaccessible to transcriptional machinery at the chromatin packs more tightly around the histones. Using the dCas9-KRAB fusion protein, the authors developed a "NOT" gate, which achieved 15-fold repression of plasmid-based EGFP, measured by fold reduction of EGFP fluorescence. This demonstrated to the field that dCas9-DNA effector domain fusions could improve the overall efficacy of CRISPR-Cas-based repression and also be applied to create user-controlled genetic circuits. Gander et al. [63] greatly expanded dCas9-based genetic circuits by synthesizing a library of "NOR" gates in *S. cerevisiae.* The authors enhanced target repression fusing dCas9 to the Mxi1 domain from human, a protein that was shown to effectively recruit histone deacetylase enzymes for chromatin remodeling. The rationale was similar to that of Gilbert et al. [62], where they repress genes of interest by driving chromatin remodeling to make target sequences sterically inaccessible. The dCas9-Mxi1 fusion achieved a maximum of 98% reduction in plasmid-based GFP expression, while standard dCas9 had a maximum of only 42% reduction in GFP expression. The "NOR" gate used two gRNA sequences simultaneously to repress one target; 20 gRNAs and target sequences were designed, allowing for a possible 400 unique gate combinations. To demonstrate the utility of this system, the authors constructed increasingly complex gates using different combinations of "NOR" gates, including "OR," "AND," "NAND," "XNOR," and "XOR" gates. Lastly, by using two of the same targeting gRNAs for the "NOR" gate, the authors converted the "NOR" gate to a simple "NOT" gate. They demonstrated seven consecutive "NOT" gates could still achieve a twofold repression. Achieving a twofold repression through a seven-tier regulatory cascade is remarkable, as this was the first instance of a CRISPR-based regulatory cascade in a eukaryotic system achieving significant target repression that consisted of more than two tiers of regulation within the circuit. Moreover, this library provided the first highly customizable system to build complex genetic circuits in eukaryotes. Yeo et al. [64] further enhanced dCas9-based repression in eukaryotes by fusing the human MeCP2 effector domain to the dCas9-KRAB fusion protein developed by Gilbert et al. [62]. The MeCP2 protein further improves chromatin remodeling, as it natively recruits deacetylase enzymes [65]. This configuration now maximizes recruitment of deacetylase enzymes to remodel chromatin around target sequences and repress transcription. The dCas9-KRAB-MeCP2 fusion showed almost twofold increased repression of endogenous genes than the previous dCas9-KRAB fusion. Next, they integrated the dCas9-KRAB-MeCP2 fusion into exogenous and endogenous gene circuits to show improved control. The authors constructed a doxycycline-inducible "NOT" gate, which achieved almost 40-fold repression of exogenous EYFP, which was four times greater than the dCas9-KRAB containing "NOT" gate. They also built a doxycycline-inducible double inverter circuit, similar to the double inverter in Kiani et al., which led to a threefold activation of exogenous EYFP. Lastly, they constructed a double inverter circuit to regulate the endogenous CXCR4 gene. The dCas9-KRAB-MeCP2 achieved an eightfold change in CXCR4 regulation, which was at least twofold greater than other dCas9 constructs used in the inverter.

In parallel to developing CRISPRi-based genetic circuits, there has been significant effort to develop CRISPR-activated (CRISPRa) control schemes in eukaryotes. CRISPR-Cas-based gene activation was first demonstrated in works from Gilbert et al.

(ref#), Cheng et al. [66], and Perez-Pinera et al. [67]. All three works utilized dCas9 fused to multiple copies of the herpes simplex virion protein (VP16) transcriptional activation domain, which natively complexes with transcription factors to drive transcriptional activation, and were able to achieve targeted gene activation of both exogenous and endogenous targets in human cell lines. The work of Nissim et al. [68] was one of the first works to develop CRISPRa-based genetic circuits. The authors utilized a dCas9-VP64 (taCas9) fusion protein to build a transcriptional genetic cascade in HEK293T cells. The transcriptional cascade utilized "gRNA1" complexed with taCas9 to activate expression of "gRNA2," which complexed with taCas9 to drive transcription of the ECFP protein.

To further improve CRISPR-based gene activation, Chavez et al. [69] rationally designed a VP64-p65-Rta (VPR) tripartite activator to fuse to dCas9. The VP64 domain, the technical term for four consecutive copies of the VP16 domain, the p65 domain, and the Rta domain are all well-established transcriptional activator domains. VP16 and Rta domains are activators from herpes simplex and Epstein-Barr viruses, while p65 is the most potent transcriptional activator found in humans [70]. Thus, the expected additive functionality of fusing the three domains is potent transcriptional activation. The dCas9-VPR fusion demonstrated a 22-fold to 320-fold increase in endogenous gene expression relative to activation by the dCas9-VP64 fusion from Gilbert et al. [62] in HEK293T cells. Of significance to the field, development of the dCas9-KRAB CRISPRi and dCas9-VPR CRISPRa platforms vastly expanded the potential for CRISPR-based RNP control, such that users could now develop powerful genetic circuits to either significantly activate or repress target genes in eukaryotic systems. Gao et al. [71] developed chemically inducible control of CRISPRa and CRISPRi genetic circuits from modified dCas9 fusion proteins. The authors modified the dCas9-VPR fusion protein from Chavez et al. [69], such that the VPR activation domain was fused via chemically inducible heterodimerization linker domains. Meaning, a proper functional fusion between the dCas9 protein and the VPR domain can only form in the presence of a specific chemical inducer, as the chemical allows the two parts of the heterodimer to successfully link to one another. After screening multiple domains, they found that an abscisic acid (ABA)-inducible and a gibberellin (GA)-inducible heterodimerization domain, previously discovered by Liang et al. [72] and Miyamoto et al. [73], could achieve a 165-fold and 94-fold activation of plasmid-based EGFP in HEK293T cells, respectively. Next, they replaced the VPR activation domain with the KRAB repression domain and observed 5.6-fold and 3.2-fold repression of EGFP only in the presence of ABA and GA, meaning the dCas9 and KRAB domains could only fuse together and achieve desired regulation via successful heterodimerization, which required the presence of either ABA or GA. The authors next created an "OR" as well as an "AND" gate by fusing both ABA-inducible and GA-inducible heterodimerization domains to a single dCas9 protein. The "OR" gate was operated by fusing the GA-inducible domain to the N-terminus of dCas9, while fusing the ABA-inducible domain to the C-terminus of dCas9. Therefore, in the presence of either or both inducers, dCas9 could fuse to the VPR domain, as the VPR domain was fused to the other half of either the ABA-inducible or GA-inducible domain. Thus, the VPR domain could properly fuse to the dCas9 protein and activate a plasmid-based EGFP target. The authors demonstrated at least 76-fold EGFP activation for any of the inducer combinations used. The "AND" gate was constructed by fusing one half of the GA-inducible domain to the N-terminus of dCas9. The other half of the GA-inducible domain was fused to one half of the ABA-inducible domains. Lastly, the remaining half of the ABA-inducible domain was fused to the VPR domain. Thus, the full dCas9-VPR fusion protein required the presence of both ABA and GA to be present for the complete fusion to be successfully constructed, as GA allowed for dCas9 to fuse to the intermediate domain and ABA allowed for VPR to fuse to the other half of the intermediate domain. The "AND" gate ultimately achieved 49-fold activation of exogenous EGFP. Lastly, the authors constructed a dCas9 fusion that could either activate or repress a desired target. To do this, they modified the "OR" gate by replacing the VPR domain that was fused to one half of the ABA-inducible domain with the KRAB repression domain. Thus, in the presence of GA, the dCas9 construct could fuse to VPR and activate its desired target, while in the presence of ABA, dCas9 would fuse to KRAB and repress its target. This is diagrammed in Fig. 1b, within the "ligand-inducible" box. Using this system, they were able to achieve 2.7-fold repression and 2.5-fold activation.

Along with engineering the Cas protein involved in the gRNA-Cas RNP regulatory complex, significant efforts have been made to engineer the gRNA sequence to improve regulatory strength as well as to provide users with novel control methods in CRISPR-driven genetic circuits. First, work from Zalatan et al. [74] developed modified gRNAs by appending specific RNA hairpins to the 3′ end of the gRNA sequence, which recruited specific RNA-binding proteins (RBPs) fused to transcriptional activator or repression domains. These engineered gRNAs were termed scaffold RNAs (scRNAs). The authors utilized the MS2, PP7, and com RNA hairpins to recruit MCP, PCP, and Com RBPs, respectively, in the scRNA constructs. First, each RBP was fused to the VP64 transcriptional activator domain and demonstrated 50-fold, 35-fold, and 20-fold upregulation of plasmid-based mVenus in

S. cerevisiae, respectively. Additionally, the dCas9-VP64 fusion from previous works [62, 66, 67] only demonstrated a two to threefold increase in expression. This system was then transfected into HEK293T cells, in which the MCP scRNA achieved 150-fold activation of plasmid-based fluorescent protein mVenus and sixfold activation of the endogenous CXCR4 gene. Activation by the MCP scRNA was three times greater than the dCas9-VP64 fusion for both the exogenous and endogenous transcripts. The authors also replaced the VP64 activator domain with the KRAB repressor domain in each scRNA construct. The Com scRNA fusion achieved the maximum of a fivefold reduction in endogenously expressed EGFP, which was consistent with the dCas9-KRAB fusion construct. Lastly, the authors demonstrated simultaneous activation and repressor of endogenous CXCR4 and B4GALNT1, respectively, thus creating a programmable bidirectional switch in eukaryotic systems using these scRNA constructs. In this case, they achieved this by expressing two sgRNAs: one that contained MS2 RNA hairpin to recruit an MCP-VP64 RBP fusion to activate the CXCR4 target and one that contained the comb hairpin to recruit a Com-KRAB fusion to repress the B4GALNT1 target. Second, Liu et al. [75] built CRISPR-based "signal conductors" by fusing gRNA sequences to ligand-responsive riboswitches, such that the CRISPR-Cas9 complex could become active in the absence of the desired ligand, the guide sequence base pairs with the antisense stem of the riboswitch. In the presence of the desired ligand, the riboswitch undergoes a conformational change and liberates the guide sequence, such that it can interact with its target DNA target, thus activating the RNP complex. The authors coupled these riboswitch-gRNA fusions with dCas9 or dCas9-VP64 fusion proteins to either selectively repress or activate endogenous targets, respectively. They first demonstrated the circuit's ability to upregulate TP53 and CDKN1A, two antitumor genes, in the presence of the oncogenic signal nucleophosmin (NPM). The authors transfected T24 bladder cells, which have elevated concentrations of NPM, and demonstrated activation of TP53 and CDKN1A, two key genes in antitumor metabolism, measured by a twofold reduction in cell growth relative to cells with mutant sgRNA aptamers. They also demonstrated the modularity of the circuit by also upregulating tumor migration suppressor, E-cadherin, in the presence of oncogenic signal Ets-1. The circuit utilized an Ets-1-sensing aptamer fused to the sgRNA; thus in the presence of Ets-1, the gRNA could rearrange into its active state. This design achieved a fivefold reduction in tumor cell migration relative to cells containing a mutant sgRNA. Lastly, the authors constructed a circuit that could upregulate and repress genes simultaneously in response to an external signal by employing the sgRNA-aptamer fusions with dCas9-MS2 and dCas9-VP64-Rev fusion proteins co-expressed in a single cell line. Using these constructs, the authors created an AND gate to repress BCL2 and upregulate BAX genes in response to NF-kB and B-catenin, respectively. These genes were selected, as repression of BCL2 with simultaneous activation of BAX drives apoptosis in cancer cells. The authors demonstrated they could selectively drive a sevenfold increase in apoptosis in three bladder cancer cell lines, while minimizing changes in apoptosis for three normal human cell lines. Ying et al. [76] developed a CRISPR activation construct that regulates transcription of a plasmid-based RNA aptamer that fluoresces via binding to a small molecule organic florigen. The gRNA sequence complexed with dCas9-VPR fusion contains a hairpin which prevents the guide sequence from binding to its target sequence. In the presence of specific target mRNA, the gRNA hybridizes with the mRNA transcript and undergoes structural switching such that the RNP complex can activate the transcription of the fluorescent RNA. This system was utilized in HeLA cells to detect *survivin* mRNA, also known as BIRC5, a gene upregulated in many cancers. Using this construct, fluorescent signals could be detected within an hour of induction of *survivin* mRNA, while protein-based sensing circuits took at least eight hours for signal detection.

With the development of CRISPR-based genetic circuits in eukaryotes, these systems have been recently applied to address human health issues. Liu et al. [77] developed an "AND" gate utilizing CRISPR-Cas9 in order to identify and respond to bladder cancer cells. The authors placed Cas9 and a lacI-targeting gRNA under the control of two promoters specifically activated in bladder cancer cells, hUPII and hTERT. Upon activation, the CRISPR-Cas9 complex would block transcription of LacI, which represses a downstream gene of interest. Additionally, there was no significant activation in seven other human cell lines tested. Next, they replaced the luciferase with hBax, p21, or E-cadherin gene, which drove apoptosis, reduced growth, and reduced motility, respectively. These CRISPR-based genetic circuits lay the foundation for potential gene therapies or cancer diagnostics. Recently, Fan et al. [78] developed a liposome-delivered CRISPR-Cas circuit to downregulate VEGFR2, Bcl-2, and *survivin* transcripts in human bladder cancer 5637 cell implanted into mice bladders. Liposomes specifically delivered target gRNAs and Cas13a, a homologue to the Cas9 protein that targets and cuts single stranded RNA transcripts, by using antibodies specific to VEGFR2 receptors on the tumor cells. Upon delivery, the CRISPR-Cas13a complex would form and target the desired transcripts; in this system, all target transcript expression levels were reduced to at least 40% of their levels prior to treatment. Additionally, overall tumor size was reduced to ~35% after 20 days of continuous treatment. This is one of the first instances of using a CRISPR-based circuit in gene therapy for the treatment of human cancers. CRISPRi regulation was also utilized to program human gut bacteria. Lastly, Mimee et al. [79] developed genetic parts to program the gut bacterium *Bacteroides thetaiotaomicron*, specifically a CRISPRi-based "NOT" gate that allowed for IPTG-inducible repression of desired exogenous and endogenous targets. They colonized the gut of mice and demonstrated IPTG-inducible repression using the CRISPRi "NOT" gate by providing IPTG to the mice through drinking water. This is one of the first examples of using CRISPRi-based genetic circuits to program bacteria directly related to the human gut microbiota.

3 Selecting an Appropriate System

One of the hallmarks of synthetic biology is the fungibility of parts and circuits, and in this vein, we believe that many of the RNP-based control systems described could be cross-utilized in a variety of settings. To select an RNP-based control system that will be of the greatest utility, we review the benefits and current limitations for each of the regulatory categories reviewed above. The benefits and limitations of each category are summarized in Table 1.

In short, both prokaryotic and eukaryotic RNP control schemes broadly fall into RBP-RNA-based and CRISPR-Cas-based categories. In general, RBP-RNA-based control provides the advantage of control via engineering RNA-RNA hybridization, while also serving as a protein interface with the remainder of metabolism. For example, a global RNA-binding protein such as Hfq or CsrA can be utilized in tandem with an sRNA to achieve posttranscriptional regulation of desired mRNA transcripts, precisely because these proteins have known functional "interpretations" in a systems context. This system-wide interpretation also allows for multiplexing to target several transcripts simultaneously. These systems have been utilized for controlling the expression of multiple genes in prokaryotic systems to globally regulate cellular metabolism; for example, Na et al. [22] and Noh et al. [24] used Hfq and target sRNAs in bioprocessing applications to regulate competing metabolic pathways.

Drawbacks to RBP-RNA-based control include the fact that systemic functional interpretations are often fixed, meaning targets can often either just be repressed or activated. That said, this drawback could likely be overcome by the simple expedient of using "ribocomputing" devices to provide RNP-based regulation, for example, by using programmed strand exchange to regulate formation of an mRNA-ribosome complex. The portability of specific regulatory systems (e.g., bacterial Hfq into eukaryotes) is

Table 1
Summary of benefits and limitations of RNP-based control schemes in prokaryotes and eukaryotes

System	Mode of control	Regulatory mechanism	System benefits	Current limitations	Circuit application
Prokaryote	Ribozyme-directed mRNA regulation	mRNA transcript degraded by RNase P and external guide RNA sequence (EGS)	Target specific mRNA transcripts for degradation with minimal off-target effects	Only represses transcripts, no way to upregulate targets	[15, 16]
	RBP-RNA regulation	Hfq-sRNA driven regulation of mRNA transcripts	Multiplexed to target several genes simultaneously Reduced metabolic burden by only overexpressing sRNAs Tunable levels of endogenous gene repression without extensive genome engineering	Only represses transcripts, no way to upregulate targets Time intensive to redesign sRNA for every mRNA target Does not work in organisms lacking Hfq	[22, 24, 25]
		Engineered CsrB sRNAs with varied affinities for CsrA to titrate cellular levels	Rapidly titrate intracellular CsrA to drive changes on the cellular level Reduced metabolic burden by only requiring overexpression of CsrB sRNAs	Can only directly modify CsrA-regulated processes Rapid depletion of intracellular CsrA may have unintentional consequences on cell health Does not work in organisms lacking CsrA/homologous systems	[30]
	Ribosome-controlled regulation	"Ribocomputing" engineered hairpins and trigger sRNAs to block/allow formation of ribosome on target mRNA	Able to construct complex genetic circuits using simple composable logic gates Circuits orthogonal to native posttranscriptional regulation Rapid response by using sRNAs and eliminating the need for translating circuit triggers	Currently only able to affect expression of exogenous genes Genes can only be activated, not repressed Genetic circuits can have inconsistent fold activation	[31, 85]
	CRISPR-based regulation	Catalytically inactive Cas9 (dCas9) and custom gRNA to block transcription of target genes (CRISPRi)	Multiplexed to target several genes simultaneously Tunable levels of gene repression without extensive genome engineering	Can only repress genes, no way to activate targets Overexpression of dCas9 protein may put metabolic stress on organism	[36, 37, 43, 47, 48]

(continued)

Table 1
(continued)

System	Mode of control	Regulatory mechanism	System benefits	Current limitations	Circuit application
		dCas9 fusion with custom gRNA to upregulate expression of target genes (CRISPRa)	Able to activate exogenous and endogenous genes using CRISPR-based regulation	Relatively low fold activation relative to other RNAP transcriptional activator systems Overexpression of dCas9-RNAP fusion may cause burden on organism	[38]
		dCas9 with modified gRNAs containing RNA-binding protein hairpins that recruit RBP-SoxS transcriptional activator fusion to upregulate targets	Titratable control of CRISPR-based activation Ability to activate and repress targets simultaneously Improved fold activation of targets relative to previous CRISPRa systems	Overexpression of dCas9 and RBP-SoxS fusion may cause a metabolic burden	[41]
	CRISPR- and RNAP-based regulation	Engineered activator sRNAs to control transcriptional elongation via RNAP of gRNAs for dCas9	Large library of composable parts for fine-tuning gene expression Operates orthogonal to the rest of the organism's metabolism Multiplexed to regulate several targets simultaneously Achieves full activation of regulated genes within an hour of induction	Only able to control expression of exogenous genes Overexpression of dCas9 can be a metabolic burden over time Only able to repress target genes, no way to activate	[49, 50]
Eukaryote	Ribozyme-directed mRNA regulation	mRNA transcript degraded by RNase P and external guide RNA sequence (EGS)	Portable method to target various mRNAs implicated in human health concerns	Limited genetic parts to provide precise user control of system	[17, 18, 54]
	RBP-RNA regulation	L7Ae-box RBP domain binding a gRNA sequence fused to the 5′ UTR of a transcript to block/allow translation	Can simultaneously activate and repress multiple targets	Can only regulate exogenous targets Must modify original transcript by fusing L7Ae binding domain to UTR of the transcript	[56]

RBP protein fusion regulation	METTL-dCas9 and ALKBH5-dCas9 fusions to selectively add or remove m6A modification to mRNA transcripts	Epitranscriptome modification of mRNA transcripts	Can only add m6A modifications to mRNA transcripts, no other modifications or regulatory effects Overexpression of dCas9 fusion proteins may cause burden on organism	[57]
	DNA adenosine methyltransferase (Dam) fused to Dnp1 to add m6A modifications, used in tandem with Dpn1 fusions to VP64 or KRAB to regulate methylated mRNA transcripts	First circuit to utilize epitranscriptome control for mRNA transcript regulatory fate Cells with "read-write" module possess regulatory "memory" of mRNA transcripts for several days	Only applied to endogenous copy of eGFP at a single locus, little studies on exogenous sequence and other loci Extensive m6A modification may cause unobserved cellular effects	[58]
CRISPR-based regulation	dCas9 or dCas9 fusions to block transcription of specific genes targeted via custom guide RNAs (CRISPRi)	Able to downregulate expression of endogenous target genes >100-fold repression of endogenous genes Large library of available complex logic gates	Overexpression of dCas9 or dCas9 fusions can be a metabolic burden on cells Circuit regulating transcription imposes lag time to achieve desired regulation Can only repress targets, no way to activate Off-target effects can occur in higher-level organisms such as mice and human cell models	[36, 62–64, 79]
	dCas9 fusions to transcriptionally activate specific genes targeted via guide RNAs (CRISPRa)	Able to activate or repress various target genes utilizing CRISPR-based circuits > 100-fold activation of endogenous targets	System is only able to activate expression of target genes Expression of dCas9 with multiple activation tags may cause metabolic burden	[62, 66–69, 71]
	Modified gRNAs for expanded regulatory control in CRISPR-Cas-based regulation	Simultaneous activation and repression of targets using CRISPR-Cas regulation Inducible control over RNP activator domain reduces leakiness of CRISPR-Cas activity CRISPR-Cas systems can respond to external metabolites or ligands	Significant time required to modify gRNA sequences to ensure desired function and maintain interaction with dCas9 Overexpression of RBP fusion proteins may cause metabolic burden	[74–76]

questionable. In this regard, there has been significantly less development of RBP-RNA-based control schemes in eukaryotes. System-level interactions have been underplayed in favor of fusing RBPs to modifying protein domains to bind and specifically modify targeted mRNA transcripts, for example, RBP protein fusions that specifically add m6A modifications to mRNA transcripts, such as in Liu et al. (2019) [57] and Park et al. [58].

CRSPR-based RNP control systems most often rely on catalytically inactive Cas9 (dCas9) fused to a transcriptional regulatory domain. These systems are highly programmable and can be used to repress (CRISPRi) or activate (CRISPRa) both endogenous and exogenous targets. Because of their modularity, CRISPRi-based systems have been integrated into genetic circuits to construct inducible "OR," "AND," as well as "NOR" logic gates. Thus, using CRISPR-based logic gates, fundamental control schemes can be constructed, such as toggle switches, feedforward loops, and oscillators, as seen in Santos-Moreno et al. [43]. Due to their programmability, CRISPR-Cas circuits can also be multiplexed to rewire cellular metabolism and improve bioprocessing product yield, although in a manner that may be significantly more complicated than relying on endogenous RBP control. Nonetheless, this feature is especially valuable in industrially relevant organisms that lack previously engineered RBP-RNA regulatory systems, such as Hfq- or CsrA-regulated systems. In this regard, dCas9 has been extensively used in eukaryotes to repress (CRISPRi) as well as activate (CRISPRa) target genes, and target genes can be regulated to greater than 100-fold wild-type levels.

A major drawback of CRISPR-based systems is the cellular burden of expressing Cas and Cas fusion proteins, which can lead in many instances to eventual circuit inactivation. In addition, there are frequently potential off-target effects due to potential mispairing of CRISPR gRNAs. Additionally, since CRISPR-based systems act through transcriptional regulation, this creates a lag time for response and can take hours to achieve the desired regulations; this is not ideal for applications that require fast-acting systems, such as those that produce toxic compounds.

Recent work suggests that it may be possible to meld the advantages of the two systems, systems integration (for RBPs) and extreme programmability (for CRISPR-Cas): guide RNA sequences have been modified to contain RBP-binding hairpins that recruit engineered RBP transcriptional regulator fusions, such as observed by Zalatan et al. [74] and Dong et al. [41]. Recruiting endogenous factors also has the advantage of allowing users to potentially simultaneously activate and repress multiple transcripts, depending on the RBP.

4 Future Perspectives

RNP-based control systems have been engineered over the last decade to provide control over major metabolic pathways and cellular responses. One potential application of RNP-based genetic control is in industrial bioprocesses, where there are already successes in producing value-added compounds such as biofuels [80] or industrial commodity chemicals [81]. Scale-up can be further improved by developing stress-responsive control schemes that quickly and continuously maximize cell viability [82]. RNP-based control systems provide unique opportunities to coordinate metabolism at the system level, providing both feedback on genetic components (transcription, replication) and on posttranscriptional regulation, ultimately reducing response lag time. Achieving complete regulation within an hour using RNP-based genetic circuits could help mitigate accumulation of toxic intermediates, which has been previously shown to improve metabolite production and cell viability [83]. Another potential avenue for RNP-based gene control is in biological therapeutics. Given that CRISPR-based methods will see increasing use in gene therapy (as a recent example, a CRISPR-Cas9 construct has been utilized to treat sickle cell disease in human patients, with edits to the exact nucleotide in 80% of the targeted alleles across multiple patients, without any observed side effects [84]), further integration of treatment with human systems biology and physiology can only further improve outcomes.

Beyond applications inside of cells, preliminary work has demonstrated potential utility of RNP-based systems in cell-free systems, such as rapid diagnostic tests for detection of antibiotics, glucose, and even specific viral strains (i.e., Ebola) [85]. In addition to diagnostic tests, RNP-based genetic circuits have the potential to be integrated into "smart materials" to detect toxic metabolites or pathogens.

Lastly, there has been a recent push to develop "minimal" or synthetic cells and organelles. These cells could serve as a chassis to create in vivo cellular systems that are entirely independent of bacterial strains and eukaryotic cell lines [86]. One key piece to improve the utility and development of these cells is finding additional ways to precisely regulate gene expression within these synthetic cells, particularly by controlling translation. Importantly, RNP control schemes (which could be easily purified and optimized for in vitro use) may serve as an excellent resource for synthetic cell systems, as they can rapidly control translation of target genes and have been utilized in vivo and in vitro systems. RNP genetic circuits have begun to be implemented into synthetic cells, such as into synthetic stress granules for spatial and temporal translational control [87]. Moving forward, combining RNP-based genetic control into synthetic cellular systems appears to be an exciting frontier in synthetic biology.

Overall, the examples of individual RNP systems that we have listed throughout highlight the opportunities for more generally engineering cells. The fact that in many instances target interactions or overall stability can be fine-tuned via nucleic acid hybridization means that well-known, quantitative thermodynamic measurements and predictions can potentially be applied to the control of gene regulation. In parallel, protein partners allow much broader systems-level integration, since proteins touch on virtually every aspect of cellular metabolism and regulation. RNPs are literally the linchpins of programmable biology, a hallmark of the ongoing revolution in synthetic biology. Our own work with CsrA shows how it is possible to allow the simplicity of nucleic acid hybridization to be parallelized and multiplexed via a single protein. Both of these features now bode well for extending interactions across the entire cellular system, where, for example, efforts to multiplex gRNAs to regulate multiple genes in parallel are mere harbingers of the possibilities available for using correlated and cross-regulated sets of nucleic acids to simultaneous regulate a wide variety of cellular processes, from intron splicing to protein translation to secretion.

Acknowledgments

We would like to thank Dr. Andrew Ellington for his written contributions to the manuscript as well as his continual guidance, input, and feedback throughout the development of this work. Additionally, we would like to thank the rest of the Contreras Lab for their helpful discussions on the manuscript. Lastly, this work was supported by grants from the National Institutes of Health [grant number R01GM135495 to LMC] and the National Science Foundation [grant numbers MCB-1932780 to L.M.C., DGE-1610403 to T.R.S].

References

1. Leistra AN, Curtis NC, Contreras LM (2019) Regulatory non-coding sRNAs in bacterial metabolic pathway engineering. Metab Eng 52:190–214. https://doi.org/10.1016/j.ymben.2018.11.013
2. Yan Q, Pfleger BF (2020) Revisiting metabolic engineering strategies for microbial synthesis of oleochemicals. Metab Eng 58:35–46. https://doi.org/10.1016/j.ymben.2019.04.009
3. Jensen MK, Keasling JD (2015) Recent applications of synthetic biology tools for yeast metabolic engineering. FEMS Yeast Res 15(1): 1–10. https://doi.org/10.1111/1567-1364.12185
4. Shivram H, Cress BF, Knott GJ, Doudna JA (2021) Controlling and enhancing CRISPR systems. Nat Chem Biol 17(1):10–19. https://doi.org/10.1038/s41589-020-00700-7
5. Khalil AS, Collins JJ (2010) Synthetic biology: applications come of age. Nat Rev Genet

11(5):367–379. https://doi.org/10.1038/nrg2775

6. Sowa SW, Gelderman G, Contreras LM (2015) Advances in synthetic dynamic circuits design: using novel synthetic parts to engineer new generations of gene oscillations. Curr Opin Biotechnol 36:161–167. https://doi.org/10.1016/j.copbio.2015.08.020
7. Hogan DJ, Riordan DP, Gerber AP, Herschlag D, Brown PO (2008) Diverse RNA-binding proteins interact with functionally related sets of RNAs, suggesting an extensive regulatory system. PLoS Biol 6(10):e255–e255. https://doi.org/10.1371/journal.pbio.0060255
8. Lee SR, Lykke-Andersen J (2013) Emerging roles for ribonucleoprotein modification and remodeling in controlling RNA fate. Trends Cell Biol 23(10):504–510. https://doi.org/10.1016/j.tcb.2013.05.001
9. Collins LJ, Kurland CG, Biggs P, Penny D (2009) The modern RNP world of Eukaryotes. J Hered 100(5):597–604. https://doi.org/10.1093/jhered/esp064
10. Akopian D, Shen K, Zhang X, Shan S (2013) Signal recognition particle: an essential protein-targeting machine. Ann Rev Biochem *82*(1):693–721. https://doi.org/10.1146/annurev-biochem-072711-164732
11. Han R, Fang J, Jiang J, Gaidamakova EK, Tkavc R, Daly MJ, Contreras LM (2020) Signal recognition particle RNA contributes to oxidative stress response in Deinococcus radiodurans by modulating catalase localization. Front Microbiol 11:3263. https://www.frontiersin.org/article/10.3389/fmicb.2020.613571
12. Esakova O, Krasilnikov AS (2010) Of proteins and RNA: the RNase P/MRP family. RNA. 16(9):1725–1747. https://doi.org/10.1261/rna.2214510
13. Cameron TA, Matz LM, De Lay NR (2018) Polynucleotide phosphorylase: not merely an RNase but a pivotal post-transcriptional regulator. PLoS Genet 14(10):e1007654. https://doi.org/10.1371/journal.pgen.1007654
14. Guerrier-Takada C, Gardiner K, Marsh T, Pace N, Altman S (1983) The RNA moiety of ribonuclease P is the catalytic subunit of the enzyme. Cell 35(3):849–857. https://doi.org/10.1016/0092-8674(83)90117-4
15. Li Y, Guerrier-Takada C, Altman S (1992) Targeted cleavage of mRNA in vitro by RNase P from Escherichia coli. Proc Nat Acad Sci 89(8):3185 LP–3189. https://doi.org/10.1073/pnas.89.8.3185
16. Guerrier-Takada C, Salavati R, Altman S (1997) Phenotypic conversion of drug-resistant bacteria to drug sensitivity. Proc Nat Acad Sci 94(16):8468 LP–8472. https://doi.org/10.1073/pnas.94.16.8468
17. Cobaleda C, Sánchez-García I (2000) In vivo inhibition by a site-specific catalytic RNA subunit of RNase P designed against the BCR-ABL oncogenic products: a novel approach for cancer treatment. Blood 95(3):731–737. https://doi.org/10.1182/blood.V95.3.731.003k28_731_737
18. Trang P, Kilani A, Kim J, Liu F (2000) A ribozyme derived from the catalytic subunit of RNase P from Escherichia coli is highly effective in inhibiting replication of herpes simplex virus 111Edited by J. Doudna. J Mol Biol *301*(4):817–826. https://doi.org/10.1006/jmbi.2000.4022
19. Yang Z, Vu G-P, Qian H, Chen Y-C, Wang Y, Reeves M, Zen K, Liu F (2014) Engineered RNase P Ribozymes effectively inhibit human cytomegalovirus gene expression and replication. Viruses 6(6):2376–2391. https://doi.org/10.3390/v6062376
20. Quendera AP, Seixas AF, dos Santos RF et al (2020) RNA-binding proteins driving the regulatory activity of small non-coding RNAs in Bacteria. Front Mol Biosci 7:78. https://www.frontiersin.org/article/10.3389/fmolb.2020.00078
21. Santiago-Frangos A, Woodson SA (2018) Hfq chaperone brings speed dating to bacterial sRNA. Wiley Interdiscip Rev RNA 9(4):e1475–e1475. https://doi.org/10.1002/wrna.1475
22. Na D, Yoo SM, Chung H, Park H, Park JH, Lee SY (2013) Metabolic engineering of Escherichia coli using synthetic small regulatory RNAs. Nat Biotechnol 31(2). https://doi.org/10.1038/nbt.2461
23. Shuo C, Aixia Z, Blyn LB, Gisela S (2004) MicC, a second small-RNA regulator of Omp protein expression in Escherichia coli. J Bacteriol *186*(20):6689–6697. https://doi.org/10.1128/JB.186.20.6689-6697.2004
24. Noh M, Yoo SM, Kim WJ, Lee SY (2017) Gene expression knockdown by modulating synthetic small RNA expression in *Escherichia coli*. Cell Syst 5(4):418–426.e4. https://doi.org/10.1016/j.cels.2017.08.016
25. Lahiry A, Stimple SD, Wood DW, Lease RA (2017) Retargeting a dual-acting sRNA for multiple mRNA transcript regulation. ACS Synth Biol 6(4):648–658. https://doi.org/10.1021/acssynbio.6b00261

26. Romeo T, Babitzke P (2018) Global regulation by CsrA and its RNA Antagonists. Microbiol Spectr 6(2). https://doi.org/10.1128/microbiolspec.RWR-0009-2017
27. Sowa SW, Gelderman G, Leistra AN, Buvanendiran A, Lipp S, Pitaktong A, Vakulskas CA, Romeo T, Baldea M, Contreras LM (2017) Integrative FourD omics approach profiles the target network of the carbon storage regulatory system. Nucleic Acids Res 45(4): 1673–1686. https://doi.org/10.1093/nar/gkx048
28. Andrew R, Stephanie P, Ying-Jung L, Archana P, Helen Y, Potts AH, Bevilacqua PC, Romeo T, Babitzke P, Cossart PF (2021) CsrA-mediated translational activation of ymdA expression in Escherichia coli. MBio 11(5):e00849–e00820. https://doi.org/10.1128/mBio.00849-20
29. Leistra AN, Gelderman G, Sowa SW, Moon-Walker A, Salis HM, Contreras LM (2018) A Canonical biophysical model of the CsrA global regulator suggests flexible regulator-target interactions. Sci Rep 8(1):1–19. https://doi.org/10.1038/s41598-018-27474-2
30. Leistra AN, Amador P, Buvanendiran A, Moon-Walker A, Contreras LM (2017) Rational modular RNA engineering based on in vivo profiling of structural accessibility. ACS Synth Biol 6(12):2228–2240. https://doi.org/10.1021/acssynbio.7b00185
31. Green AA, Kim J, Ma D, Silver PA, Collins JJ, Yin P (2017) Complex cellular logic computation using ribocomputing devices. Nature 548(7665):117–121. https://doi.org/10.1038/nature23271
32. Amalfitano E, Karlikow M, Norouzi M, Jaenes K, Cicek S, Masum F, Sadat Mousavi P, Guo Y, Tang L, Sydor A, Ma D, Pearson JD, Trcka D, Pinette M, Ambagala A, Babiuk S, Pickering B, Wrana J, Bremner R et al (2021) A glucose meter interface for point-of-care gene circuit-based diagnostics. Nat Commun 12(1):724. https://doi.org/10.1038/s41467-020-20639-6
33. Ishino Y, Shinagawa H, Makino K, Amemura M, Nakata A (1987) Nucleotide sequence of the iap gene, responsible for alkaline phosphatase isozyme conversion in Escherichia coli, and identification of the gene product. J Bacteriol 169(12):5429–5433. https://doi.org/10.1128/jb.169.12.5429-5433.1987
34. Jansen R, van Embden JDA, Gaastra W, Schouls LM (2002) Identification of genes that are associated with DNA repeats in prokaryotes. Mol Microbiol *43*(6):1565–1575. https://doi.org/10.1046/j.1365-2958.2002.02839.x
35. Jinek M, Chylinski K, Fonfara I, Hauer M, Doudna JA, Charpentier E (2012) A programmable dual-RNA-guided DNA endonuclease in adaptive bacterial immunity. Science 337(6096):816–821. https://doi.org/10.1126/science.1225829
36. Qi LS, Larson MH, Gilbert LA et al (2013) Repurposing CRISPR as an RNA-guided platform for sequence-specific control of gene expression. Cell 152(5):1173–1183. https://doi.org/10.1016/j.cell.2013.02.022
37. Nielsen AAK, Voigt CA (2014) Multi-input CRISPR/Cas genetic circuits that interface host regulatory networks. Mol Syst Biol 10(11):763. https://doi.org/10.15252/msb.20145735
38. Bikard D, Jiang W, Samai P, Hochschild A, Zhang F, Marraffini LA (2013) Programmable repression and activation of bacterial gene expression using an engineered CRISPR-Cas system. Nucleic Acids Res 41(15): 7429–7437. https://doi.org/10.1093/nar/gkt520
39. Dove SL, Joung JK, Hochschild A (1997) Activation of prokaryotic transcription through arbitrary protein–protein contacts. Nature 386(6625):627–630. https://doi.org/10.1038/386627a0
40. Dove SL, Hochschild A (1998) Conversion of the ω subunit of Escherichia coli RNA polymerase into a transcriptional activator or an activation target. Genes Dev *12*(5):745–754. http://genesdev.cshlp.org/content/12/5/745.abstract
41. Dong C, Fontana J, Patel A, Carothers JM, Zalatan JG (2018) Synthetic CRISPR-Cas gene activators for transcriptional reprogramming in bacteria. Nat Commun 9(1):2489. https://doi.org/10.1038/s41467-018-04901-6
42. Sadowski I, Ma J, Triezenberg S, Ptashne M (1988) GAL4-VP16 is an unusually potent transcriptional activator. Nature 335(6190): 563–564. https://doi.org/10.1038/335563a0
43. Santos-Moreno J, Tasiudi E, Stelling J, Schaerli Y (2020) Multistable and dynamic CRISPRi-based synthetic circuits. Nat Commun 11(1): 2746. https://doi.org/10.1038/s41467-020-16574-1
44. Gardner TS, Cantor CR, Collins JJ (2000) Construction of a genetic toggle switch in

Escherichia coli. Nature 403(6767):339–342. https://doi.org/10.1038/35002131
45. Mangan S, Itzkovitz S, Zaslaver A, Alon U (2006) The incoherent feed-forward loop accelerates the response-time of the gal system of Escherichia coli. J Mol Biol 356(5): 1073–1081. https://doi.org/10.1016/j.jmb.2005.12.003
46. Elowitz MB, Leibler S (2000) A synthetic oscillatory network of transcriptional regulators. Nature 403(6767):335–338. https://doi.org/10.1038/35002125
47. Gordon GC, Korosh TC, Cameron JC, Markley AL, Begemann MB, Pfleger BF (2016) CRISPR interference as a titratable, trans-acting regulatory tool for metabolic engineering in the cyanobacterium Synechococcus sp. strain PCC 7002. Metab Eng 38:170–179. https://doi.org/10.1016/j.ymben.2016.07.007
48. Cleto S, Jensen JVK, Wendisch VF, Lu TK (2016) Corynebacterium glutamicum metabolic engineering with CRISPR Interference (CRISPRi). ACS Synth Biol 5(5):375–385. https://doi.org/10.1021/acssynbio.5b00216
49. Chappell J, Takahashi MK, Lucks JB (2015) Creating small transcription activating RNAs. Nat Chem Biol 11(3):214–220. https://doi.org/10.1038/nchembio.1737
50. Chappell J, Westbrook A, Verosloff M, Lucks JB (2017) Computational design of small transcription activating RNAs for versatile and dynamic gene regulation. Nat Commun 8(1): 1051. https://doi.org/10.1038/s41467-017-01082-6
51. Matera AG, Wang Z (2014) A day in the life of the spliceosome. Nat Rev Mol Cell Biol 15(2): 108–121. https://doi.org/10.1038/nrm3742
52. Shay JW, Wright WE (2019) Telomeres and telomerase: three decades of progress. Nat Rev Genet 20(5):299–309. https://doi.org/10.1038/s41576-019-0099-1
53. Vazquez-Arango P, O'Reilly D (2018) Variant snRNPs: new players within the spliceosome system. RNA Biol 15(1):17–25. https://doi.org/10.1080/15476286.2017.1373238
54. Jiang X, Bai Y, Rider P, Kim K, Zhang C-Y, Lu S, Liu F (2011) Engineered external guide sequences effectively block viral gene expression and replication in cultured cells. J Biol Chem 286(1):322–330. https://doi.org/10.1074/jbc.M110.158857
55. Shotwell CR, Cleary JD, Berglund JA (2020) The potential of engineered eukaryotic RNA-binding proteins as molecular tools and therapeutics. WIREs RNA 11(1):e1573. https://doi.org/10.1002/wrna.1573
56. Saito H, Kobayashi T, Hara T, Fujita Y, Hayashi K, Furushima R, Inoue T (2010) Synthetic translational regulation by an L7Ae–kink-turn RNP switch. Nat Chem Biol 6(1): 71–78. https://doi.org/10.1038/nchembio.273
57. Liu X-M, Zhou J, Mao Y, Ji Q, Qian S-B (2019) Programmable RNA N6-methyladenosine editing by CRISPR-Cas9 conjugates. Nat Chem Biol 15(9): 865–871. https://doi.org/10.1038/s41589-019-0327-1
58. Park M, Patel N, Keung AJ, Khalil AS (2019) Engineering epigenetic regulation using synthetic read-write modules. Cell 176(1): 227–238.e20. https://doi.org/10.1016/j.cell.2018.11.002
59. Cong L, Ran FA, Cox D, Lin S, Barretto R, Habib N, Hsu PD, Wu X, Jiang W, Marraffini LA, Zhang F (2013) Multiplex genome engineering using CRISPR/Cas systems. Science 339(6121):819–823. https://doi.org/10.1126/science.1231143
60. Mali P, Yang L, Esvelt KM et al (2013) RNA-guided human genome engineering via Cas9. Science (80-) 339(6121):823 LP – 826. https://doi.org/10.1126/science.1232033
61. Kiani S, Beal J, Ebrahimkhani MR, Huh J, Hall RN, Xie Z, Li Y, Weiss R (2014) CRISPR transcriptional repression devices and layered circuits in mammalian cells. Nat Method 11(7):723–726. https://doi.org/10.1038/nmeth.2969
62. Gilbert LA, Larson MH, Morsut L, Liu Z, Brar GA, Torres SE, Stern-Ginossar N, Brandman O, Whitehead EH, Doudna JA, Lim WA, Weissman JS, Qi LS (2013) CRISPR-mediated modular RNA-guided regulation of transcription in Eukaryotes. Cell 154(2):442–451. https://doi.org/10.1016/j.cell.2013.06.044
63. Gander MW, Vrana JD, Voje WE, Carothers JM, Klavins E (2017) Digital logic circuits in yeast with CRISPR-dCas9 NOR gates. Nat Commun 8(1):15459. https://doi.org/10.1038/ncomms15459
64. Yeo NC, Chavez A, Lance-Byrne A, Chan Y, Menn D, Milanova D, Kuo C-C, Guo X, Sharma S, Tung A, Cecchi RJ, Tuttle M, Pradhan S, Lim ET, Davidsohn N, Ebrahimkhani MR, Collins JJ, Lewis NE, Kiani S, Church GM (2018) An enhanced CRISPR repressor for targeted mammalian gene regula-

tion. Nat Method 15(8):611–616. https://doi.org/10.1038/s41592-018-0048-5

65. Adkins NL, Georgel PT (2010) MeCP2: structure and functionThis paper is one of a selection of papers published in a Special Issue entitled 31st Annual International Asilomar Chromatin and Chromosomes Conference, and has undergone the Journal's usual peer review process. Biochem Cell Biol *89*(1): 1–11. https://doi.org/10.1139/O10-112
66. Cheng AW, Wang H, Yang H et al (2013) Multiplexed activation of endogenous genes by CRISPR-on, an RNA-guided transcriptional activator system. Cell Res 23(10): 1163–1171. https://doi.org/10.1038/cr.2013.122
67. Perez-Pinera P, Kocak DD, Vockley CM, Adler AF, Kabadi AM, Polstein LR, Thakore PI, Glass KA, Ousterout DG, Leong KW, Guilak F, Crawford GE, Reddy TE, Gersbach CA (2013) RNA-guided gene activation by CRISPR-Cas9–based transcription factors. Nat Method *10*(10):973–976. https://doi.org/10.1038/nmeth.2600
68. Nissim L, Perli SD, Fridkin A, Perez-Pinera P, Lu TK (2014) Multiplexed and programmable regulation of gene networks with an integrated RNA and CRISPR/Cas toolkit in human cells. Mol Cell *54*(4):698–710. https://doi.org/10.1016/j.molcel.2014.04.022
69. Chavez A, Scheiman J, Vora S, Pruitt BW, Tuttle M, Iyer EPR, Lin S, Kiani S, Guzman CD, Wiegand DJ, Ter-Ovanesyan D, Braff JL, Davidsohn N, Housden BE, Perrimon N, Weiss R, Aach J, Collins JJ, Church GM (2015) Highly efficient Cas9-mediated transcriptional programming. Nat Method *12*(4): 326–328. https://doi.org/10.1038/nmeth.3312
70. Schmitz ML, Baeuerle PA (1991) The p65 subunit is responsible for the strong transcription activating potential of NF-kappa B. EMBO J *10*(12):3805–3817. https://pubmed.ncbi.nlm.nih.gov/1935902
71. Gao Y, Xiong X, Wong S, Charles EJ, Lim WA, Qi LS (2016) Complex transcriptional modulation with orthogonal and inducible dCas9 regulators. Nat Method *13*(12):1043–1049. https://doi.org/10.1038/nmeth.4042
72. Liang F-S, Ho WQ, Crabtree GR (2011) Engineering the ABA plant stress pathway for regulation of induced proximity. Sci Signal *4*(164): rs2–rs2. https://doi.org/10.1126/scisignal.2001449
73. Miyamoto T, DeRose R, Suarez A, Ueno T, Chen M, Sun T, Wolfgang MJ, Mukherjee C, Meyers DJ, Inoue T (2012) Rapid and orthogonal logic gating with a gibberellin-induced dimerization system. Nat Chem Biol *8*(5): 465–470. https://doi.org/10.1038/nchembio.922
74. Zalatan JG, Lee ME, Almeida R, Gilbert LA, Whitehead EH, La Russa M, Tsai JC, Weissman JS, Dueber JE, Qi LS, Lim WA (2015) Engineering complex synthetic transcriptional programs with CRISPR RNA Scaffolds. Cell *160*(1):339–350. https://doi.org/10.1016/j.cell.2014.11.052
75. Liu Y, Zhan Y, Chen Z, He A, Li J, Wu H, Liu L, Zhuang C, Lin J, Guo X, Zhang Q, Huang W, Cai Z (2016) Directing cellular information flow via CRISPR signal conductors. Nat Method *13*(11):938–944. https://doi.org/10.1038/nmeth.3994
76. Ying Z-M, Wang F, Chu X, Yu R-Q, Jiang J-H (2020) Activatable CRISPR transcriptional circuits generate functional RNA for mRNA sensing and silencing. Angew Chemie Int Ed *59*(42):18599–18604. https://doi.org/10.1002/anie.202004751
77. Liu Y, Zeng Y, Liu L, Zhuang C, Fu X, Huang W, Cai Z (2014) Synthesizing AND gate genetic circuits based on CRISPR-Cas9 for identification of bladder cancer cells. Nat Commun *5*(1):5393. https://doi.org/10.1038/ncomms6393
78. Fan J, Liu Y, Liu L, Huang Y, Li X, Huang W (2020) A multifunction lipid-based CRISPR-Cas13a genetic circuit delivery system for bladder cancer gene therapy. ACS Synth Biol *9*(2): 343–355. https://doi.org/10.1021/acssynbio.9b00349
79. Mimee M, Tucker AC, Voigt CA, Lu TK (2015) Programming a human commensal bacterium *Bacteroides thetaiotaomicron* to sense and respond to stimuli in the Murine Gut Microbiota. Cell Syst *1*(1):62–71. https://doi.org/10.1016/j.cels.2015.06.001
80. Nielsen J, Keasling JD (2016) Engineering cellular metabolism. Cell 164(6):1185–1197. https://doi.org/10.1016/j.cell.2016.02.004
81. Lee JW, Na D, Park JM, Lee J, Choi S, Lee SY (2012) Systems metabolic engineering of microorganisms for natural and non-natural chemicals. Nat Chem Biol 8(6):536–546. https://doi.org/10.1038/nchembio.970
82. Wehrs M, Tanjore D, Eng T, Lievense J, Pray TR, Mukhopadhyay A (2019) Engineering

robust production microbes for large-scale cultivation. Trends Microbiol 27(6):524–537. https://doi.org/10.1016/j.tim.2019.01.006

83. Dahl RH, Zhang F, Alonso-Gutierrez J et al (2013) Engineering dynamic pathway regulation using stress-response promoters. Nat Biotechnol 31(11):1039–1046. https://doi.org/10.1038/nbt.2689

84. Frangoul H, Altshuler D, Cappellini MD et al (2020) CRISPR-Cas9 gene editing for sickle cell disease and β-Thalassemia. N Engl J Med 384(3):252–260. https://doi.org/10.1056/NEJMoa2031054

85. Pardee K, Green AA, Ferrante T et al (2014) Paper-based synthetic gene networks. Cell 159(4):940–954. https://doi.org/10.1016/j.cell.2014.10.004

86. Adamala KP, Martin-Alarcon DA, Guthrie-Honea KR, Boyden ES (2017) Engineering genetic circuit interactions within and between synthetic minimal cells. Nat Chem 9(5):431–439. https://doi.org/10.1038/nchem.2644

87. Simon JR, Eghtesadi SA, Dzuricky M, You L, Chilkoti A (2019) Engineered Ribonucleoprotein granules inhibit translation in protocells. Mol Cell 75(1):66–75.e5. https://doi.org/10.1016/j.molcel.2019.05.010

Chapter 2

Computational Design of RNA Toehold-Mediated Translation Activators

Kaiyue Wu, Zhaoqing Yan, and Alexander A. Green

Abstract

Translation activators are an important class of riboregulators that respond to nucleic acid signals by activating gene expression. Toehold switches and single-nucleotide-specific programmable riboregulators (SNIPRs) are two types of translation activators that can detect nearly any nucleic acid sequence using interactions initiated by single-stranded domains known as toeholds. Toehold switches operate with high dynamic range, orthogonality, and programmability, making them capable of detecting a variety of pathogens in paper-based cell-free diagnostic assays. SNIPRs are designed to enable the accurate detection of single-nucleotide mutations, making them valuable tools for mutation and drug-resistance assays. Here we describe the computational design process for generating toehold switches and SNIPRs active against different pathogens and mutations of interest. Such riboregulators can be deployed in paper-based diagnostic assays to enable rapid and low-cost disease detection.

Key words RNA-based regulators, toehold switches, SNIPRs, riboregulators

1 Introduction

Riboregulators are RNA molecules that respond to nucleic acid signals through Watson-Crick base pairing to control gene expression. This capability has allowed riboregulators to be used for constructing genetic circuits in vivo and in vitro that alter gene expression in response to endogenous transcripts or synthetic exogenous RNAs [1, 2]. While earlier work on riboregulators focused on their applications in living cells [3–7], in recent years, there has been a growing interest in applying riboregulators in cell-free transcription/translation systems [8]. Without the constraints of the cellular chassis, cell-free systems provide unfettered access to the translational machinery and do not raise biosafety concerns. Moreover, these systems can remain active for extended periods at room temperature by freeze-drying them onto paper substrates [8],

James Chappell and Melissa K. Takahashi (eds.), *Riboregulator Design and Analysis*,
Methods in Molecular Biology, vol. 2518, https://doi.org/10.1007/978-1-0716-2421-0_2,

raising the exciting possibility of bringing synthetic biology tools outside the lab in a convenient portable format.

Within cell-free systems, riboregulators usually serve as sensors, activating gene expression once a target nucleic acid marking the presence of a pathogen or mutation is detected. Paper-based cell-free systems powered by riboregulators have been used for detecting the genetic signature of a variety of pathogens including the Ebola virus [8], Zika virus [9], norovirus [10], and SARS-CoV-2 [11], along with gut microbiota [12] and mutations associated with cancer and drug resistance [13]. Coupling these paper-based molecular assays with isothermal amplification reactions results in diagnostic tests that cost a few dollars each, do not require expensive equipment, and provide test results that can be read out by the eye. All these features make them promising technologies for use in global health and deployment in low-resource settings.

Designing riboregulators that respond effectively to pathogens and mutations can be a challenging process. Natural nucleic acid sequences often adopt thermodynamically stable secondary structures that frustrate sensor binding. Furthermore, the structure and translational efficiency of the riboregulator itself change with each target sequence, typically requiring computational screening of hundreds of potential riboregulators for each pathogen. In this chapter, we describe the main strategies that our laboratory employs for designing riboregulators capable of detecting specific pathogens and mutations. We focus in particular on two types of de novo designed toehold-mediated translation activators: toehold switches [14], which are optimized for general pathogen detection, and single-nucleotide-specific programmable riboregulators (SNIPRs) [15], which are optimized for mutation detection. Both riboregulators employ a toehold domain, a free single-stranded sticky end, to initiate binding with their cognate target RNA, enabling them to operate with favorable thermodynamics and reaction kinetics. The following sections will cover the design of these translation activators from selection of targets and specification of design parameters through to computer-assisted generation, analysis, and selection of candidate designs.

1.1 Toehold Switches

Toehold switch riboregulators provide high dynamic range, orthogonality, and programmability [14]. They were designed from first principles based on knowledge of thermodynamic principles and the influence of secondary structures on mRNA expression. These riboregulators contain two components: a trigger RNA and a switch RNA. The switch RNA contains hairpin-based core sensing unit (*see* Fig. 1), which includes the critical toehold domain, and the sequence of the gene to be regulated. In the absence of the trigger, the hairpin conceals the ribosomal binding site (RBS) and start codon within its loop and stem regions, respectively. The branch migration process that follows trigger binding changes the

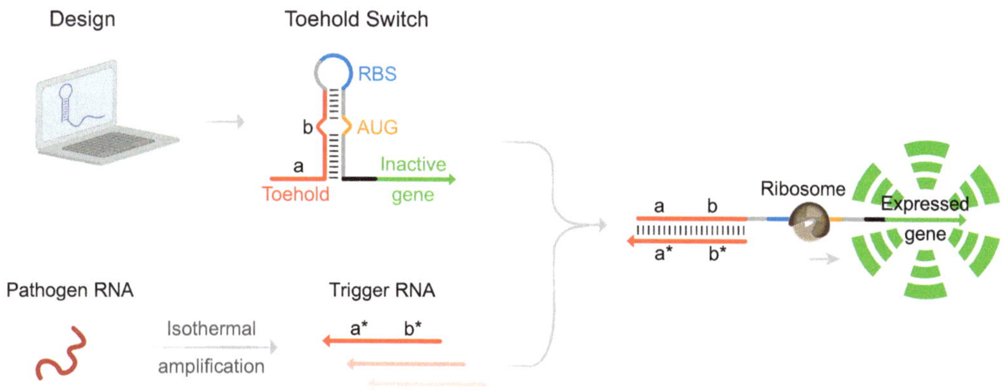

Fig. 1 Schematic of the toehold switch mechanism. In the presence of trigger RNA, the toehold domain initiates a branch migration and releases the RBS and start codon from a hairpin secondary structure to activate translation

conformation of the switch and thus exposes the RBS and the start codon to allow gene expression to occur.

Despite sharing a similar secondary structure and design principle, several different versions of toehold switches have been developed with slightly different parameters based on different optimization and selection strategies [9, 10, 14]. We will discuss the basic design principles and methods for two common versions of the toehold switch – Series A [14] and Series B [9] – and how these design principles interface with the design algorithms. These principles are generalizable to other toehold switches and riboregulator designs.

The first step of the design process involves selection of targets and specification of design parameters. The target of interest can be a fragment of a specific mRNA, rRNA, or viral RNA, preferably displaying weaker secondary structures. Even though the actual trigger binding region for toehold switches is around 30–40 nt, we recommend selecting a longer target (200–300 nt) as the input to allow the design algorithms to generate a large enough number of designs and to maximize the likelihood of getting a riboregulator with good performance.

According to the version of the toehold switch design selected, the specification of structure and sequence parameters varies and is listed as follows:

1. Series A toehold switches

As shown in Fig. 2a starting from the 5′ end of the hairpin, 5′ single-stranded toehold region is 30 nt, followed by a stem-loop structure, within which the lower and the upper stems are 9 and 6 bp, respectively, and are separated by a 3-nt bulge. The loop region is 11 nt. The trigger binding region is 36 nt, including the full toehold region and the first 6 nt in the lower stem. The upper

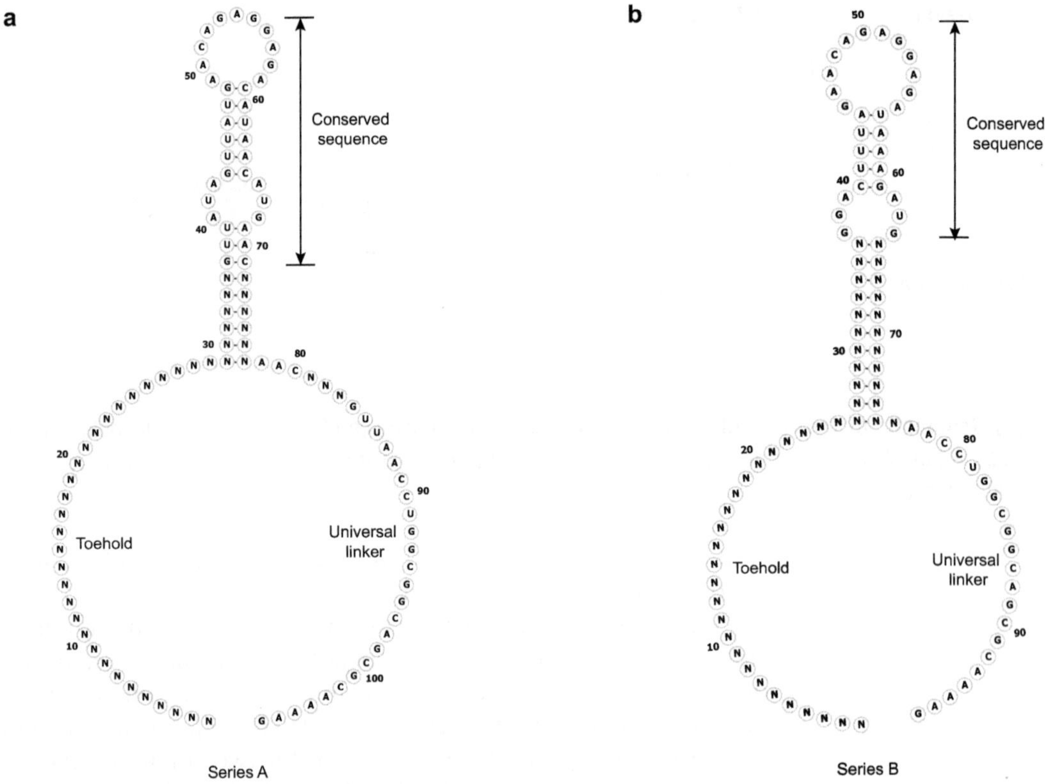

Fig. 2 Structural information for Series A and Series B toehold switches. (**a**) Structural schematic of the Series A toehold switches. (**b**) Structural schematic of the Series B toehold switches (N, undetermined nucleotide; A, adenine; G, guanine; C, cytosine; U, uracil). The diagrams were generated using Forna [17]

part of the hairpin (from position 37 to position 71) has the conserved sequence, GUUAUAGUUAUGAACAGAGGAGA CAUAACAUGAAC, which includes the RBS and the start codon sequence. A universal linker sequence is attached to the 3' end of the hairpin: AACCUGGCGGCAGCGCAAAAG.

2. Series B toehold switches

In Series B toehold switches (*see* Fig. 2b), the 5' single-stranded toehold region is 25 nt. The lower and the upper stems are 11 and 5 nt, respectively, and are separated by a 3-nt bulge. The loop region is 12 nt. The trigger binding region is also 36 nt, including the full toehold region and the 11-nt lower stem. The conserved sequence in the upper part of the hairpin 37 (from position 37 to position 64) is GGACUUUAGAACAGAGGAGAUAAAGAUG. A universal linker sequence that follows the hairpin is the same as the Series A toehold switches: AACCUGGCGGCAGCGCAAAAG.

We use the NUPACK software package [16] to assess all of the potential toehold switches for each target RNA computationally. The algorithm moves in 1-nt steps along the target RNA defining the switch RNA based on the sequence of its binding site in the target. For each candidate riboregulator, NUPACK is used to compute a variety of different parameters for the design, including the defect level of the switch hairpin, and the degree of single strandedness of the toehold, trigger, and activated switch. This process is completed in stage 1 of the design algorithm.

In stage 2, the designs for each target are analyzed to assign them scores based on a set of selection criteria. First, the variable stem region in the toehold switches can generate in-frame stop codons between the start codon and the reporter gene. Therefore, such designs are first discarded. Next, the design parameters calculated during stage 1 are combined via different weighting coefficients into a single numerical score to rank each design.

Taking Series A toehold switches as an example, the following defect levels are used to score the designs: [1] d_{sensor}, the defect for the toehold switch hairpin from 5′ toehold to the 3′ end of output gene, usually the first 29 nt; [2] $d_{toehold}$, the defect level of the toehold domain of the toehold switch hairpin compared to a fully single-stranded target structure; and [3] $d_{target\_site\_unpaired}$ predicts the availability of the binding site within the target, by calculating the probability that the bases in the binding site will be unpaired in the context of the full target transcript. Each of the defect values is first calculated and then normalized by dividing it by the maximal value in the set generated for a given target sequence.

The scoring equation with weighting coefficients is summarized as below, with lower scores corresponding to better expected performance:

$$S = 3d_{sensor} + 4d_{toehold} + 5d_{target\_site\_unpaired} \quad (1)$$

This scoring process allows us to estimate the performance of the toehold switches. However, empirical testing is still necessary to validate the performance of the toehold switches. The Series B toehold switches employ a four-parameter scoring function that was established based on experimental testing of devices for norovirus detection [10]. The Series B scoring function provides a predicted ON/OFF ratio for each candidate design.

The final stage of the algorithm selects top designs based on scores, makes final adjustments to the sequence to allow them to be ordered as single-stranded DNA oligos, and compiles them into a spreadsheet.

1.2 SNIPRs

SNIPRs are toehold-mediated translation activators designed to operate with high specificity and programmability [13]. Each SNIPR is composed of four functional elements (*see* Fig. 3): a docking site, an energy-balancing region, a translation-initiation region, and an output gene coding sequence. Translation from the SNIPR is repressed by positioning the RBS within a hairpin

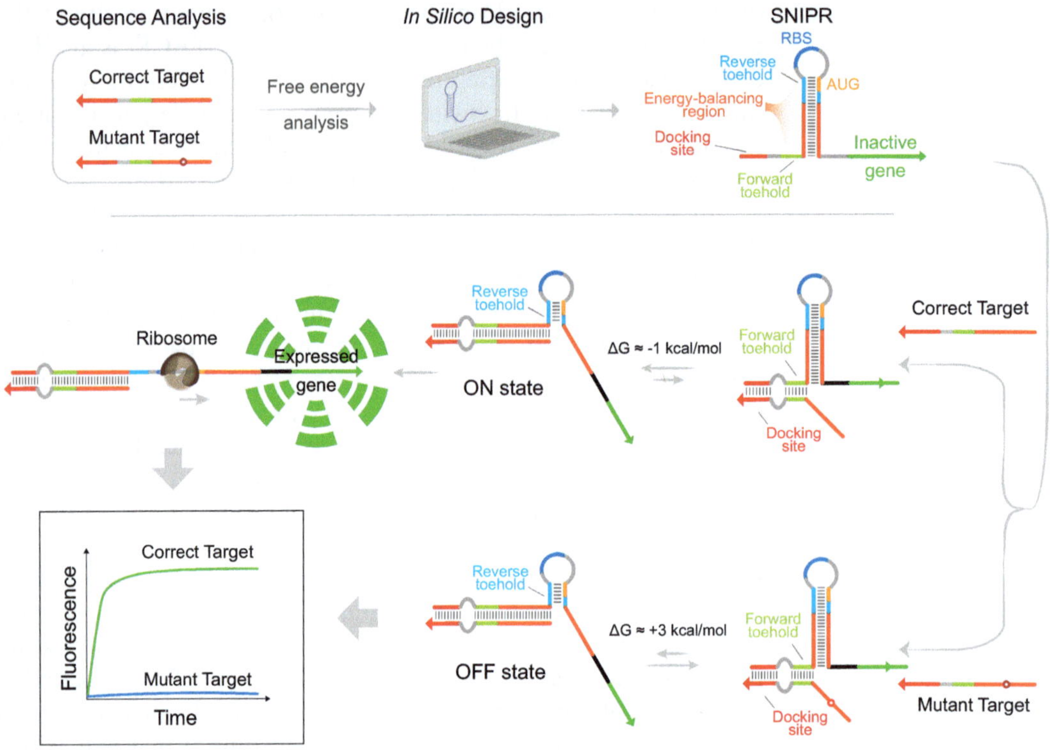

Fig. 3 Schematic of the SNIPR mechanism. The binding of trigger RNA with correct sequence yields an OFF state to ON state transition with a slightly negative free energy and hence releases the hairpin structure. A mutation in the trigger RNA results in an energy penalty leading to a positive reaction free energy, driving the SNIPR toward the OFF state

loop and the start codon (AUG) within the loop or a hairpin stem. The energy-balancing region is programmed to allow the SNIPR to operate near chemical equilibrium through a pair of forward and reverse toehold domains and a double-stranded branch-migration region. This construction enables SNIPRs to be exquisitely sensitive to small changes in target sequence that alter the reaction free energy and hence induce a shift in equilibrium.

In the presence of the trigger RNA, hybridization with a complementary single-stranded docking site drives efficient recruitment of the trigger to the SNIPR, placing it in close proximity with the energy-balancing region. The ensuing reaction between the forward toehold domain and trigger RNA leads to a branch-migration reaction that causes the SNIPR stem to unwind and reach its ON state. Importantly, this OFF-state to ON-state transition with the cognate trigger is designed to have a slightly negative free energy difference of −1 to −2 kcal/mol. On the other hand, the binding of a trigger RNA with any single-nucleotide mutation that causes a mismatch within the double-stranded branch-migration region generates a +3 kcal/mol energy penalty. This penalty leads to a positive free difference between the OFF state and ON state, thus preventing the mutated trigger from activating the SNIPR. In addition, in the case of a false activation by a mutant trigger RNA, the reverse toehold domain that is left undisturbed after a displacement reaction can initiate a reverse branch migration reaction to restore the OFF state secondary structure.

2 Materials

All computational design work for the toehold switches and SNIPRs was done using computers running MacOS or Linux. The in silico design algorithms are run using MATLAB (installed with the Bioinformatics Toolbox), and they make use of functions in the NUPACK software package version 3.2. MATLAB releases R2013a or later have been used with the software. No other materials are required.

The toehold switch design software can be found at https://github.com/AlexGreenLab/TSGEN.

The SNIPR design software can be found at https://github.com/Albert09111/SNIPR.

3 Methods

3.1 Installation of Software Packages

1. Install the NUPACK version 3.2 software package:

First, go to the http://www.nupack.org website and click "Downloads" to go to the software download page.

Second, register to download NUPACK.

Lastly, follow the NUPACK installation instructions to allow the software to run on your computer.

2. Set up the environment variables in MATLAB to enable it to call NUPACK functions (*see* **Note 1**):

First, open MATLAB.

Second, open or create a `startup.m` file in your MATLAB default user work folder. This folder is defined in the MATLAB variable `userpath` and `startup.m` automatically run each time you open MATLAB.

Third, in the `startup.m` file, set the `NUPACKINSTALL` variable to the location of your NUPACK installation. To do this, use the following line of code with the location of the package, where `[user_name]` is your username on the computer:

```
NUPACKINSTALL = '/Users/[user_name]/.../
   nupack3.2.2';
```

Lastly, add the lines of code below to set `NUPACKINSTALL` as an environment variable, and add the folder with NUPACK binaries to the `PATH`:

```
setenv('NUPACKINSTALL',NUPACKINSTALL);
setenv('PATH',[getenv('PATH'),sprintf(':%s/build/
   bin',NUPACKINSTALL)]);
```

The necessary environment variables will now be set whenever you start MATLAB.

3.2 Computational Design of Toehold Switches

The general design process is shown in Fig. 4. In brief, it can be divided into two sections, [1] selection of targets and [2] specification of design parameters and NUPACK-assisted computational design of toehold switches, which contains three stages: design generation, scoring, and sorting (*see* **Note 2**).

One of the advantages of toehold switches is their ability to target nearly any RNA sequence. The following process is often used to choose the potential toehold switch target regions that likely lead to good switch performance.

1. Determine the organism/species (e.g., viruses, bacteria, etc.) of interest, and identify the related DNA/RNA target sequence (*see* **Note 3**).
2. The targets can often be obtained from previous literature if available. Otherwise, they can be obtained by downloading sequences from different sequence databases (e.g., GenBank).

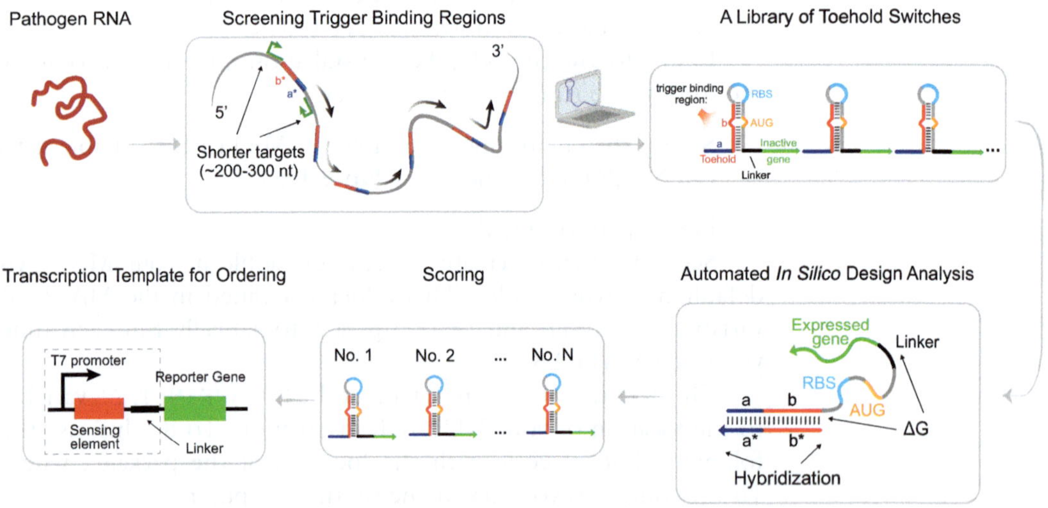

Fig. 4 Schematic of the computational design of toehold switches against user-specified target sequences. A variety of switches binding to different portions of the target (~200–300 nt) are designed and analyzed. The toehold switches are scored in silico, and top predicted devices are output in DNA form for ordering

3. Identify the regions that remain conserved from the downloaded sequences. Then compare the conserved sequences with the ones in closely related species to ensure that the sequences for the target pathogen are unique (e.g., for SARS-CoV-2, you would compare it to the other human coronaviruses). This information can be obtained by aligning the sequences using the Basic Local Alignment Search Tool (BLAST) (*see* **Note 4**).
4. For each conserved, unique sequence, generate a target input sequence that is approximately 200–300 nt in length, which is ideal for use with toehold switches (*see* **Note 5**).
5. The resulting target sequences can then be entered into the `design_input_file.csv` file in the toehold switch design software `input` folder.
6. Open the spreadsheet named `design_input_file.csv`, and specify the desired name of the target input under Column 1 as indicated in the spreadsheet.
7. Specify the target input RNA sequences in Columns 2 and 3:
   - If you have an existing set of amplification primers for your target input RNA, you can use Column 2 to specify the full expected amplicon sequence produced during amplification. Use Column 3 to specify the subsequence that does not include the sites where the amplicon binds to the amplification primers. This step ensures that the algorithm will not generate toehold switches that bind directly to one of the primers and the resulting assay interrogates three regions of the target rather than just two.
   - If you do not have amplification primers for your target sequence, you can use the same sequence in Columns 2 and 3. Alternatively, you can exclude a given number of bases on the 5′ and 3′ ends that correspond to the typical primer length of the amplification method that you plan to use.
8. Input the desired temperature in Column 4 for running the toehold switch reactions in °C (e.g., 37).
9. Input the name of the output gene in Column 5.
10. Input the sequence of the output gene in RNA format in Column 6 (*see* **Note 6**).
11. Enter more targets as needed in the rows below. Leave the rows below blank if there are no more targets.
12. Save the spreadsheet file.
13. Open MATLAB and navigate to the folder that contains the toehold switch design software.

14. Run the function:
    - toehold_switch_design_run(num_designs, input_file,options) using the parameter settings of your choice.
    - num_designs (default value = 6): Use this parameter to specify the number of designs that you want generated for each target sent to the software.
    - input_file (default value = design_input_file.csv'): Use this parameter to specify the location of the input file that provides the target sequences. The function will search for this file in an 'input' subfolder in the current working folder unless otherwise specified.
    - SeriesA (default value = 1): Defines if the function will generate Series A designs. Set to 0 to not produce these designs.
    - SeriesB (default value = 1): Defines if the function will generate Series B designs. Set to 0 to not produce these designs.
    - Parallel (default value = 1): Enables processing with multiple cores on the local computer. Set to 0 if you do not want parallelization enabled or do not have the parallel processing toolbox.
    - Antisense (default value = 0): Set to 1 if you wish to create toehold switch sensors for the antisense sequences of those provided in the input file.

For example, if you want to generate eight toehold switch designs for only Series A for targets specified in the my_sequences.csv' file in the input folder, you would specify

```
toehold_switch_design_run(8,'my_sequences.csv','SeriesB',0);
```

For the same set of targets, adding the antisense orientation and Series B designs, the function call would be

```
toehold_switch_design_run(8,'my_sequences.csv','Series-
B',1,'Antisense',1);
```

15. After the toehold switch design process is complete, information on the resulting designs will be saved in the form of csv format spreadsheets within the final_designs subfolder in your current MATLAB working folder. The files SeriesA_-sensor_DNA_sequences.csv and SeriesB_sensor_D-NA_sequences.csv will contain the DNA templates necessary to order for toehold switch construction. The files SeriesA_final_design_info.csv and

SeriesB_final_design_info.csv will contain information on the RNA sequences for each design (*see* **Note 7**).

16. The toehold switch hairpins and targets are usually ≤200 nt, so they can be ordered as individual DNA oligonucleotides. The hairpins are later assembled with the reporter gene to create fully functional toehold switches. Necessary primers should be ordered along with the hairpins for the subsequent PCR assembly process (*see* **Note 8**).

3.3 Computational Design of SNIPRs

The design process of SNIPRs (*see* Fig. 5) is also highly automated, and the software is available online for researchers at https://github.com/Albert09111/SNIPR. Follow the installation instructions provided with the SNIPR software for installation. Since the SNIPR software also uses MATLAB to call NUPACK binaries, you can also follow the instructions in Subheading 3.1 to set up the necessary packages.

1. The designs of SNIPRs require two inputs: wild-type and mutant sequences. SNIPRs are capable of distinguishing

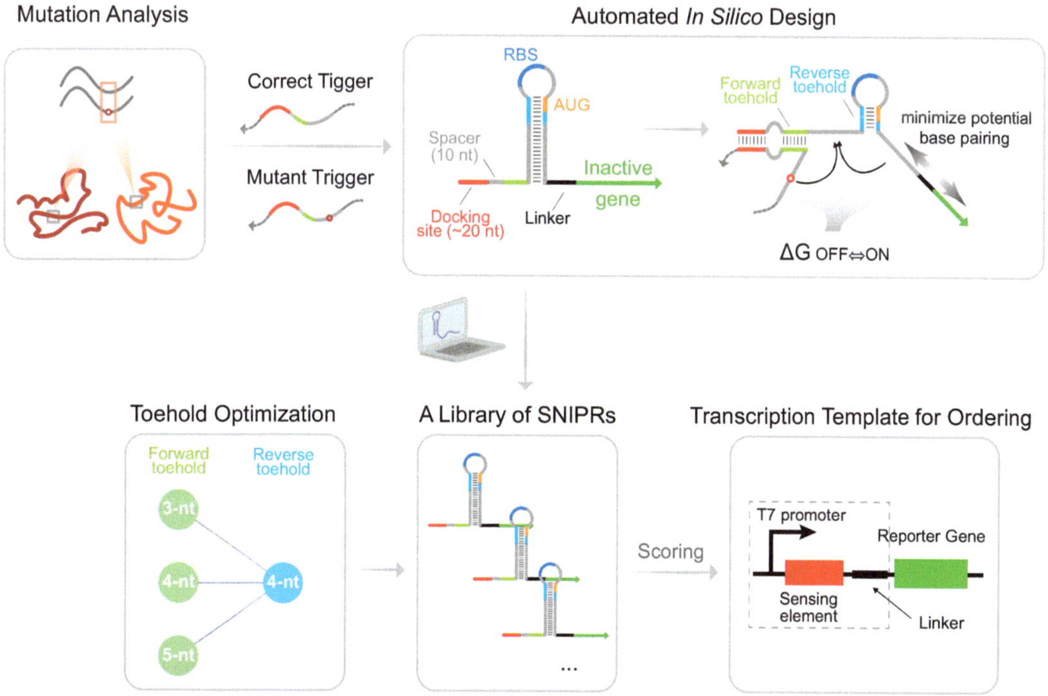

Fig. 5 Schematic of the computational design of SNIPRs. The position of the mutation to be detected is identified and positioned to coincide with the middle portion of the SNIPR stem. Three different sets of designs containing different pairs of forward (3, 4, and 5 nt) and reverse toeholds (4 nt) are generated in silico. The sequence of SNIPRs is analyzed and scored based on the free energy difference between the ON state and OFF state, the ensemble defect level, and the secondary structure of the active riboregulator

mutations with only a single-nucleotide difference between wild-type and mutant sequences.

2. Similarly, the easiest way to find the targets is to look for them in existing literature. Otherwise, they can be selected from sequence databases (e.g., GenBank). The ideal sequences should be highly conserved, except for the mutation of interest.
3. To allow the sequence to be used as the input, tailor the sequence, so the mutation position in the target is at least 15 bases away from the 5′ end of the RNA. Sequences should be specified in the 5′–3′ direction.
4. Open `target_input.csv` located in the input folder within the SNIPR master folder.
5. Specify the name of the output SNIPR sequence in Column 1.
6. Specify the wild-type and the mutated sequence in Column 2 and Column 3, respectively (*see* **Note 9**).
7. Input the first 29 nt of the output gene sequence (e.g., *GFP*, *lacZ*) in Column 4 (*see* **Note 10**).
8. Save and close the `target_input.csv` file.
9. Open the main code `SNIPR.m` in MATLAB.
10. The code contains several parameters with default values, but they can be customized. Before running the code, review the following parameters and edit as needed:
    - `library_num`: this parameter determines the number of the design libraries the code will generate (*see* **Note 11**).
    - `IS_PARALLEL`: set to 1 to enable parallel computing by harnessing multiple computer cores or 0 to disable parallel computing.
    - `select_num`: this parameter determines the final number of top designs generated and selected from the libraries (*see* **Note 12**).
    - `SNP_TARGET`: this parameter determines whether the wild-type sequence or the mutant sequence will activate the reporter gene expression. Set to 1 to allow mutant targets to activate gene expression; set to 0 to allow wild-type targets to turn on gene expression.
11. Save the parameters.
12. Run the `SNIPR.m` code to start generating SNIPR design for the specified targets.
13. The SNIPR code generates the designs automatically in four stages: [1] load sequence and generate design script, [2] run design script to generate sequence library, [3] extract the design library, and [4] analyze and screen the best design.

The progress of the algorithm can be visualized in real time on the MATLAB command window.

14. After the full running process is completed, find the SNIPR RNA sequence for each target along with other parameters in the `output` folder.
15. To order SNIPRs as DNA oligos, convert the SNIPR RNA to DNA sequence, and append the T7 promoter sequence to the 5' end (*see* **Note 13**). Necessary primer sequences should also be ordered to assemble the SNIPR hairpin with the corresponding gene.

4 Notes

1. It is also possible to link MATLAB to the NUPACK functions by modifying the bash configuration file from the command line terminal program of your choice. This procedure is described in the SNIPR software installation guide, which can be found at https://github.com/Albert09111/SNIPR. If you do choose this method, MATLAB must be started from the command line to set the necessary environment variables defined in the configuration file.
2. The toehold switch design code is available at https://github.com/AlexGreenLab/TSGEN.
3. Toehold switches have mainly been used for detection of RNA, but it is possible to detect DNA sequences. Designs may need to be adjusted for DNA detection based on the weaker binding for DNA/RNA hybridization compared to RNA/RNA hybridization.
4. We recommend selecting at least two regions from the pathogen DNA/RNA to increase the likelihood of getting a functional sensor. In addition, sense and antisense orientations can generally be detected by the toehold switch depending on whether the forward or reverse primer contains a T7 promoter.
5. To ensure the algorithms generate a sufficient number of designs in the pool, we recommend using targets that are at least 70 nt. While a longer target is generally better, it is not necessary for the initial screening process and increases the difficulty of synthesizing the target. We recommend using targets that are 200–300 nt as inputs to generate toehold switch design since targets of this length also provide many options for suitable amplification primers.
6. Only up to the first 29 nt of the output gene are considered in the algorithm in order to reduce computation times.

7. The toehold switch design algorithm also outputs to different subfolder spreadsheets containing the individual parameter values (e.g., switch defect level, toehold defect level, etc.) of all of the designs generated for each target sequence. You can use these spreadsheets to implement your own scoring functions for selecting higher-performance designs.
8. We recommend ordering and testing at least six top toehold switch designs generated from the design software to increase the likelihood of getting a toehold switch with very good performance.
9. The wild-type sequence and mutant sequence do not have to be in the same length. The algorithm will identify the mutation and proceed with designing as long as the mutation is position at least 15 nt away from the 5′ end of the target sequence.
10. Output gene sequence will be considered in the SNIPR design analysis.
11. We recommend setting the library numbers to at least ten to allow sufficient numbers of designs to be generated and screened.
12. We recommend setting this number to at least five. In experimental screening, at least five top-ranked designs should be tested to increase the likelihood of getting a functional design.
13. The 5′ GGG sequence is universally appended to all SNIPR RNA sequences generated, which represents the first three nucleotides transcribed using T7 RNA polymerases. The T7 promoter sequence we use is GCGCTAATACGACTCACTATAGGG. Alternatively, other RNA polymerases can be used. We recommend ordering and testing at least six top SNIPR designs generated from the design software to increase the likelihood of getting a SNIPR with very good performance.

References

1. Chappell J, Watters KE, Takahashi MK, Lucks JB (2015) A renaissance in RNA synthetic biology: new mechanisms, applications and tools for the future. Curr Opin Chem Biol 28:47–56. https://doi.org/10.1016/J.CBPA.2015.05.018
2. Green AA (2019) Synthetic bionanotechnology: synthetic biology finds a toehold in nanotechnology. Emerg Top Life Sci 3:507–516. https://doi.org/10.1042/ETLS20190100
3. Isaacs FJ, Dwyer DJ, Ding C et al (2004) Engineered riboregulators enable post-transcriptional control of gene expression. Nat Biotechnol 22:841–847. https://doi.org/10.1038/nbt986
4. Lucks JB, Qi L, Mutalik VK et al (2011) Versatile RNA-sensing transcriptional regulators for engineering genetic networks. Proc Natl Acad Sci 108:8617–8622. https://doi.org/10.1073/PNAS.1015741108
5. Callura JM, Cantor CR, Collins JJ (2012) Genetic switchboard for synthetic biology applications. Proc Natl Acad Sci 109:5850–5855. https://doi.org/10.1073/PNAS.1203808109

6. Callura JM, Dwyer DJ, Isaacs FJ et al (2010) Tracking, tuning, and terminating microbial physiology using synthetic riboregulators. Proc Natl Acad Sci 107:15898–15903. https://doi.org/10.1073/PNAS.1009747107
7. Chappell J, Takahashi MK, Lucks JB (2015) Creating small transcription activating RNAs. Nat Chem Biol 11:214–220. https://doi.org/10.1038/nchembio.1737
8. Pardee K, Green AA, Ferrante T et al (2014) Paper-based synthetic gene networks. Cell 159:940–954. https://doi.org/10.1016/j.cell.2014.10.004
9. Pardee K, Green AA, Takahashi MK et al (2016) Rapid, low-cost detection of Zika virus using programmable biomolecular components. Cell 165:1255–1266. https://doi.org/10.1016/j.cell.2016.04.059
10. Ma D, Shen L, Wu K et al (2018) Low-cost detection of norovirus using paper-based cell-free systems and synbody-based viral enrichment. Synth Biol 3:ysy018. https://doi.org/10.1093/SYNBIO/YSY018
11. Amalfitano E, Karlikow M, Norouzi M et al (2021) A glucose meter interface for point-of-care gene circuit-based diagnostics. Nat Commun 12:1–10. https://doi.org/10.1038/s41467-020-20639-6
12. Takahashi MK, Tan X, Dy AJ et al (2018) A low-cost paper-based synthetic biology platform for analyzing gut microbiota and host biomarkers. Nat Commun 9:1–12. https://doi.org/10.1038/s41467-018-05864-4
13. Hong F, Ma D, Wu K et al (2020) Precise and programmable detection of mutations using ultraspecific riboregulators. Cell 180:1018–1032.e16. https://doi.org/10.1016/j.cell.2020.02.011
14. Green AA, Silver PA, Collins JJ, Yin P (2014) Toehold switches: de-novo-designed regulators of gene expression. Cell 159:925–939. https://doi.org/10.1016/j.cell.2014.10.002
15. Hong F, Ma D, Wu K et al (2020) Precise and programmable detection of mutations using ultraspecific riboregulators. Cell 180:1018–1032.e16. https://doi.org/10.1016/j.cell.2020.02.011
16. Zadeh JN, Wolfe BR, Pierce NA (2011) Nucleic acid sequence design via efficient ensemble defect optimization. J Comput Chem 32:439–452. https://doi.org/10.1002/JCC.21633
17. Kerpedjiev P, Hammer S, Hofacker IL (2015) Forna (force-directed RNA): simple and effective online RNA secondary structure diagrams. Bioinformatics 31:3377–3379. https://doi.org/10.1093/BIOINFORMATICS/BTV372

Chapter 3

Design of RNA-Based Translational Repressors

Seongho Hong, Dongwon Park, Soma Chaudhary, Griffin McCutcheon, Alexander A. Green, and Jongmin Kim

Abstract

The toehold switch is an RNA-based riboregulator that activates translation in response to a cognate trigger RNA and provides high ON/OFF ratios, excellent orthogonality, and logic capabilities. Riboregulators that provide the inverse function – turning off translation in response to a trigger RNA – are also versatile tools for sensing and efficiently implementing logic gates such as NAND or NOR. Toehold and three-way junction (3WJ) repressors are two de novo designed translational repressors devised to provide NOT functions with an easily programmable and intuitive structural design. Toehold and 3WJ repressors repress translation upon binding to cognate trigger RNAs by forming strong hairpin and three-way junction structures, respectively. These two translational repressors can be incorporated into multi-input NAND and NOR gates. This chapter provides methods for designing these translational repressors and protocols for in vivo characterization in *E. coli*.

Key words RNA-based regulator, Toehold switch, Toehold repressor, 3WJ repressor, Ribocomputing

1 Introduction

Synthetic biology is a nascent field aimed at designing biological systems with novel functionality at the genetic level [1, 2]. A wide range of synthetic circuits have been designed including bistable switches, oscillators, logic gates, and memory devices [3–5]. Historically, many protein regulators, including transcription factors or enzymes, have been used to construct genetic circuits, but the narrow dynamic range and unwanted cross talk displayed by many components has hindered the task of implementing more complex circuits [7]. In this respect, nucleic acid-based regulatory circuits provide an alternative to constructing genetic circuits with design flexibility and high programmability due to predictable Watson-Crick base pairing rules. In particular, RNA-based regulators have the advantage of requiring much smaller encoding space compared to protein regulators, which helps to overcome the limited ability to carry out genetic programs in a single cell [8–10].

James Chappell and Melissa K. Takahashi (eds.), *Riboregulator Design and Analysis*,
Methods in Molecular Biology, vol. 2518, https://doi.org/10.1007/978-1-0716-2421-0_3,

A representative example of an RNA-based regulator is the toehold switch, which controls gene expression at the translational level [6]. The toehold switch inhibits translation initiation by hiding the ribosome binding site (RBS) and start codon (AUG) inside the RNA hairpin stem-loop structure, precluding access of ribosomes (*see* Fig. 1a). In the presence of the cognate trigger RNA, the toehold switch can unwind the stem-loop structure as the trigger RNA binds to the toehold region and branch migration occurs. Through this, the RBS and AUG that were hidden in the stem loop can be exposed, allowing access to ribosomes and translation initiation. The toehold switch can be implemented in NOT logic gates through direct hybridization of a third species, the deactivating RNA, to a trigger RNA to silence the trigger and prevent trigger-switch binding. Alternatively, if the trigger-switch complex has formed, extended single-stranded domains of the trigger RNA also provide a binding region for the deactivating RNA, leading to the separation of the trigger RNA from the switch RNA.

We reported that the programmable toehold switch design can be implemented as a multi-input AND and OR logic processing ribocomputing devices with good performance [11]. However, NOT logic gates composed of only two strands, a trigger and a switch RNA, enable versatile logic gates like NAND or NOR to be implemented more intuitively and efficiently. Thus, we developed two de novo designed translational repressor models: the toehold repressor and the three-way junction (3WJ) repressor [12]. The toehold repressor has a strong hairpin structure and a downstream exposed RBS and start codon that enables translation to be initiated (*see* Fig. 1b). In the presence of the trigger RNA, the hairpin structure is released, and a secondary hairpin is formed downstream, isolating the RBS and start codon and turning off translation. On the other hand, the 3WJ repressor has an unstable hairpin structure, which makes it easy to unwind the stem-loop and exposes RBS and AUG to enable downstream gene translation (*see* Fig. 1c). In the presence of the 3WJ trigger RNA, the trigger RNA binds to a pair of single-stranded domains on either side of the hairpin, forming a three-way junction structure to inhibit translation.

These two translational repressors can be used for implementing NAND and NOR logic gates through RNA-RNA interactions alone. To implement the NAND gate with toehold repressors, two different triggers were designed to hybridize to each other through complementary bridging domains to unwind the repressor hairpin structure (*see* Fig. 2a, b). In the case of 3WJ repressor, two orthogonal repressor RNAs are connected in a single RNA transcript such that translation can be inhibited only when both input triggers are present (*see* Fig. 2c, d). The NOR gate was implemented by including multiple sequestered trigger RNAs upstream of the

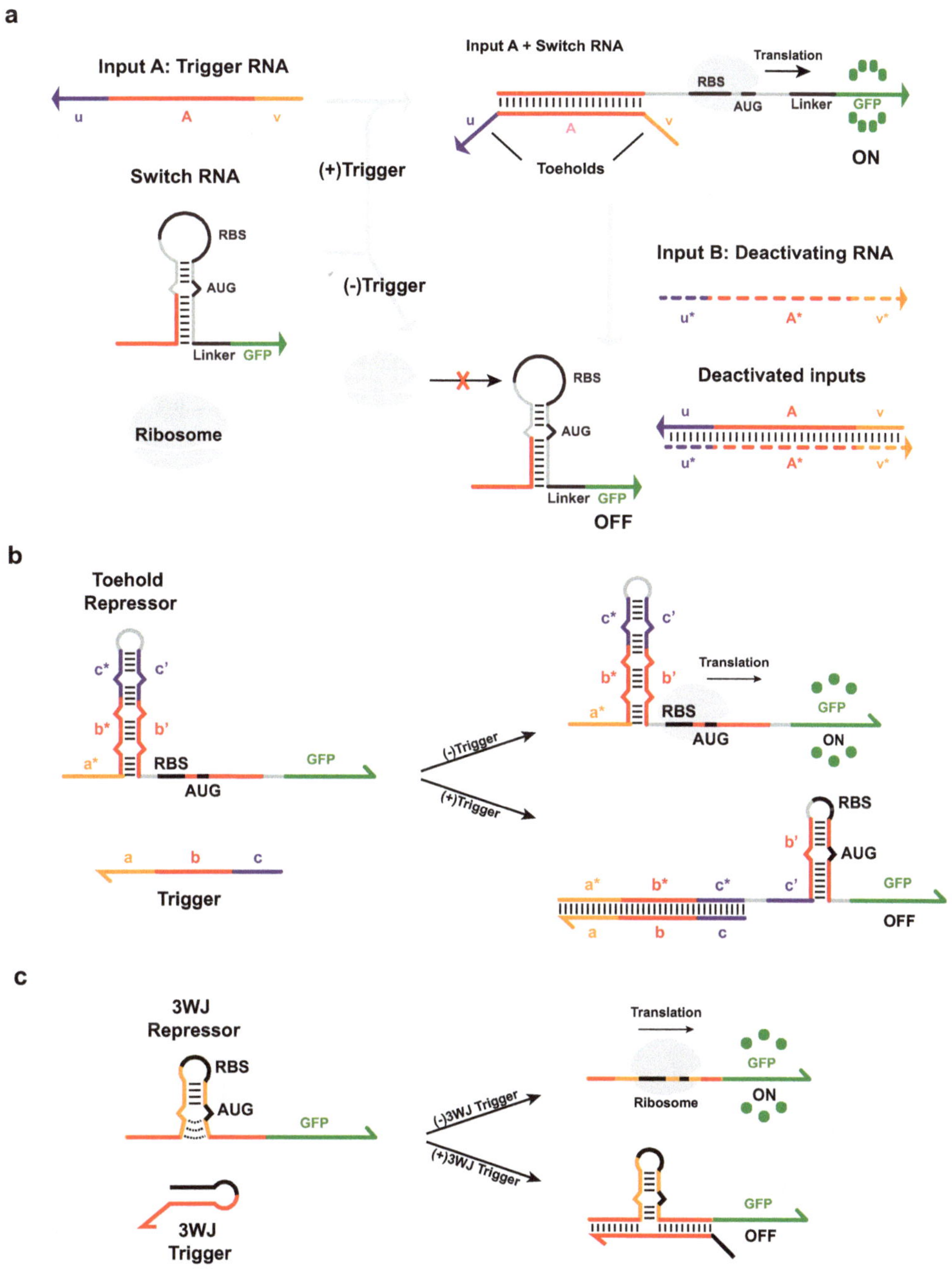

Fig. 1 De novo designed synthetic RNA translational regulators. (**a**) Schematic of the toehold switch mechanism. The ribosomal binding site (RBS) and start codon (AUG) of the switch RNA are exposed only when the cognate trigger RNA (input A) disrupts the secondary structure. A AND (NOT B) logic can be implemented by a deactivating RNA (input B), which induces strand displacement to abolish trigger RNA (input A) activity. (**b**) Schematic of the toehold repressor mechanism. In the presence of the trigger RNA, the hairpin structure is released, and a secondary hairpin is formed, isolating the RBS and start codon. (**c**) Schematic of the 3WJ repressor mechanism. In the presence of the 3WJ trigger RNA, the trigger RNA binds to the domains around the bottom of repressor hairpin, forming a three-way junction structure that inhibits translation

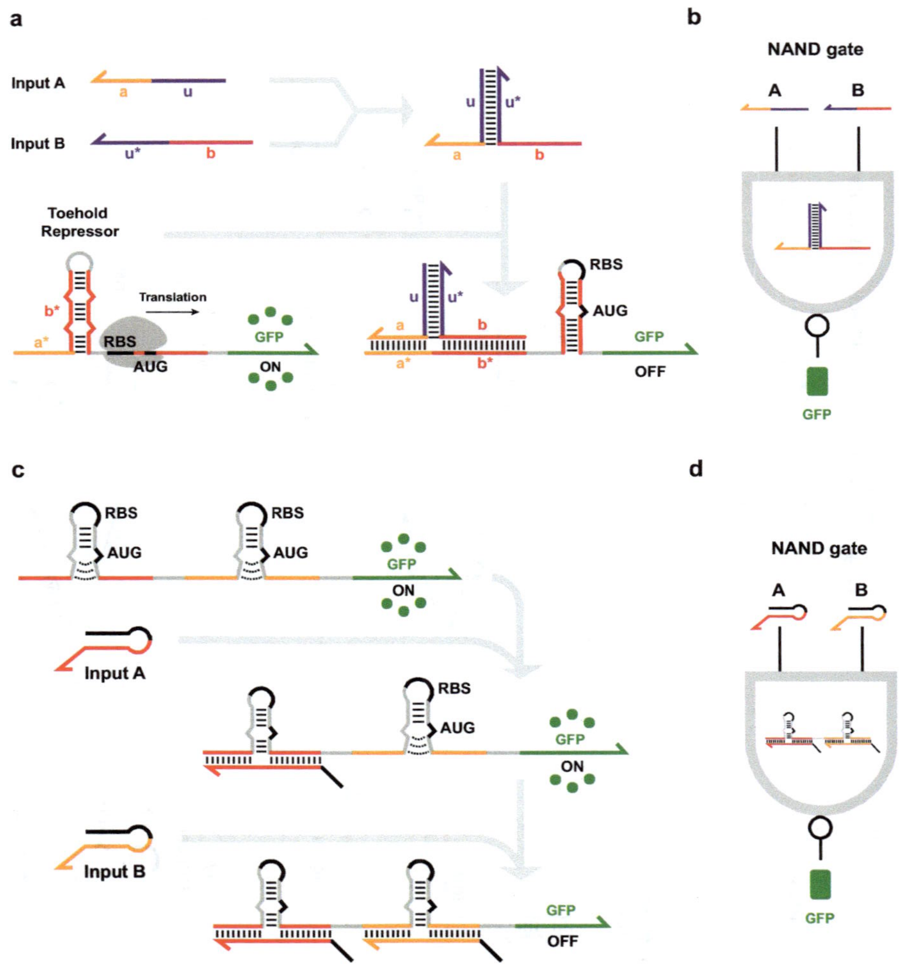

Fig. 2 Two-input NAND logic gates using toehold and 3WJ repressor. (**a**, **b**) Design of a toehold repressor NAND gate where two input RNAs hybridize to each other to unwind the repressor hairpin. The downstream gene is repressed only if both input A and input B are present. (**c**, **d**) Design of a 3WJ repressor NAND gate. In the gate RNA, two orthogonal repressor RNAs are inserted in-frame and upstream of the reporter gene so that translation can be inhibited only when both input triggers are present

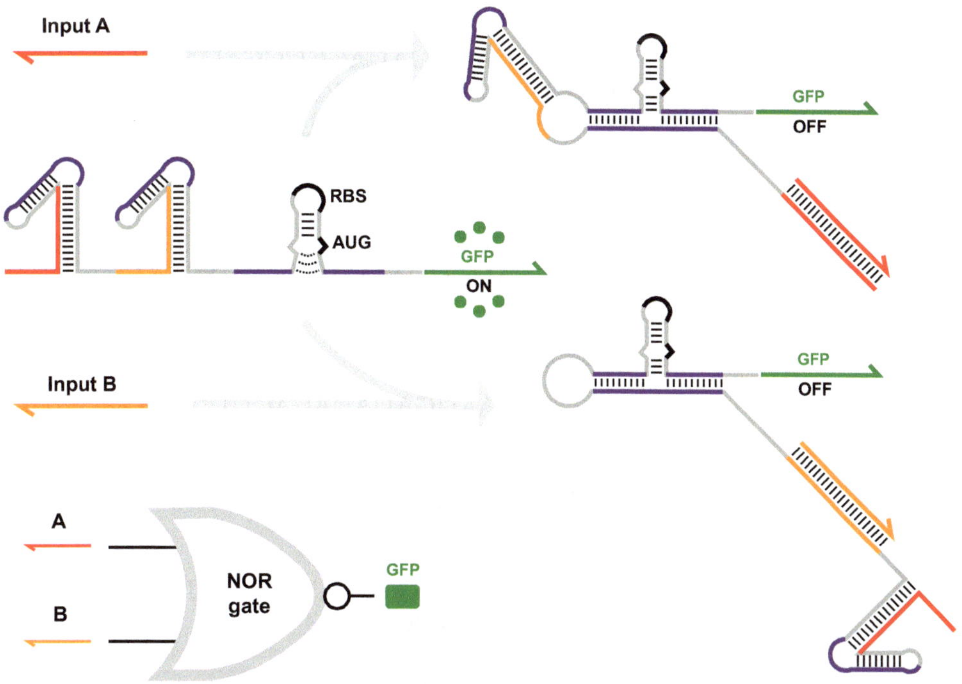

Fig. 3 NOR logic gate implemented by 3WJ repressor. The gate RNA contains a repressor module that regulates the output gene and two sequestered trigger RNAs upstream of the repressor. Binding of either input RNA causes the corresponding hairpin to unwind, inducing trigger binding to the 3WJ repressor module and inhibiting translation

3WJ repressor to induce intramolecular interactions only in the presence of the input RNAs (*see* Fig. 3). These NAND and NOR gates could be extended to multi-input ribocomputing devices that recognize up to four independent input RNAs.

This chapter briefly introduces how to design toehold and 3WJ repressors, how to clone them, and how to identify their repressor activity in vivo by measuring GFP fluorescence.

2 Materials

Prepare all solutions using distilled water (DW) and analytical grade reagents. Prepare and store all reagents at room temperature unless indicated otherwise.

2.1 Toehold Repressor and Three-Way Junction (3WJ) Repressor

1. An example of toehold repressor is as follows:

   Switch sequence:

   5′- GGGAAAGUGAAAGUAAGAAUAAAGUGGAAAGAGA CGAGAAUAAGAAGAGACAUCAUAUCUUCUUAUU CUCGUCUCUUUCUACUUUAUAAAC **AGAGGA-GAAUAAAGAUG**AAAGAGACGAACCUGGCGGCAG CGCAAAAGAUGCGUAAA-3′ (*see* **Note 1**)

Trigger sequence:

5′-GGGCGAUAGCUAUACCUGGAUAGCUAUCGCAUU CUUCUUAUUCUCGUCUCUUUCCACUUUAUUC UUACUUUCACUUUACU-3′

Decoy sequence:

5′-GGGUCUCACGCCCUCAGCUGGGCGUGAGAUGAG CCUCGUCUCCAGAUGACGAGGCAACGUAGGAUC UGACUGAUCCUACUAU-3′ (*see* **Note 2**)

2. An example of 3WJ repressor

   Switch sequence:

   5′- GGGAUGAAUGAUAUACACUUGUUAUAGUUAUG AAC**AGAGGAGA**CAUAAC**AUG**AACAAGCACGAAU UGACUACACUAAACCUGGCGGCAGCGCAAAAGA UGCGUAAA-3′

   Trigger sequence:

   5′- GGGACGAAUUGAUUUGUCAAUUCGUGCGUGUA UAUCAUUCAUCAU-3′

   Decoy sequence:

   5′-GGGUCUCACGCCCUCAGCUGGGCGUGAGAUGAG CCUCGUCUCCAGAUGACGAGGCAACGUAGGAUC UGACUGAUCCUACUAU-3′

2.2 Suitable Vectors

1. Switch RNA expressed from a relatively low copy vector is suitable for proper switch RNA-trigger RNA interaction (*see* **Note 3**).
2. Trigger RNA is expressed from a relatively high copy vector (*see* **Note 3**).

2.3 Cloning

1. 14 mL round bottom tube.
2. Plasmid miniprep kit.
3. 2× Pfu PCR master mix, store at −20 °C.
4. Thermal cycler.
5. 1% agarose gel: agarose, 1× TAE buffer.
6. 1 kb DNA ladder.
7. SYBR™ Safe DNA Gel Stain.
8. Gel imaging system.
9. 20,000 units/mL DpnI, store at 20 °C.
10. 1× rCutsmart™ buffer: 50 mM potassium acetate, 20 mM Tris-acetate, 10 mM magnesium acetate, 100 μg/mL recombinant albumin, pH 7.9, store at −20 °C.
11. Gibson assembly master mix, store at −20 °C.
12. 10 units/μL T4 polynucleotide kinase (T4 PNK), store at −20 °C.

13. Type IIS restriction enzyme (BsaI-HFv2, BsmBI-v2), store at −20 °C.
14. 350 units/μL T4 DNA ligase, store at −20 °C.
15. 10× T4 DNA ligase buffer: 660 mM Tris-HCl, pH 7.6, 66 mM $MgCl_2$, 100 mM DTT, 1 mM ATP, store at −20 °C.

2.4 Chemically Competent E. coli

1. Inoue transformation buffer: 10 mM pipes, pH 6.7, 55 mM $MnCl_2$, 15 mM $CaCl_2$, 250 mM KCl.
2. Dimethyl sulfoxide (DMSO).
3. Liquid nitrogen.

2.5 Strains and Growth Condition

1. BL21 DE3 (F− ompT hsdSB (rB−mB−) gal dcm (DE3)).
2. BL21 Star DE3 (F− ompT hsdSB (rB−mB−) gal dcm rne131 (DE3)) (*see* **Note 4**).
3. BL21-AI™ (F-ompT hsdSB (rB- mB-) gal dcm araB:: T7RNAP-tetA) (*see* **Note 5**).
4. DH5α (endA1 recA1 gyrA96 thi-1 glnV44 relA1 hsdR17 (rK−mK+) λ−).
5. Lysogeny Broth (LB) medium: 5 g/L yeast extract, 10 g/L tryptone, 10 g/L NaCl.
6. 30 mg/mL kanamycin in DW, filter sterilize and store 1 mL aliquots at −20 °C.
7. 50 mg/mL ampicillin in DW, filter sterilize and store 1 mL aliquots at −20 °C.
8. 25 mg/mL spectinomycin in DW, filter sterilize and store 1 mL aliquots at −20 °C.
9. 1 M isopropyl β-D-1-thiogalactopyranoside (IPTG) in DW, filter sterilize and store 1 mL aliquots at −20 °C.
10. 20% w/v arabinose in DW, filter sterilize and store 1 mL aliquots at −20 °C.
11. 96-well deep plate.
12. Breathable sealing film.

2.6 Plate Reader

1. 96-well black/clear plate.
2. Microtiter plate reader with monochromator or lasers and filters for simultaneous measurement of green, red, yellow, and cyan fluorescent proteins.

2.7 Flow Cytometry

1. Flow cytometer with appropriate lasers and filters for simultaneous measurement of green, red, yellow, and cyan fluorescent proteins.
2. 1× phosphate buffered saline (PBS).
3. Paraformaldehyde powder.

3 Methods

3.1 Design of RNA Translational Repressors

To generate de novo RNA translational repressors, use the NUPACK sequence design package (http://nupack.org/design/new) [13].

1. Enter the design code, select an appropriate RNA energy parameter, and write structural information and sequence information to NUPACK (*see* **Note 6**, Figs. 4 and 5).
2. Click the update button.
3. Check that the structure generated within the preview window is the correct one.
4. Click the design button.
5. Choose candidates by considering normalized ensemble defects and ensuring that they lack in-frame stop codons downstream of the translation start site.
6. Use NUPACK's analysis function to examine whether candidates form the intended structures (*see* **Note 7**).

3.2 Cloning of RNA Translational Repressor Expression Vectors

1. Choose an optimal cloning method based on the following recommendations. If your sequence has a strong structure (like a toehold repressor), or repeating sequences, choose CRATES [14] as described in **steps 2–6**. If you change a small part (few bases deletion, insertion, substitution) of existing sample, choose site-directed plasmid mutagenesis protocol [15] as described in **steps 7–10**. Otherwise, choose Gibson assembly [16] as described in **steps 11–15**.
2. For easy colony selection, prepare a backbone plasmid containing a constitutively expressed reporter gene cassette with each end having type IIS restriction sites. Determine the concentration of plasmid by measuring the absorption at 260 nm according to Beer-Lambert's law using sequence-specific extinction coefficients.
3. Design DNA blocks using compatible 4 nt junctions. For detailed instructions, see [14].
4. Phosphorylate the single stranded DNA oligo: Combine 2 μL of 20 nM single-stranded DNA oligo with 5 μL of 10× T4 DNA ligase buffer, 1 μL of T4 PNK, and 42 μL of DW to a total volume of 50 μL. Incubate for 30 min at 37 °C and heat inactivate by incubating at 65 °C for 20 min.
5. Anneal the single-stranded DNA oligo to double-stranded DNA block with 4 nt junctions: add 25 μL of phosphorylated (+) strand oligo to 25 μL of phosphorylated (−) strand oligo. Anneal in a PCR machine by heating up to 95 °C and cooling down to 15 °C at rate of about −1 °C/min.

```
# Sample NUPACK code for a toehold repressor

# specify RNA energy parameters, values = rna1995 or rna1999
# number of designs, values = 1 to 10
material = rna1999
temperature = 37.0
trials = 10

# The GGG sequence is only required when using T7 promoter. If there is
# a required sequence for other promoters, design with that sequence.
domain preG = GGG
domain a = N15
domain b = N18
domain c = N12

# cp_star domain specifies the top right portion of the switch stem.
# It is not fully complementary to domain c.
# bp_star domain specifies the bottom right portion of the switch
# stem. It is not fully complementary to domain b.
domain cp_star = N12
domain bp_star = N18

# The first 6 nt in this bpp domain specifies the 6-nt spacer between
# RBS and AUG that is required for efficient translation. This domain is
# not fully complementary to bp_star due to the start codon bulge
domain bpp = N6 AUG N9
domain loop1 = N9

# AGAGGAGA sequence is the RBS sequence to allow efficient translation.
domain loop2 = N4 AGAGGAGA

# Common linker sequence from Green et al.[1]
domain linker = AACCUGGCGGCAGCGCAAAAG

# target switch structure
structure Tr = U3 U45
structure Sw = U3 U15 D7 (U1 D7 (U1 D7 (U1 D6 (U9) U1) U1) U1) U12 U18
        U21
structure Sw_Tr = U3 D45 (U9 U12 D9 (U3 D6 U12 U3) U21 + U3)

Tr.seq = preG c* b* a*
Sw.seq = preG a b c loop1 cp_star bp_star loop2 bpp linker
Sw_Tr.seq = preG a b c loop1 cp_star bp_star loop2 bpp linker preG c* b*
        a*

# prevent sequence patterns
prevent = AAAA, CCCC, GGGG, UUUU, KKKKKK, MMMMMM, RRRRRR, SSSSSS, WWWWWW,
        YYYYYY
```

Fig. 4 Example of NUPACK code for toehold repressor

```
# Sample NUPACK Code for a 3WJ repressor

# specify RNA energy parameters, values = rna1995 or rna1999
# number of designs, values = 1 to 10
material = rna1999
temperature = 37.0
trials = 10

# target structures
structure Sw = U3 U15 U5 U3 D6 U11 U3 U5 U12 U7 U21
structure Tr = U3 D8 U6 U4 U13
structure Sw_Tr = U3 D15 (D5 (U3 D6 U11 U3) D12 (U7 U21 + U3 U12))

# sequence domains
domain preG = GGG
domain targ_stem = N8
domain targ_loop = N6
domain toehold = N15
domain stem1 = UUGUU
domain stem2 = AUA
domain stem3 = GUUAUG

# AGAGGAGA sequence is the RBS sequence to allow efficient translation.
# bases added after binding site to ensure the 3WJ repressor and gene
# are in the same reading frame
domain loop = AACAGAGGAGA
domain stem2_star = AUG
domain post_hairpin1 = N2
domain post_hairpin2 = N10
domain post_hairpin3 = N7

# Common linker sequence from Green et al.[6]
domain linker = AACCUGGCGGCAGCGCAAAAG

Sw.seq = preG toehold stem1 stem2 stem3 loop stem3* stem2_star stem1*
        post_hairpin1 post_hairpin2 post_hairpin3 linker
Tr.seq = preG targ_stem targ_loop targ_stem* post_hairpin1* toehold*
Sw_Tr.seq = preG toehold stem1 stem2 stem3 loop stem3* stem2_star stem1*
        post_hairpin1 post_hairpin2 post_hairpin3 linker preG targ_stem
        targ_loop targ_stem* post_hairpin1* toehold*

# prevent sequence patterns
prevent = AAAA, CCCC, GGGG, UUUU, KKKKKKK, MMMMMM, RRRRRR, SSSSSS,
        WWWWWW, YYYYYY
```

Fig. 5 Example of NUPACK code for 3WJ repressor

6. Digest backbone plasmid and assemble with double-stranded DNA blocks in one-pot reaction: mix 20 fmol backbone plasmid, 400 fmol each double stranded DNA blocks, 2 μL of T4 DNA ligase buffer, 2000 units of T4 DNA ligase, 10 units of type IIS restriction enzyme, and DW to a total of 20 μL. Incubate for 2 min at 42 °C and 5 min at 16 °C for 25 cycles. Incubate for 10 min at 60 °C and heat inactivate by incubating at 80 °C for 10 min.
7. Design forward and reverse primers (*see* **Note 8**).
8. Polymerase chain reaction: Combine 10 μL of Pfu master mix with 2 μL of forward primer, 2 μL of reverse primer, 1 μL of template plasmid, and 5 μL of DW to a total of 20 μL. Denature template plasmid for 5 min at 95 °C. Incubate for 1 min at 95 °C, for 1 min at T_m of nonoverlapping region – 5 °C, and for 1 min/kb at 72 °C for 12–15 cycles. Incubate for 1 min at T_m of overlapping region – 5 °C and for 2 min/kb at 72 °C.
9. Check the size of the PCR product by gel electrophoresis (*see* **Note 9**).
10. Digest template plasmid: Add 1 μL of DpnI and 2 μL of CutSmart Buffer to 17 μL of PCR product. Incubate 45–60 min at 37 °C.
11. Design forward and reverse primers to make insert DNA and backbone DNA (*see* **Note 10**).
12. Polymerase chain reaction: Combine 10 μL of Pfu master mix with 2 μL of forward primer, 2 μL of reverse primer, 1 μL of template plasmid, and 5 μL of DW to a total of 20 μL. Denature template plasmid for 2 min at 95 °C. Incubate for 20 sec at 95 °C, for 40 sec at T_m – 5 °C, and for 1 min/kb + 1 min at 72 °C for 30 cycles. Incubate for 2 min/kb – 1 min at 72 °C.
13. Check the size of the PCR product by gel electrophoresis (*see* **Note 11**).
14. Digest template plasmid: Add 1 μL of DpnI and 2 μL of CutSmart Buffer to 17 μL of PCR product. Incubate 45–60 min at 37 °C.
15. Gibson assembly: Combine 5 μL of Gibson assembly master mix with 1 μL of insert PCR product, 1 μL of backbone PCR product, and 3 μL of DW to a total of 10 μL. Incubate 45–60 min at 50 °C.
16. Transform 50 μL of chemically competent *E. coli* DH5a cells with the 5 μL of CRATE product, site-directed plasmid mutagenesis product, or Gibson assembly product using heat shock.
17. Incubate transformed bacteria with LB medium without antibiotics for 1–2 h at 37 °C in a shaker.
18. Plate on selective LB agar plates.

3.3 Making of Chemically Competent E. coli for Cloning and Experimental Transformant

1. Plate *E. coli* (DH5a, BL21 DE3) on LB agar plates followed by incubation at 37 °C overnight (*see* **Note 12**).
2. Select a single colony from the plate to inoculate 7 mL of LB medium. Incubate overnight at 37 °C with shaking.
3. Inoculate 5 mL of overnight cell culture to 500 mL of fresh LB in a 2 L narrow mouth Erlenmeyer flask. Incubate at 37 °C with shaking until OD600 of 0.35 (*see* **Note 13**).
4. Aliquot the cells ino 50 mL conical tubes (50 mL per tube) and chill on ice for 10 min (*see* **Note 14**).
5. Centrifuge the cells at 2012 × *g* for 10 min at 4 °C.
6. Resuspend each pellet in 6 mL of ice-chilled Inoue transformation buffer and chill on ice for 10 min.
7. Centrifuge the cells at 2012 × *g* for 10 min at 4 °C.
8. Resuspend all pellets in 20 mL of ice-chilled Inoue transformation buffer with 7% v/v DMSO and chill on ice for 10 min.
9. Aliquot the cells into 200 μL PCR tubes (50 μL per tube) and flash freeze with liquid nitrogen. Store at −70 °C freezer.

3.4 Experiments with RNA Translational Repressors

1. Thaw 50 μL of competent *E. coli* on ice, add 150 ng of each plasmid for expressing the switch RNA and trigger RNA to the cells, and incubate for 10 min on ice. To apply a heat shock, incubate the reaction tube for 30 s at 42 °C using a water bath. Immediately after the heat shock, chill the cells on ice for 2 min. Recover the cells with 1 mL of LB media without antibiotics for 1 h (in case of single transformant) to 2 h (in case of double transformant) in shaker. Plate on selective LB agar plates.
2. Select three single colony from the plate to inoculate 1 mL of LB media with antibiotics in 96-well deep plate with shaking at 800 rpm and 37 °C (*see* **Note 15**).
3. Dilute overnight cultures by 100-fold into fresh LB medium with antibiotics and return to the shaker (800 rpm, 37 °C). After 80 min, cells are induced using IPTG at a final concentration of 1 mM from a 100 mM stock solution in DW (in case of BL21 DE3 and BL21 Star DE3 strain) or L-arabinose at a final concentration of 0.2% (w/v) from a 20% stock solution in DW (in case of BL21 AI strain) for 3 h and 30 min or 6 h (in case of two-input logic gate test).
4. After IPTG or L-arabinose induction, transfer 200 μL of induced cultures to a 96-well black/clear plate. Measure the fluorescence level and OD600 in the plate reader. Calculate normalized fluorescence level as follows:

   In the case of GFP,

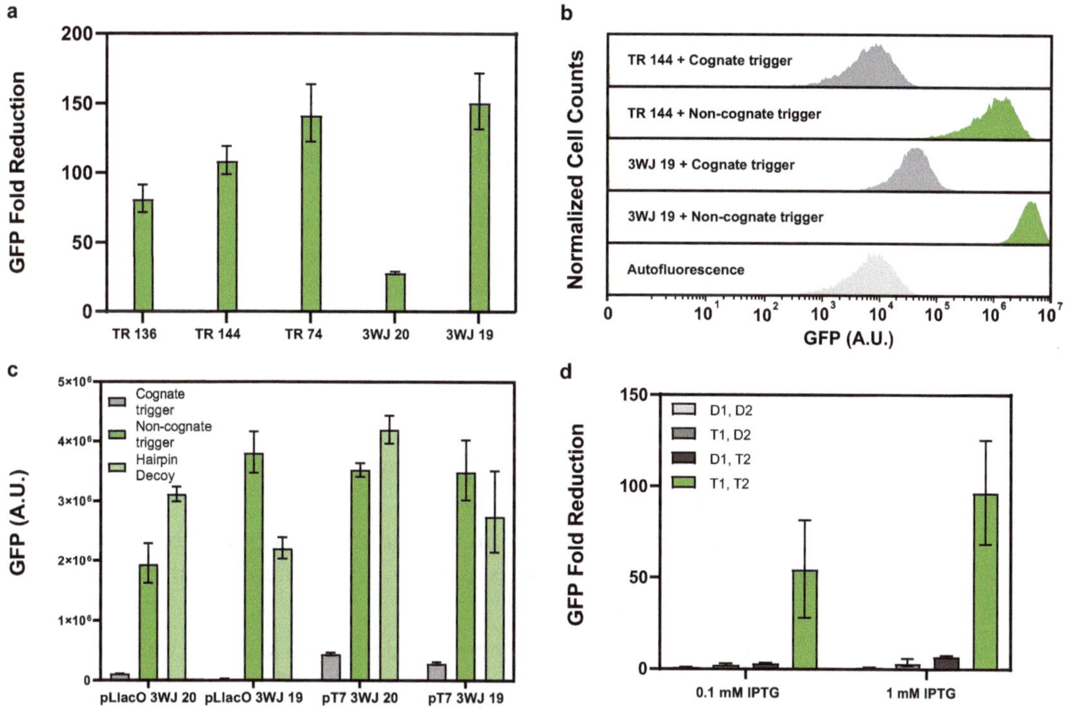

Fig. 6 GFP fluorescence measurements of toehold and 3WJ repressor. (**a**) GFP fold reduction of toehold repressors (TR 136, TR144, TR 74) and 3WJ repressors (3WJ 20, 3WJ 19) [12]. *E. coli* cells were induced with 0.1 mM IPTG. (**b**) Flow cytometry GFP fluorescence histograms for TR 144 and 3WJ 19 compared to *E. coli* autofluorescence. Autofluorescence level was measured in cells transformed with a control plasmid that does not contain the GFP sequence. (**c**) GFP fluorescence result of 3WJ repressors with the promoters pLlac0 and pT7 (*see* **Note 17**). *E. coli* cells were induced with 0.1 mM IPTG. (**d**) GFP fold reduction of 3WJ NAND gate with two different IPTG concentrations: 0.1 mM and 1 mM (*see* **Note 18**). Error bars represent standard deviations of three biological replicates

$$\text{Normalized GFP} = \frac{\textit{GFP of test sample} - \textit{GFP of LB with antibiotics}}{\textit{OD600 of test sample} - \textit{OD600 of LB with antibiotics}}.$$

5. Flow cytometry can be used for more precise analysis of cell fluorescence (*see* **Note 16**). Figure 6 shows the results of the flow cytometry assay.

4 Notes

1. The first GGG sequence is a conserved sequence for proper transcription via the T7 promoter. The bold sequences are ribosome binding site and initiation codon.
2. The decoy is an RNA that minimally binds to the switch RNA. The experimental result using the decoy represents the output

of the noninteracting state (in the case of repressors, this is the ON state). The most suitable decoy sequence may be different for each switch, but in general, hairpin RNA can be used which forms a strong secondary structure by itself.

3. In these experimental conditions, the copy number differences in the plasmids expressing switch and trigger RNAs led to an approximately 6- to 8-fold excess of trigger compared to switch molecules as determined by fluorescence measurements of the GFP reporter expressed separately from each plasmid [6]. pCOLADuet (kanamycin resistance, ColA origin) for switches and pET15b (ampicillin resistance, ColE1 origin) and pCDFDuet (spectinomycin resistance, CDF origin) for triggers are used.
4. The BL21 Star DE3 strain with an IPTG inducible genomic T7 RNA polymerase and decreased RNase activity is used for repressor characterization.
5. The BL21-AI™ strain with an arabinose inducible genomic T7 RNA polymerase is used in experiments that require tight expression control.
6. Online NUPACK sequence design package does not recognize the wobble base pair, so be careful. Alternatively, you can use the python NUPACK library or Linux software, which offers many of the latest functions. The example of NUPACK scripts for the repressors in Figs. 4 and 5 can also be found at https://github.com/AlexGreenLab/DesignofRNA-basedTranslationalRepressors.
7. If you want more accurate results, include the first ~100 nt of the reporter gene after the switch RNA, and include the T7 terminator sequence after the trigger RNA. Short hairpin regions with stems 6–10 bp long can also be added to the 5′ end of the transcripts to increase their stability.
8. Each primer contains a nonoverlapping region at the 5′ end and overlapping region at the 3′ end. The melting temperature of nonoverlapping region is 5–10 °C higher than the melting temperature of overlapping region (we use parameter of SantaLucia et al. [17]). This difference minimizes forward and reverse primer dimerization and enables the primers to use the PCR products as the template. The mutation site (insertion, deletion, substitution) can be placed in any region.
9. If you get smeared or unwanted bands, increase annealing temperature, or use more dilute template DNA.
10. The length of Gibson region, which overlaps the insert and backbone DNA, is 15–30 bp. We usually use a T7 promoter and its upstream sequence [for the forward primer] and the linker sequence or T7 terminator sequence [for the reverse primer] to design primers.

11. If you get smeared or unwanted bands, increase the annealing temperature, or use more dilute template DNA. However, if the size of the unwanted backbone band is under 1500 bp, it can be ignored, and you can proceed to the next step. DNA bands under 1500 bp cannot form an intact plasmid that contains the antibiotic resistance gene and the origin of replication, so this product is simply removed at the antibiotic selection step.
12. If you need to make a triple transformant for NAND or NOR gate experiments, it helps to prepare a chemically competent *E. coli* that contains a gate RNA. Plate a single transformant on a selective agar plate, and all the next steps are carried out in selective LB medium.
13. If you need a small number of competent cells, inoculate 300 μL of overnight cell culture to 30 mL of fresh LB in 50 mL conical tube. And incubate at 37 °C with shaking with a slightly opened lid until reaching an OD600 of 0.35. You might get approximately 40 vials of competent cells (50 μL per vial).
14. Performing aliquot, chilling, and resuspension steps in a cold room is recommended for better competency.
15. If your gene is cytotoxic, adding 0.2% glucose is recommended to prevent recombination.
16. After **step** 3.4, 3, dilute cells with 1× PBS (adjusting OD600 to 0.1) to make a final volume of 100 μL. Detect cells with a flow cytometer using a forward scatter (FSC) trigger, recording at least 15,000 cell events for each measurement. Analyze acquired data using CytExpert software as follows: cell populations were gated according to their FSC and side scatter (SSC) distributions using auto polygon tool (gated cells should be at least 80% of the total populations). Calculate geometric mean of GFP fluorescence histograms to extract the average fluorescence from three biological replicates. If the flow cytometer is not available due to instrument availability issues, IPTG-induced cells can be fixed with paraformaldehyde and measured later. Collect IPTG-induced cells by centrifugation, resuspend the pellet in 1% paraformaldehyde solution, and wait for 30 min. When stored in a 4 °C refrigerator, fixed cells are guaranteed to be stable for up to 7 days.
17. The 3WJ repressor system can work well enough with promoters other than the T7 promoter (*see* Fig. 6c). Also, pLlacO 3WJ 19 and pT7 3WJ 19 with non-cognate trigger and hairpin decoy data show that the hairpin decoy does not necessarily have a minimum interaction with switches.

18. When three or more input RNAs are required, such as a NAND gate, it can be better to perform induction with a higher concentration of IPTG (*see* Fig. 6d). This is presumably due to the requirement of higher concentrations of RNA components for multicomponent complexes.

Acknowledgments

This work was supported by the National Research Foundation of Korea (NRF- 2019R1A2C1086830) grant funded by the Korean government (MSIT). This work was also supported by an NIH Director's New Innovator Award (1DP2GM126892), an NIH U01 award (1U01AI148319-01), and an NSF RAPID award (2029532) to A.A.G.

References

1. Cameron DE, Bashor CJ, Collins JJ (2014) A brief history of synthetic biology. Nat Rev Microbiol 12(5):381–390
2. Cheng AA, Lu TK (2012) Synthetic biology: an emerging engineering discipline. Annu Rev Biomed Eng 14:155–178
3. Elowitz MB, Leibler S (2000) A synthetic oscillatory network of transcriptional regulators. Nature 403(6767):335–338
4. Kim J, White KS, Winfree E (2006) Construction of an in vitro bistable circuit from synthetic transcriptional switches. Mol Syst Biol 2(1):68
5. Siuti P, Yazbek J, Lu TK (2013) Synthetic circuits integrating logic and memory in living cells. Nat Biotechnol 31(5):448–452
6. Green AA, Silver PA, Collins JJ, Yin P (2014) Toehold switches: de-novo-designed regulators of gene expression. Cell 159(4):925–939
7. Grünberg R, Serrano L (2010) Strategies for protein synthetic biology. Nucleic Acids Res 38(8):2663–2675
8. Chappell J, Watters KE, Takahashi MK, Lucks JB (2015) A renaissance in RNA synthetic biology: new mechanisms, applications and tools for the future. Curr Opin Chem Biol 28:47–56
9. Liang JC, Bloom RJ, Smolke CD (2011) Engineering biological systems with synthetic RNA molecules. Mol Cell 43(6):915–926
10. Hong S, Kim J, Kim J (2021) Multilevel gene regulation using switchable transcription terminator and toehold switch in Escherichia coli. Appl Sci 11(10):4532
11. Green AA, Kim J, Ma D, Silver PA, Collins JJ, Yin P (2017) Complex cellular logic computation using ribocomputing devices. Nature 548(7665):117–121
12. Kim J, Zhou Y, Carlson PD, Teichmann M, Chaudhary S, Simmel FC, Silver PA, Collins JJ, Lucks JB, Yin P (2019) De novo-designed translation-repressing riboregulators for multi-input cellular logic. Nat Chem Biol 15(12): 1173–1182
13. Zadeh JN, Steenberg CD, Bois JS, Wolfe BR, Pierce MB, Khan AR, Dirks RM, Pierce NA (2011) NUPACK: analysis and design of nucleic acid systems. J Comput Chem 32(1): 170–173
14. Liao C, Ttofali F, Slotkowski RA, Denny SR, Cecil TD, Leenay RT, Keung AJ, Beisel CL (2019) Modular one-pot assembly of CRISPR arrays enables library generation and reveals factors influencing crRNA biogenesis. Nat Commun 10(1):2948
15. Liu H, Naismith JH (2008) An efficient one-step site-directed deletion, insertion, single and multiple-site plasmid mutagenesis protocol. BMC Biotechnol 8(1):91
16. Gibson DG, Young L, Chuang R-Y, Venter JC, Hutchison CA, Smith HO (2009) Enzymatic assembly of DNA molecules up to several hundred kilobases. Nat Methods 6(5):343–345
17. SantaLucia J (1998) A unified view of polymer, dumbbell, and oligonucleotide DNA nearest-neighbor thermodynamics. Proc Natl Acad Sci 95(4):1460–1465

Chapter 4

Design of Ribocomputing Devices for Complex Cellular Logic

Griffin McCutcheon, Soma Chaudhary, Seongho Hong, Dongwon Park, Jongmin Kim, and Alexander A. Green

Abstract

The ability to control cell function is a critical goal for synthetic biology and motivates the development of ever-improving methods for precise regulation of gene expression. RNA-based systems represent powerful tools for this purpose since they can take full advantage of the predictable and programmable base pairing properties of RNA to control gene expression. This chapter is focused on the computational design of RNA-only biological circuits that can execute complex Boolean logic expressions in living cells. These ribocomputing devices use toehold switches as building blocks for circuit construction, integrating sensing, computation, and signal generation functions within a gate RNA transcript that regulates expression of a gene of interest. The gate RNA in turn assesses the assembly state of networks of interacting input RNAs to execute AND, OR, and NOT operations with high dynamic range in *E. coli*. Harnessing in silico tools for device design facilitates scaling of the circuits to complex logic expressions, including four-input AND, six-input OR, and disjunctive normal form expressions with up to 12 inputs. This molecular architecture provides an intuitive and modular strategy for devising logic systems that can be readily engineered using RNA sequence design software and applied in vivo and in vitro. In this chapter, we describe the process for designing ribocomputing devices from the generation of orthogonal toehold switch libraries through to their use as building blocks for AND, OR, and NOT circuitry.

Key words RNA-based regulator, Riboregulator, Toehold switch, Complex logic, Ribocomputing

1 Introduction

In recent years, much work in synthetic biology has focused on developing novel biological functions that replicate the complex logical operations and capabilities of electronic systems [1, 2]. Biochemical circuits have been created to replicate bistable switches [3, 4], oscillators [5, 6], logic gates [7–9], and memory devices [10], but these systems often rely on a limited number of protein-based regulators whose unpredictability and crosstalk hamper our ability to implement biochemical circuits at larger scales [11]. By comparison, the predictability and selectivity of Watson-Crick base

James Chappell and Melissa K. Takahashi (eds.), *Riboregulator Design and Analysis*,
Methods in Molecular Biology, vol. 2518, https://doi.org/10.1007/978-1-0716-2421-0_4,

pairing in DNA and RNA mean that numerous orthogonally interacting nucleic acid devices can be implemented within a single circuit without the crosstalk that often affects protein-based systems [12–15]. As electronic circuits utilize large numbers of identical logical gates and components to function, the orthogonality, scalability, and programmability of RNA-based devices provide an analogous starting point for constructing complex biological circuits.

In particular, the riboregulator known as the toehold switch [16] demonstrates how such orthogonal and programmable devices are possible using purely RNA-based components. The toehold switch is designed to form a hairpin stem-loop structure upstream of a regulated gene. This hairpin hides a ribosome binding site (RBS) within the loop and the start codon (AUG) along its stem. By blocking the ribosome access, it controls gene expression at the translational level [16] (*see* Fig. 1a). The hairpin of the toehold switch also features a single-stranded toehold region that is used to initiate binding to a cognate trigger RNA. Once binding occurs, the trigger RNA competitively binds through the rest of the hairpin stem, unwinding it and allowing ribosomal access to initiate translation. The design of the toehold switch enables these riboregulators to detect a vast array of trigger sequences while providing a wide dynamic range. Moreover, further developments have led to toehold switches optimized for AND logic [17] (*see* Fig. 1b), programmable repressor systems [18], and ultraspecific riboregulators termed SNIPRs (single-nucleotide-specific programmable riboregulators) capable of discerning point mutations and base modifications [19]. Toehold switches and SNIPRs can reliably discriminate triggers with lengths of 21 nt or more, providing billions of unique potential sequences. While in practice the number of functional and noninteracting sequences is lower, the capacity of riboregulators to interface with networks of numerous modular, noninterfering devices has enabled the creation of multi-input logic devices utilizing AND, OR, and NOT logic.

The architecture of RNA-based computing or ribocomputing devices is illustrated in Fig. 1c [17, 20]. The central processing unit of the network is the gate RNA, an extended transcript consisting of one or more toehold switch hairpins upstream of an output gene. The gate RNA integrates the core sensing and output functions of the network into a single molecule, simplifying expression of the circuit. A network of programmed RNAs provides input signals to the circuit, and the output signal is a protein that is translated upon activation of the gate RNA. RNA inputs interact with themselves and the gate through predictable and programmable base pairing. Inputs that bind to individual sensor domains on the gate can independently activate protein production to enable OR logic operations. Inputs can bind cooperatively to each other to form structures that activate the sensor domains for AND logic.

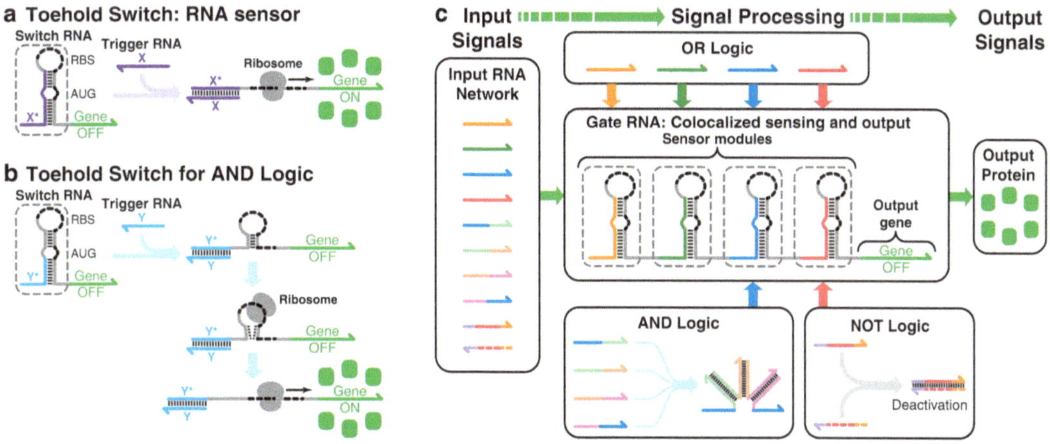

Fig. 1 In vivo computation using synthetic ribocomputing devices. (**a**) Schematic of the toehold switch riboregulator. The ribosomal binding site (RBS) and start codon (AUG) of the switch RNA are exposed when the cognate trigger RNA (input X) disrupts the secondary structure, allowing translation. (**b**) Schematic of toehold switches optimized for AND logic. These toehold switches retain a weak hairpin upon activation by the trigger RNA that still allows efficient translation by the ribosome while reducing translational leakage. (**c**) Schematic of the ribocomputing device architecture. Signal processing is carried out by the gate RNA, which contains sensing and output modules and is comprised of toehold switch elements. AND, OR, and NOT logic occurs through the programmable interactions established between the input and gate RNAs. (Figure adapted from Ref. [17])

Alternatively, they can bind competitively to inhibit sensor binding for NOT logic. Sensors utilizing AND logic are optimized to function with multi-input RNAs (*see* Fig. 1b).

The design of these systems is handled using computational tools, notably the NUPACK software suite [21]. To design a multi-input logic device, a user employs NUPACK's coding language to specify the structures of the sensor domains, the input RNAs, and the desired interactions between all the components. After testing individual devices for compatibility and orthogonality, they can be assembled into full logic circuits.

This chapter provides an introduction to using the NUPACK software package for ribocomputing device design. It covers the process for generating libraries of orthogonal toehold switches. It further shows how these libraries can be applied for the computational design of AND, OR, and NOT ribocomputing devices using validated toehold switch modules. For more information on construction and testing of these RNA-based systems, refer to Chap. 3 [22] or [23].

2 Materials

An example of a single-input toehold switch element (*see* **Note 1**). Sequences are listed below:

2.1 Toehold Switch Elements and Sample Two-Input Ribocomputing Devices

1. Gate sequence: GGGACUACUAUUGAUUACACGCUUUA CUUCGAAAUUCAUAAUGAACAG **AGGAGA**UAU-GAA**AUG** CGAAGUAAAGCGAACCUGGCGGCAGCG CAAAAG (*see* **Note 2**).
2. Trigger sequence: GGGCGUGAGAUAAGCACAUCUCAC GAGACGAAGUAAAGCGUGUAAUCAAUAGUAGUAAG.

An example of a two-input AND ribocomputing device. Sequences are listed below:

1. Gate sequence: GGGAUUGAAUAUGAUAGAAGUUUA GUAGUAGACAAUAGAACAG **AGGAGA**UAUUG**AUG** A CUACUAAACUAAACCUGGCGGCAGCGCAAAAG.
2. Input A1 sequence: GGGCCUAGCGCAGUGCAGCG CUAGGAUAACUACUAAACUUCACAAUAACUAA GAACGACGAUGACAAC.
3. Input A2 sequence: GGGAGCGCCUCUGAUAUGAGGCG CUAUAGUCAUCGUCGUUCUUAGUUAUACAUAU CAUAUUCAAUUCU.

An example of a two-input OR ribocomputing device. Sequences are listed below:

1. Gate sequence: GGGAUGAACUUAAGAUCGCG CUUUAUGGUUAUGAAACAG **AGGAGA**CAUAAC**AUG** AAAGCGCGAAACAGAACAAGAAUAAGUGCAAAUG CUUGUUCUAUCGUCUUGAAACAG**AGGAGA**CAAGA-C**AUG**AGAACAAGCAACCUGGCGGCAGCGCAAAAG.
2. Input A1 sequence: GGGACCUAUUGGACCCGUUUC CAAUAGGUGAACAAGACGAUAGAACAAGCAUUUG CACUUAUAGA.
3. Input A2 sequence: GGGAGUCGGCGUGGACUCG CACGCCGACUAAACAUAACCAUAAAGCGCGAU CUUAAGUUCAUACC.

An example of a two-input A AND (NOT B) ribocomputing device. Sequences are listed below:

1. Gate sequence: CUUGGGCCUAGAAUCGCUAGGCCC GUACUGAUGUAUUGUAGUGUCGUUGAUUCGAA GUUUAGAACAG **AGGAGA**UAAAC**AUG** GAACCAACGACUAACCUGGCGGCAGCGCAAAAG.
2. Input A (activating) sequence: GGGCCACUGACAACUCC GUCAGUGGGAGCACAAUAACAGAAACAGAAUCAAC GACACUACAAUACAUCAGCCAAAUACAAUAACA CUAC.
3. Input B (deactivating) sequence: GGGCCAUCUAGUGA

GAUCUAGAUGGUAAGUGUUAUUGUAUUUGUCU
GAUGUAUUGUAGUGUCGUUGAUUCCGUUUCU
GUUAUUGUGAGC.

2.2 Suitable Vectors

Gate RNA expressed from a relatively low copy vector is suitable for proper switch RNA-trigger RNA interaction, with input RNA expressed from a relatively high copy vector.

3 Methods

3.1 Introduction to NUPACK Nucleic Acid Sequence Design

Efficient design of the ribocomputing circuits described here requires familiarity with the NUPACK software suite, which is available at http://www.nupack.org. Here we describe the concepts and structure used to code RNA devices in NUPACK's web portal (*see* **Note 3**) by walking through the components for a script that codes a basic minimal toehold switch as described in Fig. 2 (*see* **Note 4**).

To begin, let's describe what is happening at a conceptual level when using NUPACK. In general terms, the NUPACK software requires users to define DNA/RNA structures and assign these structures arbitrary or strict sequence values. A collection of these definitions, along with any additional parameters, is what makes up a NUPACK script. From there, the software begins optimizing the sequence such that its predicted structure minimizes the predicted normalized ensemble defect compared to the structures defined by the user.

Using Fig. 2 as reference, we now discuss the script components. At the top of the script, we set a few parameters: what our device is made of, the temperature we expect it to be used at, and how many simulations to evaluate. Our devices are made of RNA, and we intend to test them at 37 °C. We specify this using the following lines:

```
material = rna (see Note 5)

temperature = 37.0
```

We typically attempt to form as many different sequences as possible to increase the likelihood of obtaining designs with minimal defect levels or to obtain libraries of orthogonal designs. So, we request the maximum possible trials in the web portal, ten, using the following line:

```
trials = 10
```

From here, we can specify the structures we want using dot-parens-plus or DU+ notation (*see* **Note 6**). To receive a functional

```
# Sample NUPACK Code for a Minimal Toehold Switch

material = rna1999
temperature = 37.0
trials = 10

# target structures
structure trigger = U30
structure switch  = U12 D9 (U3 D6 U12 U3)
structure complex = D30 + U30

# sequence domains
domain toehold = N12
domain b = N9
domain c = N3
domain d = N6

# minimal RBS and start codon required for translation
domain start_codon = AUG
domain loop   = N4 AGAGGAGA

trigger.seq  = d* c* b* toehold*
switch.seq = toehold b c d loop d* start_codon b*
complex.seq = d* c* b* toehold* toehold b c d loop d* start_codon b*

# prevent sequence patterns
prevent = AAAA, CCCC, GGGG, UUUU, KKKKKK, MMMMMM, RRRRRR, SSSSSS,
        WWWWWW, YYYYYY

# specify stop conditions for normalized ensemble defect
switch.stop = 5
trigger.stop = 5
complex.stop = 5
```

Fig. 2 Sample NUPACK code for a minimal toehold switch design

design from the software, you need to specify the structure of the individual components and how they interact. For example, a single-input toehold switch requires three structures: the switch RNA, the trigger RNA, and the complex formed when the two interact. The following code examples show the minimal NUPACK code required for defining a toehold switch. More detailed code examples can be found in the sections below. These structures can be established using the following lines:

```
structure switch = U12 D9 (U3 D6 U12 U3)
structure trigger = U30
structure complex = D30 + U30
```

The structure of each RNA species and any structure of interest formed by one or more interacting RNAs can be defined.

In addition to establishing structures, we must define the sequences of the strands themselves. This is typically done by defining different sequence "domains" which can be concatenated into "strands." These domains can be defined with arbitrary assignments such as.

```
domain toehold = N12
```

in which N specifies an arbitrary nucleotide and the number immediately following specifies the number of bases to be defined as arbitrary. NUPACK also supports IUPAC nucleotide codes for setting less arbitrary sequence requirements. These definitions can be combined with strict definitions such as in the following domain:

```
domain loop = N4 AGAGGAGA
```

Multiple domains can then be concatenated into strands. Consider the trigger domains described in Fig. 2 (*see* **Note 7**), which collectively describe **30** arbitrary nucleotides. These can be consolidated into a single strand with the line, making use of the "*" operator to define the following reverse complement:

```
strand trigger_str = d* c* b* toehold*
```

The sequence of our structures can be established using either domains, strands, or a combination of the two. For example, the following two lines are identical in function to the NUPACK interpreter if you had defined `trigger_str` properly in the previous line:

```
trigger.seq = d* c* b* toehold*
trigger.seq = trigger_str
```

What we have covered is the minimum information required to generate potential designs on NUPACK, but there are a few additional parameters that can be added to increase the likelihood of returning successful designs. To start, you can prevent excessive base repetition by specifying "pattern prevent constraints" using IUPAC nucleotide codes. The software will avoid incorporating these sequence patterns into its designs. The prevent constraint used in our sample code is as follows:

```
prevent = AAAA, CCCC, GGGG, UUUU, KKKKKKKK, MMMMMM, RRRRRR,
SSSSSS, WWWWWW, YYYYYY
```

Additionally, you can specify a target ensemble defect that you want the software to reach. The default normalized ensemble defect the software optimizes toward is 1%, but larger structures may face exceptional difficulty in reaching such a goal, requiring you to set higher acceptable defects. For example, setting a cutoff ensemble defect of 5% for the principal structures in the system is accomplished with the following lines:

```
switch.stop = 5
trigger.stop = 5
complex.stop = 5
```

While not a parameter, it is recommended that your designs include at least 30–50 bases of the output protein's RNA sequence downstream from your logic gate when defining your design structure and sequence. Doing so will reduce the likelihood of interference or crosstalk in your final designs.

Once the script has been defined, place it in the web portal's "Target Structure" field and click "Update." After confirming in the preview windows that the designs are correct, click the "Design" button to submit the job to NUPACK's servers. Evaluate the returned designs to select for those with low normalized ensemble defect, and screen out those that contain stop codons in frame with the RBS. NUPACK's analysis function can be used to examine whether the candidates form their intended structures.

3.2 Design of Orthogonal Toehold Switch Libraries

Before designing ribocomputing devices, it is necessary to first have a library of toehold switch modules known to display wide dynamic range and good orthogonality. Such libraries have been reported in previous work [16, 17]. However, it is often useful to have modules specifically designed for a particular purpose, which can be used to achieve new functions [18]. In this section, we detail the process for computational design of new orthogonal toehold switch libraries. Two types of toehold switches are provided as examples: the original toehold switch [16] and a modified toehold switch optimized for performing AND logic [17, 24] (*see* **Note 8**). The first-generation toehold switch provides high ON/OFF ratios and employs a 12-nt toehold region and 30-nt trigger RNAs (*see* Fig. 1a). The AND computing toehold switch (ACTS) type I version employs a 16-nt toehold and a shifted binding site that prevents the trigger RNA from binding to the upper part of the switch stem (*see* Fig. 1b). The latter modification still allows the switch to be activated when the full RNA trigger binds, but it reduces the likelihood that translational leakage will occur when half of the trigger is expressed in AND logic circuits.

Sample NUPACK scripts for the first-generation toehold switches and ACTS type I [17] systems are provided in Figs. 3 and 4, respectively. These NUPACK scripts have several key differences from the minimal design provided in Fig. 2. First, the full toehold switches now include additional conserved sequences necessary for in vivo expression, namely, the transcriptional terminator (`T7_term`) and output gene sequences (`linker21`, `output_gene`). Importantly, we typically set these conserved domains to be completely single stranded in the NUPACK web portal to ensure that the script runs without errors. Since the domains are not single stranded, we in turn increase the stop conditions for the designs since their intrinsic secondary structure will lead to an increase in the calculated ensemble defect. Second, we append a short hairpin sequence to the 5′ end of the trigger RNA to enhance its stability against RNases and increase its intracellular lifetime [25]. Third, we add a GGG sequence to the start of the transcripts to increase the efficiency of transcription when using T7 RNA polymerase. This leader sequence may not be necessary if you are using a different RNA polymerase. A 3-nt short spacer is also added between the switch-binding sequence and the terminator to increase the separation between these two functional domains. To generate a library of orthogonal toehold switches, do the following:

1. Paste your toehold switch design script into the NUPACK target structure text entry box, and click "Update" to verify that it is correctly interpreted. Set the number of trials to 10.
2. Click "Design" to begin the toehold switch design process on the server.
3. Determine the number of toehold switches that are required for the ribocomputing devices that you wish to implement. We will typically generate ~20-fold more designs than are needed to accumulate a sufficiently large library for experimental validation. In practice, this number will depend on the expected fraction of designs that will have the desired output characteristics.
4. For each design, evaluate the switch RNA sequence to ensure that it does not have any in-frame stop codons. Since ribocomputing devices can translate through the full switch RNA during OR operations, it is important to screen for stop codons before and after the start codon in the hairpin.
5. Repeat the NUPACK design process until you reach the desired number of stop codon-free designs (*see* **Note 9**).
6. Select an orthogonal set of toehold switch designs for experimental testing. We recommend using the following **steps** (7 through **9**) for the selection process so as to provide high-performance orthogonal devices.

```
#Sample NUPACK Code for the First-Generation Toehold Switch

material = rna1999
temperature = 37.0
trials = 10

# target structures

# The T7 terminator contains G-U wobble base pairs that can sometimes
# cause errors during NUPACK design. As a result, we assert that
# the terminator is fully single stranded. This modification will lead
# to an increased normalized ensemble defect for the designs but still
# allow good riboregulators to be generated.
structure trigger = U3 D8 U6 U3 U30 U3 U47

#We similarly assert that the linker and output gene are single-stranded
#to avoid design errors.
structure switch = U3 U12 D9 (U3 D6 U12 U3) U50

structure complex = U3 D8 U6 U3 D30 (U3 U47 + U3) U12 U18 U50

# sequence domains
domain T7_term = UAGCAUAACCCCUUGGGGCCUCUAAACGGGUCUUGAGGGGUUUUUUG
domain linker21 = AACCUGGCGGCAGCGCAAAAG

#first 29 nt of GFP
domain output_gene = AUGCGUAAAGGAGAAGAACUUUUCACUGG

domain preG = GGG
domain trig_stem = N8
domain trig_loop = N6
domain trig_spacer1 = N3
domain trig_spacer2 = N3

domain toehold = N12
domain stem1 = N9
domain stem2 = N3
domain stem3 = N6

#using a 12-nt loop domain to make full switch a multiple of 3 and
# in-frame in gate RNAs
domain loop   = N4 AGAGGAGA
domain stem2_star = AUG #start codon

switch.seq = preG toehold stem1 stem2 stem3 loop stem3* stem2_star
        stem1* linker21 output_gene
trigger.seq = preG trig_stem trig_loop trig_stem* trig_spacer1 stem3*
        stem2* stem1* toehold* trig_spacer2 T7_term
complex.seq = preG trig_stem trig_loop trig_stem* trig_spacer1 stem3*
        stem2* stem1* toehold* trig_spacer2 T7_term preG toehold stem1
        stem2 stem3 loop stem3* stem2_star stem1* linker21 output_gene

# specify stop conditions for normalized ensemble defect
# Stop condition is increased to 10 to accommodate terminator
# and output gene non-ideal secondary structure

switch.stop = 10
trigger.stop = 10
complex.stop = 10

# prevent sequence patterns
prevent = AAAA, CCCC, GGGG, UUUU, KKKKKK, MMMMMM, RRRRRR, SSSSSS, WWWWWW,
        YYYYYY
```

Fig. 3 Sample NUPACK code for the first-generation toehold switch

```
#Sample NUPACK Code for the AND-Computing Toehold Switch (ACTS) Type I

material = rna1999
temperature = 37.0
trials = 10

# target structures

structure switch = U3 U16 D12 (U3 D6 U14 U3) U50
structure target = U3 D8 U6 U3 U28 U3 U47
structure complex = U3 D8 U6 U3 D28 (U3 U47 + U3) U6 U12 U14 U12 U50

# sequence domains
domain T7_term = UAGCAUAACCCCUUGGGGCCUCUAAACGGGUCUUGAGGGGUUUUUUG
domain linker21 = AACCUGGCGGCAGCGCAAAAG

#first 29 nt of GFP
domain output_gene = AUGCGUAAAGGAGAAGAACUUUUCACUGG

domain preG = GGG
domain targ_stem = N8
domain targ_loop = N6
domain targ_spacer1 = N3
domain targ_spacer2 = N3

domain toehold = N16
domain stem1 = N12
domain stem2 = N3
domain stem3 = NNNNNN
domain loop  = NNNAACAGAGGAGA
domain stem2_star = AUG #start codon

switch.seq = preG toehold stem1 stem2 stem3 loop stem3* stem2_star stem1*
        linker21 output_gene
target.seq  = preG targ_stem targ_loop targ_stem* targ_spacer1 stem1*
        toehold* targ_spacer2 T7_term
complex.seq = preG targ_stem targ_loop targ_stem* targ_spacer1 stem1*
        toehold* targ_spacer2 T7_term preG toehold stem1 stem2 stem3 loop
        stem3* stem2_star stem1* linker21 output_gene

# specify stop conditions for normalized ensemble defect
# Stop condition is increased to 10 to accommodate terminator
# and output gene non-ideal secondary structure

switch.stop = 10
target.stop = 10
complex.stop = 10

# prevent sequence patterns
prevent = AAAA, CCCC, GGGG, UUUU, KKKKKK, MMMMMM, RRRRRR, SSSSSS,
        WWWWWW, YYYYYY
```

Fig. 4 Sample NUPACK code for the AND computing toehold switch (ACTS) type I

7. Take the full set of designs, and sort them by the normalized ensemble defect output calculated by NUPACK. Eliminate the bottom 50% of designs with the highest defect levels.
8. Use pairwise comparisons to predict computationally the potential for crosstalk between different devices. For a library of N toehold switches, this will usually entail computing the full N^2 comparisons. We have used three different pairwise measurements to assess crosstalk with varying levels of computation time and test stringency (*see* **Note 10**):

   Binding fraction. Use NUPACK to predict the fraction of switch molecules that will bind to a non-cognate trigger RNA. This is the most stringent test since triggers can readily bind to any exposed single-stranded region of the switch RNA nonspecifically to increase this measure (*see* **Note 11**).

   Hairpin disruption. Use NUPACK to compute the interaction between the trigger and switch RNAs. Use the pair probability table or minimum free energy structures of the predicted switch-trigger complex to assess the stability of the switch RNA hairpin. This metric is not as stringent as computing the binding fraction since it disregards interactions that induce switch-trigger binding but are unlikely to generate translational leakage (*see* **Note 11**).

   Edit distance. Use the portion of the trigger RNA that binds to switch RNA (i.e., exclude the 5′ hairpin and terminator), and compute the edit distance, also known as the Levenshtein distance, for each pair of triggers. The computation time for this operation is considerably smaller and is essentially half the number of comparisons needed (*see* **Note 12**).
9. Select the subset of designs with the minimum overall expected crosstalk. We use a Monte Carlo approach for this process. Random subsets of the library are selected, and their orthogonality is assessed based on the desired crosstalk metric. The subset of selected designs is modified iteratively through random design exchanges to continually reduce the crosstalk until an optimum sub-library is identified.
10. Perform experimental testing of the toehold switch library to validate device performance and crosstalk. We typically test a library that is at least five-fold larger than the highest number of expected switch modules in the final ribocomputing device. This library size helps ensure that there is an adequate number of high-performance toehold switches for the final circuit.

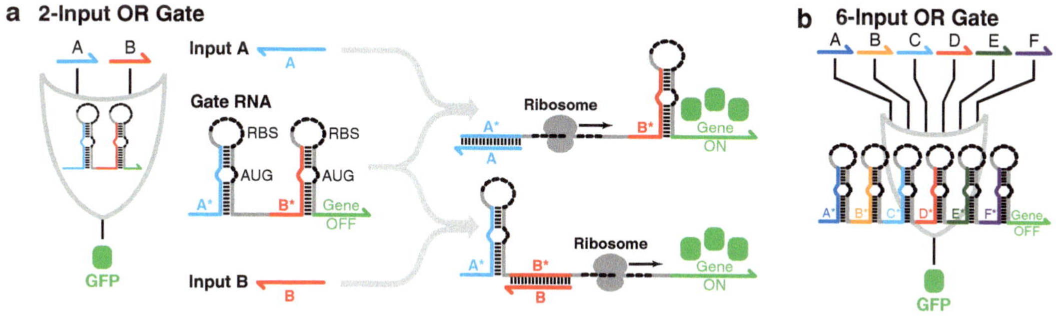

Fig. 5 (**a**) Design of a two-input AND gate RNA using two RNA strands that hybridize to form a complete trigger RNA. Each input has half of the full toehold switch trigger sequence. (**b**) Design of a four-input AND gate ribocomputing device. Inputs A1 and A4 each contain half of the trigger RNA sequence. Inputs A2 and A3 are designed to bring A1 and A4 into close proximity for toehold switch activation. (Figure adapted from Ref. [17])

3.3 Design of Multi-input AND Logic Devices

Ribocomputing AND logic devices make use of input RNAs that hybridize to one another to form a full trigger sequence that activates the corresponding toehold switch in the gate RNA. Operation of a two-input AND gate is illustrated in Fig. 5a. The two inputs each contain half of the trigger sequence for the gate RNA and are unable to unwind the ACTS hairpin on their own. Complementary domains u and u* are used to bring the two inputs together to form the complete trigger, which can then unwind the gate RNA and activate translation. This general AND logic approach can be extended to *N*-input operation using the hybridization of *N*-input species to form the trigger, as illustrated for the four-input system in Fig. 5b. Use the following procedure and the NUPACK script in Fig. 6 to design the input RNA network for an AND logic system (*see* **Note 13**):

1. Identify the validated ACTS device that you wish to use for your AND ribocomputing circuit.
2. Obtain the portion of trigger sequence that binds to the switch RNA. For ACTS type I and type II systems, this sequence should be 28 and 26 nt long, respectively.
3. Split this sequence at the midpoint to generate your half trigger sequences. For the system in Fig. 6, the whole trigger sequence is GAGACGACGAAUAGAAUUGAACAUAGAU , which leads to the 5′ half trigger GAGACGACGAAUAG and the 3′ half trigger AAUUGAACAUAGAU.
4. In the NUPACK script, use the reverse complement of the 5′ half trigger to help define the domain L00 (*see* **Note 14**).
5. In the NUPACK script, use the 3′ half trigger to help define the last input domain L04 or more generally the L*N* domain for an *N*-input AND gate (*see* **Note 14**).

```
#Sample NUPACK Code for a Four-Input AND RNA Network

material = rna1999
temperature = 37.0
trials = 10

# ACTS_TypeI_N2 switch with extended loop domain, which does not include
# GFP
domain switch = GGGAUCUAUGUUCAAUUCUAUUCGUCGUCUCAUGAGUAUAGGAGCCGAAACAG
        UAAACAGAGGAGAUAUACUAUGGAGACGACGAAUAACCUGGCGGCAGCGCAAAAG
# ACTS_TypeI_N2 trigger:   GGGACCGGUCUUUGCAGAGACCGGUGUA
# GAGACGACGAAUAGAAUUGAACAUAGAU AAG
#              structure:   ...((((((((((....))))))))))...
# [   switch binding region   ] ...

# L00 contains the reverse complement of the 5' half of the trigger. The
# sequence should appear in the switch.
domain L00 = NNN CUAUUCGUCGUCUC
domain L01 = NNNNNNNNNNNNNNNNNNNNN
domain L02 = NNNNNNNNNNNNNNNNNNNNN
domain L03 = NNNNNNNNNNNNNNNNNNNNN
domain L04 = NNN AAUUGAACAUAGAU #This is the 3' half of the trigger
# 3-nt spacers are added on to the half trigger domains in L00 and L04
# to facilitate binding of multi-input multi-arm junction complex to
# the switch

structure input_01 = U38
structure input_02 = U42
structure input_03 = U42
structure input_04 = U38
structure complex_01_04 = U38+U38
structure complex_01_02 = U17 D21 + U21
structure complex_02_03 = U21 D21 + U21
structure complex_03_04 = U21 D21 + U17
structure hcomp_02 = U18 D13 (U3 D6 U26 U3) U20 + U42
structure hcomp_03 = U18 D13 (U3 D6 U26 U3) U20 + U42

input_01.seq = L00* L01
input_02.seq = L01* L02
input_03.seq = L02* L03
input_04.seq = L03* L04
complex_01_04.seq = L00* L01 L03* L04
complex_01_02.seq = L00* L01 L01* L02
complex_02_03.seq = L01* L02 L02* L03
complex_03_04.seq = L02* L03 L03* L04

#add switch+input complexes as necessary to specify that particular
# inputs should not interact with the switch
hcomp_02.seq = switch L01* L02
hcomp_03.seq = switch L02* L03

input_01.stop = 1
input_02.stop = 1
input_03.stop = 1
input_04.stop = 1
complex_01_04.stop = 1
complex_01_02.stop = 1
complex_02_03.stop = 1
complex_03_04.stop = 1
hcomp_02.stop = 1
hcomp_03.stop = 1

prevent = AAAA, CCCC, GGGG, UUUU, KKKKKK, MMMMMM, RRRRRR, SSSSSS, WWWWWW,
        YYYYYY
```

Fig. 6 Sample NUPACK code for a four-input AND RNA network

6. Define the desired structures of each of the inputs, and specify the structures of the two-strand complexes. In particular, you can specify complexes where strands do not bind to one another (e.g., `complex_01_04`, `hcomp_02`, `hcomp_03`) to define strands that should not interact.
7. Run the resulting NUPACK script on the web portal.
8. Select the top resulting design(s).
9. To ensure optimal AND gate activity, perform a final round of NUPACK sequence design to add 5′ and 3′ functional domains onto each input RNA. To the 5′ end, append a GGG leader sequence, if using T7 RNA polymerase expression, and a short hairpin to increase transcript stability. Add a 3-nt spacer and a transcriptional terminator to the 3′ end of the input RNA.
10. Construct the plasmids necessary for expressing the input RNAs, and proceed with performance evaluation.

3.4 Design of Multi-input OR Logic Devices

Ribocomputing OR logic devices use a concatenated switch RNA hairpin array upstream and in the same frame as the output gene. Each switch RNA is tasked with responding to a specific input RNA or a combination of input RNAs. Operation of a two-input OR gate is illustrated in Fig. 7a. Binding of input A causes unwinding of switch A* and allows recruitment of a ribosome to the exposed RBS. Translation can then proceed through the B* switch region to the output gene. The general approach can be extended to *N*-input OR logic using a gate RNA with *N* switch hairpins, as can be seen with the schematic of the six-input OR gate RNA. Use the following procedure and the NUPACK script in Fig. 8 to design the gate RNA for a six-input OR logic system:

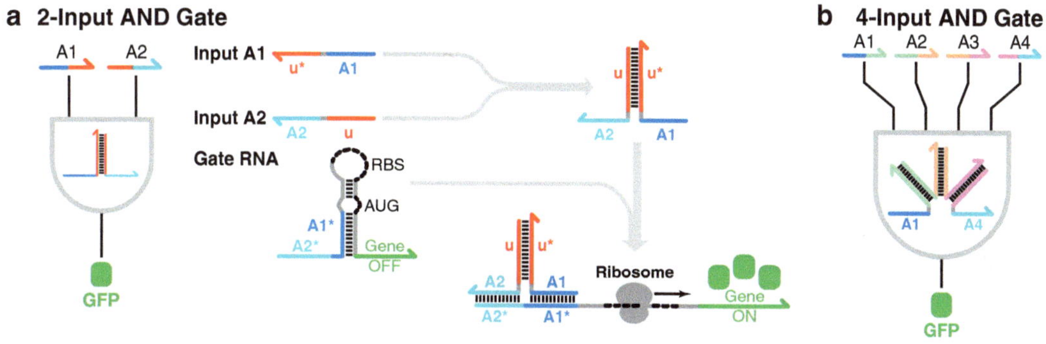

Fig. 7 (**a**) Design of a two-input OR gate RNA composed of two switch RNA hairpins. Each sensor domain has an input RNA recognition site, RBS, and start codon for independent signal activation. (**b**) Schematic of a six-input OR gate RNA featuring six independent switch RNAs concatenated along the same transcript and upstream of an output gene of interest. (Figure adapted from Ref. [17])

```
# Sample NUPACK Code for a Six-Input OR Gate RNA

material = rna1999
temperature = 37.0
trials = 10

# target structure: six toehold switch structures separated by 12-nt
# spacers followed by the linker and output gene
structure gate = U3 U12 D9 (U3 D6 U12 U3) U12 U12 D9 (U3 D6 U12 U3) U12
        U12 D9 (U3 D6 U12 U3) U12 U12 D9 (U3 D6 U12 U3) U12 U12 D9 (U3
        D6 U12 U3) U12 U12 D9 (U3 D6 U12 U3) U50

domain switch_A = GUAAUGAAUUGUAGGCUUGUUAUAGUUAUGAAACAGAGGAGACAUAACAUGAAC
        AAGCCU
domain switch_B = AUAGAUUUGAUGGCUAUUUGUACGUGUUCGAAACAGAGGAGACGAACAAUGACA
        AAUAGC
domain switch_C = CGAAUAAGACUGAGGCUGGGUAUGGUUAGAAAACAGAGGAGAUCUAACAUGACC
        CAGCCU
domain switch_D = AUCACUUAUUGUCGUCUUUGUAUGUCUGUAAAACAGAGGAGAUACAGAAUGACA
        AAGACG
domain switch_E = ACAAAGAUUGGUCGUUUCAUUACCGUUAGAAAACAGAGGAGAUCUAACAUGAAU
        GAAACG
domain switch_F = AAUUGGAUGAAGGCGGUAAGUAUGAUUGUAAGACAGAGGAGAUACAAUAUGACU
        UACCGC

#linkers between switch modules
domain linkerA1 = NNNNNNNNNNNN
domain linkerA2 = NNNNNNNNNNNN
domain linkerA3 = NNNNNNNNNNNN
domain linkerA4 = NNNNNNNNNNNN
domain linkerA5 = NNNNNNNNNNNN

domain preG = GGG
domain linker21 = AACCUGGCGGCAGCGCAAAAG
domain output_gene = AUGCGUAAAGGAGAAGAACUUUUCACUGG #first 29 nt of GFP

# thread sequence domains onto target structures
gate.seq = preG switch_A linkerA1 switch_B linkerA2 switch_C linkerA3
        switch_D linkerA4 switch_E linkerA5 switch_F linker21 output_gene

# specify stop condition for normalized ensemble defect
gate.stop = 10

#
# prevent sequence patterns
#
prevent = AAAA, CCCC, GGGG, UUUU, KKKKKK, MMMMMM, RRRRRR, SSSSSS, WWWWWW,
        YYYYYY
```

Fig. 8 Sample NUPACK code for a six-input OR gate RNA

1. Identify a set of *N* high-performance toehold switches for use in your *N*-input OR gate RNA. Ensure that their switch RNA sequences do not encode any in-frame stop codons.
2. Truncate the *N*-switch RNA sequences to remove any GGG leaders, 21-nt linkers, and output gene sequences since these will not be required for all the switch modules in the gate RNA.
3. Determine the length of the spacer(s) you wish to include between each switch module within the gate RNA. Ensure that the length of the spacer allows each switch RNA to remain in the same reading frame.
4. Thread the switch RNA sequences into the gate RNA sequence in the NUPACK script in Fig. 8.
5. Specify the target structure for the gate RNA using the optimal hairpin structure for each switch RNA. Assert that each linker between the switch modules should be single stranded to maximize accessibility of the toehold domains.
6. Run the resulting NUPACK script on the web portal.
7. Select the top resulting design(s).
8. Construct the plasmids necessary for expressing the input RNAs, and proceed with performance evaluation.

3.5 Design of A AND (NOT B) Logic Devices

Circuits that perform A AND (NOT B) logic can be constructed by using pairs of nearly perfectly complementary transcripts. As shown in Fig. 9, the trigger RNA for a toehold switch is adapted to such logic systems by appending single-stranded binding sites on its 5′ and 3′ ends. A cognate deactivating RNA is then designed to bind and block the trigger RNA from interacting with the gate RNA. Such deactivating transcripts can be implemented against any other input RNA to suppress their function in the circuit. Use the following steps to design the input RNAs for an A AND (NOT B) circuit:

1. Identify a high-performance toehold switch to use for the logic circuit, and isolate the trigger sequence region that binds to the switch RNA. Use this sequence to design the `act_core` domain in the NUPACK script in Fig. 10.
2. Run the NUPACK script using the NUPACK web server.
3. Perform a final round of NUPACK sequence design to add 5′ and 3′ functional domains onto each input RNA. To the 5′ end, append a GGG leader sequence, if using T7 RNA polymerase expression, and a short hairpin to increase transcript stability. Add a 3-nt spacer and a transcriptional terminator to the 3′ end of the input RNA.
4. Construct the plasmids necessary for expressing the input RNAs, and proceed with performance evaluation.

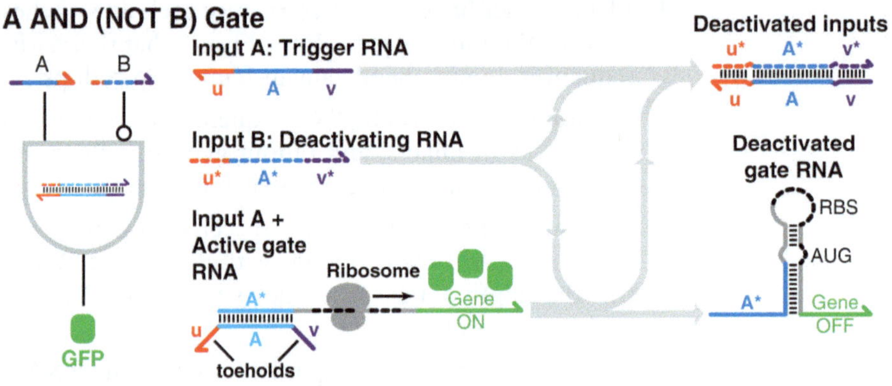

Fig. 9 Design of an A AND (NOT B) logic circuit. The RNA inputs utilize a deactivating RNA (input B) that prevents sensor domain activation through competitive hybridization and strand displacement to inactivate a trigger RNA (input A). (Figure adapted from Ref. [17])

```
# Sample NUPACK Code for an Activating/Deactivating RNA Pair

material = rna1999
temperature = 37.0
trials = 10
prevent = AAAA, CCCC, GGGG, UUUU, KKKKKK, MMMMMM, RRRRRR, SSSSSS,
        WWWWWW, YYYYYY

domain act_core = GAAUCAACGACACUACAAUACAUCAG
domain act_pre  = N16
domain act_post = N16
domain deact_pre  = N16
domain deact_post = N16

structure activator = U58
structure deactivator = U58
structure act_deact_complex = D15 (U1 D26 (U1 D15 + U1) U1)

activator.seq = act_pre act_core act_post
deactivator.seq = deact_pre act_core* deact_post
act_deact_complex.seq = act_pre act_core act_post deact_pre act_core*
        deact_post

activator.stop = 1
deactivator.stop = 1
act_deact_complex.stop = 1
```

Fig. 10 Sample NUPACK code for an activating/deactivating RNA pair

4 Notes

1. All sequences are written in the 5′ to 3′ direction.
2. The bold sequences are ribosome binding sites and initiation codon.
3. NUPACK is also available in several standalone versions for Unix-based systems at http://www.nupack.org. These standalone versions have more features and allow for NUPACK to be used in conjunction with other scripting languages such as Python and MATLAB, which can accelerate the design and evaluation process.
4. All of the code figures here can be downloaded for easier input at https://github.com/AlexGreenLab/DesignOfRibocomputingDevices.
5. NUPACK allows for two different RNA modeling algorithms, the default `rna` (based on the algorithm of Serra and Turner (1995) [26]) and `rna1999` (based on Mathews et al. (1999) [27]). Note that the rna1999 modeling parameter is only valid at 37 °C.
6. The DU+ and dot-parens-plus notations can be explained as follows. In dot-parens-plus, an unpaired base is represented as a dot, a paired base as matching parentheses, and a nick as a plus symbol. When using DU+ notation, a plus also indicates a nick. A length *x* of unpaired bases is represented as U*x*, while a duplex of length *y* is represented as D*y*, with whatever substructure that is at the end of the duplex defined immediately after the D*y*. More elaborate substructures at the end of duplexes are encapsulated in parentheses. For example, a ten base pair hairpin with a six base loop would be defined as D10 U6. If we wanted to add a five base-pair hairpin with its own four base loop into the center of that loop, we would need to encapsulate it as D10 (U3 D5 U4 U3). In dot-parens-plus notation, this would be ((((((((((...(((((....)))))...)))))))))). The reader is advised to seek out the "Demos" and "Help" sections of the NUPACK design web portal for further examples.
7. When defining strands or structure sequences, the asterisk symbol (*) following a domain or strand indicates that the sequence is the reverse complement of the referenced domain or strand.
8. Gate RNA should be expressed from a relatively low copy vector for proper switch RNA-trigger RNA interaction, while input RNA should be expressed from a relatively high copy vector. Past work has used the pCOLADuet (kanamycin resistance, ColA origin) for the gate RNA and pET15b (ampicillin resistance, ColE1 origin) for input RNA.

9. If you need to generate large numbers of riboregulator designs, we strongly advise you to use the local implementation of NUPACK for the design process. This will avoid overburdening the NUPACK web portal and likely generate your designs faster. The local NUPACK implementation uses a slightly different script syntax, but it is straightforward to convert the code. The local NUPACK software package can be downloaded from http://www.nupack.org.
10. To reduce the computation time required, you can truncate the switch RNA sequences to remove the linker and output gene sequences when doing the pairwise predictions. Since all switches have these conserved sequences, in most cases, they will not affect orthogonality.
11. If you need to screen large numbers of riboregulator designs, we strongly encourage you to use the local implementation of NUPACK for computing the expected interactions between switch and trigger RNAs.
12. Even though this edit distance is much less stringent, we have found that it can generate libraries with good orthogonality. Code for computing the edit distance should be readily available and not require a separate software installation.
13. It is worth noting that the design approach that we employ for generating input RNAs for AND logic systems does not use NUPACK to simulate the full N-strand complex required for N-input logic, where $N \geq 3$, nor does it consider unwinding of the switch RNA. Despite this, we have found that functional ribocomputing AND gates can be developed by specifying only two-strand complexes as detailed in Fig. 6. Specifying only two-strand complexes also greatly decreases the time required for NUPACK to generate the input RNA sequences.
14. The L00 domain has a short NNN spacer to the 5′ end of the half trigger domain to facilitate binding between the gate RNA and the multi-input complex. A corresponding NNN spacer is also present in the LN domain to improve binding.

Acknowledgments

This work was supported by an NIH Director's New Innovator Award (1DP2GM126892), an NIH U01 award (1U01AI148319-01), and an NSF RAPID award (2029532) to A.A.G. This work was also supported by the National Research Foundation of Korea (NRF- 2019R1A2C1086830) grant funded by the Korean government (MSIT) to J.K.

References

1. Cameron DE, Bashor CJ, Collins JJ (2014) A brief history of synthetic biology. Nature Reviews Microbiology 12(5):381–390
2. Cheng AA, Lu TK (2012) Synthetic biology: An emerging engineering discipline. Annual Review of Biomedical Engineering 14:155–178
3. Gardner TS, Cantor CR, Collins JJ (2000) Construction of a genetic toggle switch in Escherichia coli. Nature 403(6767):339–342. https://doi.org/10.1038/35002131
4. Kim J, White KS, Winfree E (2006) Construction of an in vitro bistable circuit from synthetic transcriptional switches. Molecular Systems Biology 2:68–68. https://doi.org/10.1038/msb4100099
5. Elowitz MB, Leibler S (2000) A synthetic oscillatory network of transcriptional regulators. Nature 403(6767):335–338. https://doi.org/10.1038/35002125
6. Potvin-Trottier L, Lord ND, Vinnicombe G, Paulsson J (2016) Synchronous long-term oscillations in a synthetic gene circuit. Nature 538(7626):514–517. https://doi.org/10.1038/nature19841
7. Siuti P, Yazbek J, Lu TK (2013) Synthetic circuits integrating logic and memory in living cells. Nature Biotechnology 31(5):448–452. https://doi.org/10.1038/nbt.2510
8. Nielsen AAK, Der BS, Shin J, Vaidyanathan P, Paralanov V, Strychalski EA, Ross D, Densmore D, Voigt CA (2016) Genetic circuit design automation. Science 352(6281): aac7341. https://doi.org/10.1126/science.aac7341
9. Weinberg BH, Pham NTH, Caraballo LD, Lozanoski T, Engel A, Bhatia S, Wong WW (2017) Large-scale design of robust genetic circuits with multiple inputs and outputs for mammalian cells. Nature Biotechnology 35(5):453–462. https://doi.org/10.1038/nbt.3805
10. Yang L, Nielsen AAK, Fernandez-Rodriguez J, McClune CJ, Laub MT, Lu TK, Voigt CA (2014) Permanent genetic memory with >1-byte capacity. Nature Methods 11(12): 1261–1266. https://doi.org/10.1038/nmeth.3147
11. Grünberg R, Serrano L (2010) Strategies for protein synthetic biology. Nucleic Acids Research 38(8):2663–2675. https://doi.org/10.1093/nar/gkq139
12. Chappell J, Watters KE, Takahashi MK, Lucks JB (2015) A renaissance in RNA synthetic biology: New mechanisms, applications and tools for the future. Current Opinion in Chemical Biology 28:47–56. https://doi.org/10.1016/j.cbpa.2015.05.018
13. Green AA (2019) Synthetic bionanotechnology: Synthetic biology finds a toehold in nanotechnology. Emerging Topics in Life Sciences 3:507–516
14. Liang JC, Bloom RJ, Smolke CD (2011) Engineering biological systems with synthetic RNA molecules. Molecular Cell 43(6):915–926. https://doi.org/10.1016/j.molcel.2011.08.023
15. Hong S, Kim J, Kim J (2021) Multilevel gene regulation using switchable transcription terminator and toehold switch in Escherichia coli. Applied Sciences 11(10):4532
16. Green AA, Silver PA, Collins JJ, Yin P (2014) Toehold switches: De-novo-designed regulators of gene expression. Cell 159(4):925–939
17. Green AA, Kim J, Ma D, Silver PA, Collins JJ, Yin P (2017) Complex cellular logic computation using ribocomputing devices. Nature 548(7665):117–121
18. Kim J, Zhou Y, Carlson P, Teichmann M, Chaudhary S, Simmel FC, Silver PA, Collins JJ, Lucks JB, Yin P, Green AA (2019) De novo-designed translation-repressing riboregulators for multi-input cellular logic. Nature Chemical Biology 15:1173–1182. https://doi.org/10.1101/501783
19. Hong F, Ma D, Wu K, Mina LA, Luiten RC, Liu Y, Yan H, Green AA (2020) Precise and programmable detection of mutations using ultraspecific riboregulators. Cell 180(5): 1018–1032.e1016
20. Kim J, Yin P, Green AA (2018) Ribocomputing: Cellular logic computation using RNA devices. Biochemistry 57(6):883–885
21. Zadeh JN, Steenberg CD, Bois JS, Wolfe BR, Pierce MB, Khan AR, Dirks RM, Pierce NA (2011) NUPACK: Analysis and design of nucleic acid systems. Journal of Computational Chemistry 32(1):170–173. https://doi.org/10.1002/jcc.21596
22. Hong S, Park D, Chaudhary S, McCutcheon G, Green AA, Kim J (2022) Design of RNA-based translational repressors. Methods in Molecular Biology
23. Green AA (2017) Construction and in vivo testing of prokaryotic riboregulators. Methods in Molecular Biology 1632:285–302. https://doi.org/10.1007/978-1-4939-7138-1_19
24. Pardee K, Green AA, Takahashi MK, Braff D, Lambert G, Lee JW, Ferrante T, Ma D, Donghia N, Fan M (2016) Rapid, low-cost

detection of Zika virus using programmable biomolecular components. Cell 165(5): 1255–1266

25. Zhang Q, Ma D, Wu F, Standage-Beier K, Chen X, Wu K, Green AA, Wang X (2021) Predictable control of RNA lifetime using engineered degradation-tuning RNAs. Nature Chemical Biology 17(7):828–836. https://doi.org/10.1038/s41589-021-00816-4

26. Serra MJ, Turner DH (1995) Predicting thermodynamic properties of RNA. Methods in Enzymology 259:242–261. https://doi.org/10.1016/0076-6879(95)59047-1

27. Mathews DH, Sabina J, Zuker M, Turner DH (1999) Expanded sequence dependence of thermodynamic parameters improves prediction of RNA secondary structure. Journal of Molecular Biology 288(5):911–940. https://doi.org/10.1006/jmbi.1999.2700

Chapter 5

Computational Design of Small Transcription Activating RNAs (STARs)

Baiyang Liu and James Chappell

Abstract

A major goal of synthetic biology has been to develop libraries of versatile genetic regulators that enable the precise control of gene expression. In recent years, the creation of novel RNA design motifs has allowed for the bottom-up, computational design of large libraries of high-performing and orthogonal RNA regulator systems. One example of this is Small Transcription Activating RNAs (STARs), which function through the conditional formation of terminator hairpins to activate the transcription of targeted genes. STARs have found broad utility for creating synthetic gene circuits, engineering metabolic pathways, and creating new types of diagnostics. Here we describe the method to computationally design, build, and characterize STAR regulators.

Key words Synthetic RNA, sRNA regulators, RNA synthetic biology

1 Introduction

A major goal of synthetic biology is to develop versatile and programmable genetic regulators that enable the precise control of gene expression [1]. In recent years, there has been increased interested in developing synthetic small RNA regulators (sRNAs) for this purpose [2–4]. These regulatory systems are composed of a *trans*-acting sRNA that exerts regulation via direct RNA-RNA interactions with a target RNA sequence contained within an mRNA [2, 4]. These interactions result in structural rearrangements of the mRNA that regulate gene expression, typically by occluding or exposing regulatory features, such as ribosome binding sites (RBS) in the case of translation or intrinsic terminator hairpins in the case of transcription. Motivating interest in sRNAs for synthetic gene control is their versatility of regulation. For example, natural and synthetic sRNAs are available that can control almost every aspect of gene expression (e.g., transcription, translation, RNA processing, degradation). Additionally, sRNA

James Chappell and Melissa K. Takahashi (eds.), *Riboregulator Design and Analysis*,
Methods in Molecular Biology, vol. 2518, https://doi.org/10.1007/978-1-0716-2421-0_5,

regulation is typically achieved through the formation of simple RNA structures (e.g., hairpin motifs, RNA:RNA duplexes) that are governed by the predictable rules of Watson-Crick base pairing. As a result, there is great potential for the bottom-up molecular-level design of these genetic control systems.

While creation of synthetic sRNAs was an early success of synthetic biology [5], their wide spread has been limited by their low dynamic range (i.e., *difference between on/off states*) and limited numbers of orthogonal variants (i.e., *regulators with no cross-regulation)*. This has recently been addressed by combining synthetic sRNA mechanisms dependent upon simple design motifs, with computational RNA design algorithms [6–9]. For example, Small Transcription Activating RNAs (STARs) are an entirely synthetic regulatory mechanism that work through conditional formation of a terminator hairpin encoded within a target RNA placed upstream of a gene to be controlled (Fig. 1a) [10]. While STARs were initially created by rational engineering of natural terminator sequences [10, 11], it was shown they can also be computationally designed [6]. This approach used a design motif composed of the highly efficient target RNA terminator scaffold from the AD1 STAR system [10]. Then, an RNA structure design algorithm called NUPACK [12] was used to design and optimize the key recognition sequence of this switch, the linear region and corresponding linear binding region of the target RNA and STAR pair, all while maintaining the intrinsic terminator sequence (Fig. 1b).

Importantly, NUPACK can be used to optimize the recognition region of this switch, to create STAR-target pairs that efficiently interact, resulting in higher dynamic range of regulation, or to create sequences that do not cross-react, allowing for orthogonal control. For example, previously a library of 100 STARs was computationally designed which led to the identification of variants with high dynamic ranges of 600-fold and creation of orthogonal libraries [6]. Once created, these regulatory systems can be used in a variety of applications, including genetic circuits [6, 10, 13, 14], metabolic engineering [15], and diagnostics [16].

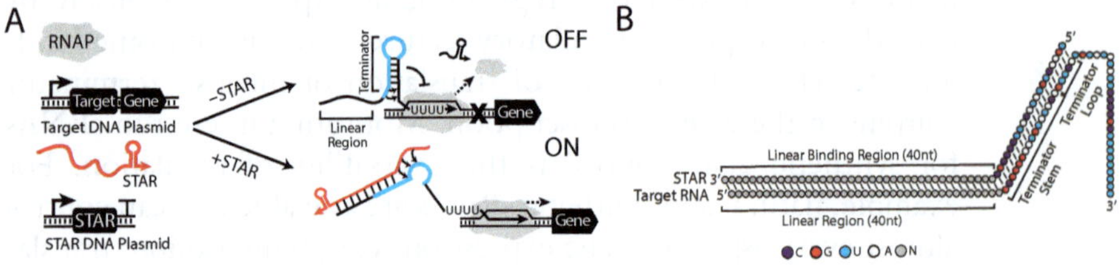

Fig. 1 Design of Small Transcription Activating RNAs (STARs). (**a**) The transcription of a target gene is regulated by an intrinsic terminator (target RNA) inserted upstream of the coding sequence. The RNA polymerase (RNAP) will only transcribe the target gene if the STAR binds to the target RNA and inhibits terminator formation. (**b**) Schematic of the sequence and structure constraints for the STAR and target RNA used in the NUPACK design algorithm. Nucleotides colored according to identity with N representing an unconstrained nucleotide that is designed by NUPACK

2 Materials

2.1 Computational STAR Design

1. NUPACK design web application [12] or locally installed version.

2.2 Construction of STAR Plasmids Using Inverse PCR

1. Target forward/STAR reverse primer: 5′ NNNNNNNNNN NNNNNNNNNNNNNNNNNNNNNNNNNNNNNNGC GGGGAATGTATACAGTTC 3′. N represents a NUPACK-designed nucleotide.
2. Target RNA reverse primer: 5′ ACTAGTATTATACCTAG GACTGAGCTAGCTGTCAAAGATCTTTAGAATT CAGCCGGAAG 3′.
3. STAR forward primer: 5′ GGATCTCAAAGCCCGCC GAAAGGC 3′.
4. Suitable DNA template for STAR/target RNA PCR reactions (e.g., STAR plasmid pJBL2807 (Addgene plasmid #71203) and target RNA plasmid pJBL2801 (Addgene plasmid #71207)).
5. 10,000 units/mL T4 polynucleotide kinase and suitable buffer (e.g., 10× ligase buffer).
6. 37 °C static incubator.
7. Phusion DNA polymerase and 5× Phusion buffer.
8. 10 mM dNTPs.
9. Molecular grade water.
10. Thermocycler.
11. 20,000 units/mL DpnI.
12. PCR purification kit.
13. DNA gel electrophoresis reagents and apparatus.
14. 400,000 units/mL T4 DNA ligase and 10× ligase buffer.
15. *E. coli* NEB Turbo competent cells.
16. LB agar plates with 100 μg/mL carbenicillin (carb).
17. LB agar plates with 100 μg/mL chloramphenicol (cam).
18. LB media with 100 μg/mL carbenicillin (carb).
19. LB media with 100 μg/mL chloramphenicol (cam).
20. 37 °C shaking incubator.
21. DNA plasmid miniprep kit.

Table 1
M9 minimal media recipe. We recommend sterilizing each component separately by autoclaving, before combining. For thiamine hydrochloride, we recommend filter sterilization

Component	Volume
5× M9 minimal salts	10 mL
10 mg/mL thiamine hydrochloride	1.7 mL
40% glycerol	500 μL
10% casamino acids	1 mL
1 M $MgSO_4$	100 μL
1 M $CaCl_2$	5 μL
Sterilized ddH_20	Up to 50 mL

22. Sanger sequencing primers. STAR seq primer: 5′ TTACCGC CTTTGAGTGAGCTGATACCGCTCGC 3′. Target RNA sequencing primer: 5′ AAATGTAGCACCTGAAGT CAGCCCC 3′.

2.3 GFP Activation Assays

1. LB media with 100 μg/mL carb and 100 μg/mL cam.
2. LB agar plates with 100 μg/mL carb and 100 μg/mL cam.
3. M9 minimal media with 100 μg/mL carb and 100 μg/mL cam (Table 1).
4. Deep 96-well plates and breathable sealing membrane.
5. Transparent 96-well plates.
6. 1× PBS solution.
7. Shaking well plate incubator.
8. Fluorescence plate reader.
9. Purified DNA plasmids: Sequence-verified STAR-encoding plasmid, sequence-verified target RNA-encoding plasmid, and suitable blank plasmids (e.g., pJBL001 (Addgene plasmid #71211) and pJBL002 (Addgene plasmid #71202)).

3 Methods

In this section, we will describe the design, construction, and characterization of computationally designed STAR regulators.

3.1 Computational STAR Design

To computationally design STAR regulators, we suggest using the NUPACK RNA design algorithm [17], specifically, the NUPACK design web application [12]. NUPACK can also be run locally using the NUPACK script provided. Other RNA design programs can also be used, following the general design principles outlined previously [6]:

1. Open the NUPACK RNA online design tool (http://www.nupack.org/).
2. Upload a NUPACK design script (*see* **Note 1**).
3. Click the update button and check the structure generated within the preview window (*see* **Note 2**).
4. Click the design button to initiate design (*see* **Note 3**).
5. Save the NUPACK-designed sequences for the STAR and target pair(s) in an appropriate file (e.g., text file, DNA sequence file, excel).

3.2 Construction of the STAR Plasmids Using Inverse PCR

This method will describe construction of a STAR and target RNA-encoding plasmid using inverse PCR. In our experience, this is the most cost-effective and straightforward way to clone these relatively short sequences; however, other methods are available and can be used. We will describe construction of two plasmids encoding one STAR/target RNA pair, which can easily be scaled for the desired application.

1. Design primers for inverse PCR. A single primer is used to encode the NUPACK-designed sequence. This primer serves as both the forward primer for the target RNA-encoding plasmid and the reverse primer for the STAR-encoding plasmid. A universal reverse primer and forward primer are used for the target RNA-encoding plasmid and STAR-encoding plasmid, respectively. These primers are considered universal because they can be used to construct any STAR/target RNA plasmid regardless of the NUPACK-designed sequence. Thus, for a single STAR/target plasmid pair, only one unique primer is required (Fig. 2).
2. For each primer (3 total), set up a PNK phosphorylation reaction: 1 μL of 100 μM primer, 1 μL of 10× ligase buffer, 0.5 μL of 10,000 units/mL T4 PNK, and water for a total of 10 μL (*see* **Note 4**).
3. Incubate at 37 °C for 1 h.
4. Once incubation is complete, set up two separate PCR reactions: 10 μL 5× Phusion buffer, 2.5 μL phosphorylated forward primer (from **Step 2**), 2.5 μL phosphorylated reverse primer (from **Step 2**), 1–10 ng DNA plasmid template, 1 μL 10 mM dNTPs, and 0.5 μL Phusion DNA polymerase and water for a total of 50 μL.
5. Add the PCR tube to thermocycler, and run the following program, as seen in Table 2 (*see* **Note 5**).

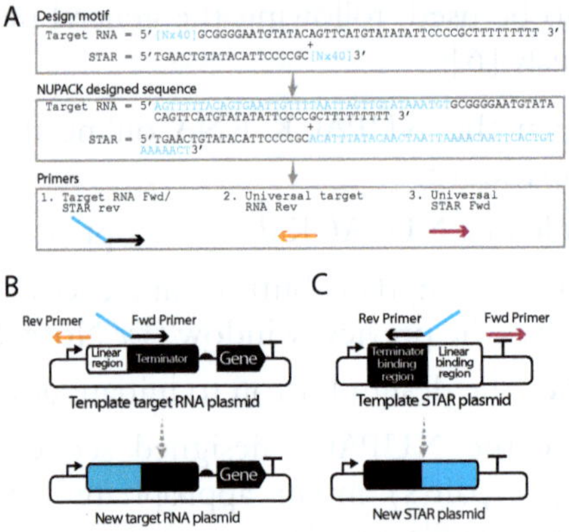

Fig. 2 Cloning strategy for STAR system. (**a**) To create a new STAR variant, a pair of complementary 40 nt linear regions is first generated in NUPACK, which are added to the 5′ of target and the 3′ of STAR, respectively. Three primers are needed to clone a STAR-target RNA pair, including two universal primers and one unique primer. (**b**) The unique primer is designed to be used in both target RNA and STAR cloning, which contains the NUPACK-generated 40 nt linear region and a part of terminator for annealing. One universal primer is used as a reverse primer (rev primer) for the target RNA plasmid. One universal primer is used as a forward primer (fwd primer) for the STAR plasmid

Table 2
Thermocycler protocol

Step	
1	98 °C, 30 s
2	98 °C, 15 s
3	Tm 1, 45 s
4	72 °C, 45s–1 min
5	Repeat **Steps 2–4**, 10×
6	98 °C, 15 s
7	Tm 2, 45 s
8	72 °C, 45s–1 min
9	Repeat **Steps 6–8**, 20×
10	72 °C, 5 min
11	12 °C, hold

6. Analyze the PCR products by performing DNA gel electrophoresis using an appropriate DNA ladder (e.g., NEB quick load 1 kb DNA ladder). For an efficient and specific PCR amplification, expect a band at ~3400 bp and ~2200 bp for the target RNA and STAR PCR reactions, respectively.
7. If unsuccessful, repeat PCR using alternative conditions (*see* **Note 6**).
8. If successful, add 0.5 μL of DpnI to each PCR reaction, mix by pipetting, and incubate at 37 °C for 1 h.
9. Perform a column-based PCR purification according to the manufacturer's instructions, and elute in molecular grade water.
10. Quantify DNA using preferred method. A suitable yield is 10–100 ng/μL.
11. Using these purified PCR products, set up two ligation reactions: 50–150 ng of PCR product, 1 μL 10× ligase buffer, and 0.5 μL 400,000 units/mL T4 ligase and water for a total of 10 μL.
12. Incubate for 2 hours (h) at room temperature.
13. Transform competent *E. coli* cells separately with the two ligation reactions. For transformation of STAR-encoding plasmids, plate cells on LB agar with 100 μg/mL carb, and for transformation of target RNA-encoding plasmids, plate cells on LB agar with 100 μg/mL cam.
14. After growth, pick colonies to grow cultures using LB media with appropriate antibiotics, and isolate plasmids using a DNA miniprep kit according to the manufacturer's instructions.
15. Sanger sequence DNA plasmids. Example sequencing primers are given (*see* Subheading 2.2).

3.3 GFP Activation Assays

1. For a given STAR-target RNA pair, transform *E. coli* TG1 (*see* **Note 7**) using the following combinations of plasmids, as seen in Table 3 (*see* **Note 8**).
2. Plate transformed cells onto LB agar plates with 100 μg/mL carb and 100 μg/mL cam, and grow overnight at 37 °C.
3. Pick ≥3 colonies from each transformation condition into a deep 96-well plate containing 300 μL of LB media with appropriate antibiotics. Include three wells with only media (i.e., no cells). Seal with a breathable sealing membrane.
4. Grow in a shaking well plate incubator at 1000 rpm at 37 °C.
5. After overnight growth, take 4 μL from each well, and inoculate a deep 96-well plate containing 196 μL of M9 minimal media (*see* **Note 9**).

Table 3
Example of transformation conditions to be used to characterize a STAR-target RNA pair

Condition	Plasmid 1 (chloramphenicol resistant)	Plasmid 2 (ampicillin resistant)
Blank/empty cells	Empty target RNA plasmid (e.g., pJBL001 (Addgene plasmid #71211))	Empty STAR plasmid (e.g., pJBL002 (Addgene plasmid #71202))
No STAR condition (off state)	Target RNA plasmid	Empty STAR plasmid (e.g., pJBL002 (Addgene plasmid #71202))
+ STAR condition (on state)	Target RNA plasmid	STAR plasmid

6. Include three wells with only media (i.e., no cells). Seal with a breathable sealing membrane.
7. Grow in a shaking well late incubator at 1000 rpm at 37 °C.
8. After 6–8 h, take 50 μL of culture from each well of the deep 96-well plate, and add to a transparent 96-well plate. Then, add 50 μL of PBS to each well.
9. Measure the GFP fluorescence (FL) (485 nm excitation and 520 nm emission) and optimal density at 600 nm (OD) in the plate reader.
10. Calculate the normalized FL/OD values. First, subtract the average media FL and OD values from each culture's FL and OD values. Second, divide FL values by corresponding OD for each culture. Normalize the FL/OD for each culture by subtracting FL/OD from cells transformed with blank plasmids.

4 Notes

1. An example of design script for the NUPACK web application is provided below:

```
# STAR computational design script for NUPACK web-application
material = rna
temperature = 37.0
trials = 3
sodium[M] = 1.0
dangles = some
allowmismatch = true
structure STAR =
..............................................................
structure Target=
.............................................................-
```

```
.........................
s t r u c t u r e                 O N                          =
(((((((((((((((((((((((((((((((((((((((((((((((((((((((((((-
(+)))))))))))))))))))))))))))))))))))))))))))))))))))))))))-
))).........................
domain a = N40
domain b = GCGGGGAATGTATACAGTTCATGTATATATTCCCCGCTTTTTTTTT
domain c = TGAACTGTATACATTCCCCGC
STAR.seq = c a*
Target.seq = a b
ON.seq = c a* a b
prevent = AAAA, CCCC, GGGG, UUUU, KKKKKK, MMMMMM, RRRRRR,
SSSSSS, WWWWWW, YYYYYY
STAR.stop = 30.0
```

2. For the provided design script, you should expect to see a linear STAR and target RNA structure. For the ON structure, you should see the STAR (a* c) fully hybridized to the target RNA (a b).
3. Designs can take several minutes to perform, and it is recommended that you provide an email address for NUPACK.
4. The protocol provided is based upon reagents from New England Biolabs (NEB). If other manufacturers' reagents are used, the protocols can be adjusted as needed.
5. We typically perform a 30-cycle PCR protocol in which a lower melting temperature (Tm 1) is used for the first 10 cycles and then a higher melting temperature (Tm 2) is used for the last 20 cycles. For Tm 1, we use a temperature that is around the predicted melting temperature of the non-variable region from the target forward/STAR reverse primer (*see* Subheading 2.2, **Step 1**). For Tm 2, we use a temperature that is around the predicted melting temperature of the entire target forward/STAR reverse primer. We strongly recommend splitting each PCR reaction into smaller aliquots and performing a temperature gradient for Tm 1 annealing step (i.e., **Step 3** in the thermocycler protocol). For example, you can take the 50 μL PCR reaction and split it into two 25 μL reactions. Each reaction can be run at a different temperature. We have found best success running at the predicted melting temperature (Tm) and +3 °C above the predicted Tm.
6. Small RNAs often form strong secondary structures (e.g., hairpins) that can reduce efficiencies of PCR. We have found that addition of DMSO (5–10% total volume) into the PCR reaction can address this. We recommend performing PCRs without DMSO and then repeating PCRs with DMSO for any reactions that fail to yield a PCR product.

7. We have performed STAR measurements in a diversity of different *E. coli* strains and have observed consistent performance.
8. Blank plasmids are plasmids containing the antibiotic resistance gene but no RNA expression cassette. These plasmids are used in place of STAR and target RNA-encoding plasmids for conditions in which these RNAs should be absent (e.g., blank cells and no STAR conditions). We provide the name of blank plasmids used previously (*see* Subheading 2.3).
9. We have performed STAR measurements in a diversity of different media and have observed consistent performance.

Acknowledgments

Welch Foundation [C-1982–20190330 to J.C.]; Alfred P. Sloan Research Fellowship [FG-2018-10500 to J.C.]; J.C. is an Alfred P. Sloan Research Fellow.

References

1. Nielsen AA, Segall-Shapiro TH, Voigt CA (2013) Advances in genetic circuit design: Novel biochemistries, deep part mining, and precision gene expression. Current Opinion in Chemical Biology 17:878–892. https://doi.org/10.1016/j.cbpa.2013.10.003
2. Ameruoso A, Gambill L, Liu B, Villegas Kcam MC, Chappell J (2019) Brave new 'RNA' world – Advances in RNA tools and their application for understanding and engineering biological systems. Current Opinion in Systems Biology 14:32–40. https://doi.org/10.1016/j.coisb.2019.02.005
3. Chappell J, Takahashi MK, Meyer S, Loughrey D, Watters KE, Lucks J (2013) The centrality of RNA for engineering gene expression. Biotechnology Journal 8:1379–1395. https://doi.org/10.1002/biot.201300018
4. Chappell J, Watters KE, Takahashi MK, Lucks JB (2015) A renaissance in RNA synthetic biology: New mechanisms, applications and tools for the future. Current Opinion in Chemical Biology 28:47–56. https://doi.org/10.1016/j.cbpa.2015.05.018
5. Isaacs FJ, Dwyer DJ, Ding C, Pervouchine DD, Cantor CR, Collins JJ (2004) Engineered riboregulators enable post-transcriptional control of gene expression. Nature Biotechnology 22:841–847. https://doi.org/10.1038/nbt986
6. Chappell J, Westbrook A, Verosloff M, Lucks JB (2017) Computational design of small transcription activating RNAs for versatile and dynamic gene regulation. Nature Communications 8:1051. https://doi.org/10.1038/s41467-017-01082-6
7. Kim J, Zhou Y, Carlson PD, Teichmann M, Chaudhary S, Simmel FC, Silver PA, Collins JJ, Lucks JB, Yin P, Green AA (2019) De novo-designed translation-repressing riboregulators for multi-input cellular logic. Nature Chemical Biology 15:1173–1182. https://doi.org/10.1038/s41589-019-0388-1
8. Green AA, Silver PA, Collins JJ, Yin P (2014) Toehold switches: De-novo-designed regulators of gene expression. Cell 159:925–939. https://doi.org/10.1016/j.cell.2014.10.002
9. Lee YJ, Kim S-J, Amrofell MB, Moon TS (2019) Establishing a multivariate model for predictable antisense RNA-mediated repression. ACS Synthetic Biology 8:45–56. https://doi.org/10.1021/acssynbio.8b00227
10. Chappell J, Takahashi MK, Lucks JB (2015) Creating small transcription activating RNAs. Nature Chemical Biology 11:214–220. https://doi.org/10.1038/nchembio.1737
11. Meyer S, Chappell J, Sankar S, Chew R, Lucks JB (2016) Improving fold activation of small transcription activating RNAs (STARs) with rational RNA engineering strategies: Improving Small Transcription Activating RNAs. Biotechnology and Bioengineering 113:216–225. https://doi.org/10.1002/bit.25693
12. Zadeh JN, Steenberg CD, Bois JS, Wolfe BR, Pierce MB, Khan AR, Dirks RM, Pierce NA

(2011) NUPACK: Analysis and design of nucleic acid systems. Journal of Computational Chemistry 32:170–173. https://doi.org/10.1002/jcc.21596

13. Lee YJ, Kim S-J, Moon TS (2018) Multilevel regulation of bacterial gene expression with the combined STAR and antisense RNA system. ACS Synthetic Biology 7:853–865. https://doi.org/10.1021/acssynbio.7b00322
14. Westbrook A, Tang X, Marshall R, Maxwell CS, Chappell J, Agrawal DK, Dunlop MJ, Noireaux V, Beisel CL, Lucks J, Franco E (2019) Distinct timescales of RNA regulators enable the construction of a genetic pulse generator. Biotechnology and Bioengineering 116:1139–1151. https://doi.org/10.1002/bit.26918
15. Glasscock CJ, Biggs BW, Lazar JT, Arnold JH, Burdette LA, Valdes A, Kang M-K, Tullman-Ercek D, Tyo KEJ, Lucks JB (2021) Dynamic control of gene expression with riboregulated switchable feedback promoters. ACS Synthetic Biology 10:1199–1213. https://doi.org/10.1021/acssynbio.1c00015
16. Verosloff M, Chappell J, Perry KL, Thompson JR, Lucks JB (2019) PLANT-Dx: A molecular diagnostic for point-of-use detection of plant pathogens. ACS Synthetic Biology 8:902–905. https://doi.org/10.1021/acssynbio.8b00526
17. Wolfe BR, Porubsky NJ, Zadeh JN, Dirks RM, Pierce NA (2017) Constrained multistate sequence design for nucleic acid reaction pathway engineering. Journal of the American Chemical Society 139:3134–3144. https://doi.org/10.1021/jacs.6b12693

Chapter 6

Design and Assembly of Multilevel Transcriptional and Translational Regulators for Stringent Control of Gene Expression

F. Veronica Greco, Thea Irvine, Claire S. Grierson, and Thomas E. Gorochowski

Abstract

Precise control of gene expression is crucial when reprogramming the behavior of living cells. However, common inducible systems often lack the ability to stringently control gene expression due to the use of a single type of regulator that can be susceptible to unavoidable biomolecular fluctuations. In contrast, multilevel controllers (MLCs) employ several forms of regulation simultaneously to overcome this issue, ensuring a reduced basal expression while minimally affecting the maximum induced expression level that can be achieved. Here, we show how our publicly available genetic toolkit can be used to simplify the assembly of MLCs for the stringent control of gene expression. We demonstrate how new compatible parts can be designed and explain the rapid end-to-end assembly procedure.

Key words Gene regulation, Genetic circuits, Transcription, Translation, Synthetic biology

1 Introduction

Advances in synthetic biology over the past few decades have generated an array of well-characterized genetic parts that can be assembled to create more complex circuits able to dynamically regulate gene expression [1–5]. A fundamental requirement in many of these genetic circuits is the ability to externally control the expression of a gene through the use of small molecules [6] or environmental factors such as temperature [7] or light [8, 9]. In these cases, it is common to use transcription factors (TFs) that are sensitive to a particular signal (i.e., inducer) such that the presence or absence of the signal controls binding of the TF to a DNA region in or near the promoter of a gene of interest. Such binding is used to either activate or repress transcription of the gene of interest by stimulating or sterically blocking initiation of RNA polymerase

James Chappell and Melissa K. Takahashi (eds.), *Riboregulator Design and Analysis*,
Methods in Molecular Biology, vol. 2518, https://doi.org/10.1007/978-1-0716-2421-0_6,

(RNAP). A commonly used example is the P_{tac} system [10], which consists of a constitutively expressed LacI repressor protein. Under normal conditions, LacI strongly binds to operator sites within P_{tac} promoter sequences, hampering initiation of RNAP and thus switching off expression of any downstream genes. To turn on expression, the inducer isopropyl β-d-1-thiogalactopyranoside (IPTG) is added, which abolishes the DNA binding activity of LacI, allowing for transcription from the P_{tac} promoter to occur and the expression of target genes.

In many cases, such regulation is sufficient. However, some applications require more precise control over gene expression that is difficult to achieve with these basic inducible systems [11]. Specifically, stochastic biomolecular fluctuations in a cell can lead to low levels of basal ("leaky") gene expression that are difficult to irradicate [12, 13]. For systems expressing genes that are toxic to the cell or which drive highly sensitive downstream processes [14], this low-level expression can result in reduced cell viability (and thus production), as well as cause the failure of complex circuits through unwanted triggering of subcomponents [15–17].

One approach for more stringent control of gene expression is to use multiple forms of regulation simultaneously to create what we call multilevel controllers (MLCs) (*see* Fig. 1) [18]. These can be constructed by supplementing an existing inducible system that relies on transcriptional regulation by a TF with secondary control of transcriptional or translational processes. Examples of such secondary RNA-based regulators include (1) toehold switches (THS) where the ribosome binding site (RBS) of a target gene is sequestered into a strong RNA secondary structure, halting translation initiation until the expression of a cognate small "trigger" RNA (sRNA) that disrupts the secondary structure and enables translation initiation to occur [19, 20], (2) small transcription activating RNAs (STARs) that are used to inhibit additional transcriptional terminators that are placed in the 5′ untranslated region (UTR) of a gene of interest [21, 22], and (3) what has been referred to as a dual control (DC) system where control of both transcription and translation is achieved through hybrid RNA regulators [23].

In this chapter, we describe how multilevel controllers (MLCs) can be assembled using our publicly available genetic toolkit (Addgene: #166922–166,936) and describe how compatible parts can be created to implement new types of MLC. The protocols and toolkit presented have been designed to ensure the rapid and accurate assembly of MLCs through the careful selection of assembly sequences that minimize unwanted cross-reactivity [24, 25] and the use of a reporter gene "dropout" to simplify the screening of properly assembled designs [26]. Furthermore, a key feature of the MLCs is the alternating orientation of internal components, which ensures any internal transcriptional read-through from one element to the next does not cause unwanted triggering

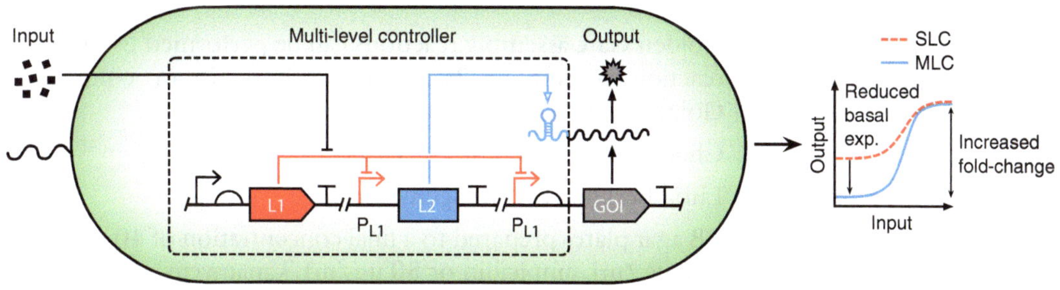

Fig. 1 Overview of a multilevel controller (MLC). Level 1 (L1) and level 2 (L2) regulation shown in red and blue, respectively. Compared to a single-level controller (SLC) (e.g., the P_{tac} system used alone), an MLC is able to reduce basal expression, leading to a larger fold change in activation and the suppression of low-level gene expression noise. Diagram shown using Synthetic Biology Open Language Visual (SBOLv) symbols [30, 31] and created using DNAplotlib [32, 33]

of the system [15, 27]. We anticipate that this step-by-step guide will support those requiring precise control of gene expression and help to improve the design of high-performance synthetic genetic circuits.

2 Materials

2.1 Creation of New Toolkit Compatible Parts

1. Computational design of DNA sequences and assessment of primer suitability (i.e., GC content, secondary structure propensity, and off-target binding analysis) can be performed using the Benchling website (https://benchling.com) or other plasmid editor tools.
2. Chemically competent DH10 beta *E. coli* cells.
3. Plasmid purification kit.
4. High-fidelity DNA polymerase kit.
5. PCR/DNA cleanup kit.
6. TAE buffer: 40 mM Tris, 20 mM acetate, 1 mM ethylenediaminetetraacetic acid, and 0.5 M sodium EDTA.
7. 1% agarose gels for DNA electrophoresis are prepared using molecular biology grade agarose and TAE buffer containing DNA gel stain.
8. DNA ladder: 1 kb + DNA ladder.
9. Annealing buffer: 10 mM Tris base, pH 7.5–8.0, 50 mM NaCl, and 1 mM EDTA.
10. Oligo phosphorylation: Performed using the phosphokinase enzyme in T4 ligase buffer following manufacturer's instructions.
11. Antibiotics: 50 μg/mL kanamycin and 100 μg/mL ampicillin.

2.2 Reagents for MLC Assembly

1. Golden Gate assembly reactions can be performed using a final reaction volume of 20 μL using BsaI-HF or commercial Golden Gate assembly kit using BsaI-HF.
2. Chemically competent DH10-beta *E. coli* cells.
3. Luria Bertani (LB) lysogeny broth.
4. LB agar plates prepared to a final concentration of 40 g/L with 100 μg/mL ampicillin or 50 μg/mL kanamycin.
5. Plasmid purification kit.

3 Methods

An MLC consists of eight different types of building blocks that must be assembled using a one-pot Golden Gate reaction [25, 28] to create a functional design (*see* Fig. 2). Each of these blocks can take the form of a pre-cloned plasmid (*see* Subheading 3.1), amplicon (*see* Subheading 3.2), or annealed pair of complementary DNA oligos (*see* Subheading 3.3). The use of pre-cloned plasmids is the preferred method as they reduce the chance of part mutations. Amplicons are easy to generate by PCR and allow for the rapid addition of required sequences at the 5′- and 3′-end of each block. Annealed complementary DNA oligos are quick to synthesize and are ideal for smaller sequences less than 60 nt long. However, this approach tends to lead to even higher error rates than when using amplicons due to the fidelity of oligo synthesis. In all cases, every new part must be flanked by a specific sequence that is used to create necessary 4 bp overhangs which are generated by BsaI digestion and used during assembly (Table 1).

Our toolkit comprises of plasmids (pA–pH) encoding each of the basic parts and a backbone/destination vector (pMLC-BB1) in which the final MLC will be inserted (Table 2; *see* Fig. 2b). This backbone plasmid contains a constitutively expressed orange fluorescent protein (*ofp*) that is removed upon successful insertion of an MLC, allowing for the absence of fluorescence in transformed cells to be used for screening of successful assembly. Although the toolkit is tailored to produce MLCs, it is also possible to create single-level controllers (SLCs) (e.g., a P_{tac} inducible system) by substituting several blocks with predesigned spacer elements. Specifically, blocks B and C can be replaced by a single inert spacer element that is able to connect with block A at its 5′-end and block D at its 3′-end. Furthermore, the removal of the secondary regulator also requires the use of an additional RBS part for block G that can easily be made by annealing complementary oligos containing the required RBS sequence (*see* Subheading 3.3). Another element of our MLC design is the ability to further tune the turnover

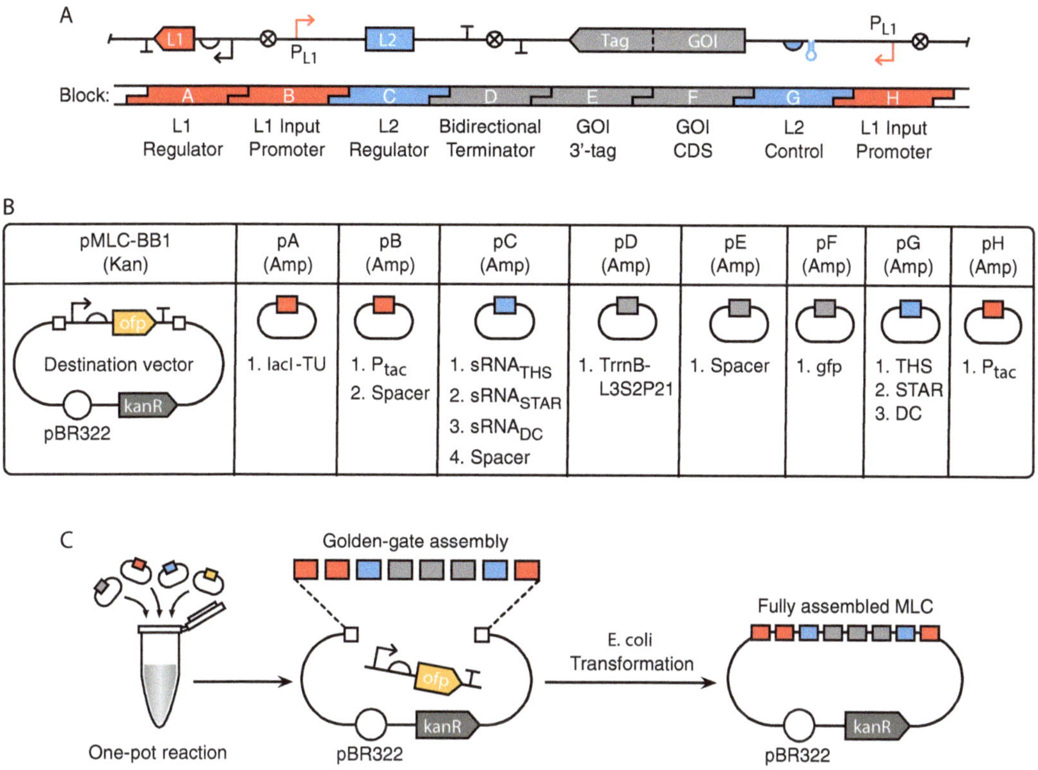

Fig. 2 Overview of the genetic design for MLCs. (**a**) Core design of an MLC. Level 1 (L1) and level 2 (L2) regulation shown in red and blue, respectively. (**b**) Genetic toolkit and parts available from Addgene. See Table 1 for further details regarding each plasmid. (**c**) Key steps in the assembly of an MLC. Antibiotic to be used for selection shown below each plasmid name in parentheses (Kan, 50 μg/mL kanamycin; Amp, 100 μg/mL ampicillin)

Table 1
Flanking sequences required for assembly of each MLC block

Block	5′ sequence[a]	3′ sequence[a]
A	NNNNNNGGTCTCA**gtga**	**ggag**TGAGACCNNNNNN
B	NNNNNNGGTCTCA**ggag**	**aggt**TGAGACCNNNNNN
C	NNNNNNGGTCTCA**aggt**	**gctt**TGAGACCNNNNNN
D	NNNNNNGGTCTCA**gctt**	**cgct**TGAGACCNNNNNN
E	NNNNNNGGTCTCA**cgct**	**ccgc**CGAGACCNNNNNN
F	NNNNNNGGTCTCA**ggcg**	**gtag**CGAGACCNNNNNN
G	NNNNNNGGTCTCA**catc**	**tact**TGAGACCNNNNNN
H	NNNNNNGGTCTCA**tact**	**tact**TGAGACCNNNNNN

[a]Each sequence contains a 4 bp overhang (shown in lowercase and bold), a BsaI restriction site used for Golden Gate assembly (underlined), and six undefined bases that improve BsaI digestion efficiency

Table 2
MLC toolkit plasmids

Plasmid	Description	References	AddGene #
pMLC-BB1	Destination vector for MLC	[18]	166922
pA01	LacI – L1 regulator	[29]	166923
pB01	P_{tac} promoter – L1 input promoter	[29]	166924
pC01	sRNA (STAR pT1810 Dual Control) – L2 regulator	[23]	166925
pC02	sRNA (STAR AD1.A5) – L2 regulator	[21]	166926
pC03	sRNA (STAR AD1.A6) – L2 regulator	[21]	166927
pC04	sRNA (Toehold Switch) – L2 regulator	[20]	166928
pD01	TrrnB-L3S2P21 – bidirectional terminator	[3]	166929
pE01	Spacer with TAA stop codon – GOI 3′-tag	[18]	166930
pF01	Green fluorescent protein – GOI CDS	[18]	166931
pG01	Target pT181-DC + RBS – L2 control	[23]	166932
pG02	Target AD1.S5 + RBS – L2 control	[21]	166933
pG03	Target AD1.S6 + RBS – L2 control	[21]	166934
pG04	Target – THS – L2 control	[20]	166935
pH 01	P_{tac} promoter – L1 input promoter	[29]	166936

dynamics of a gene of interest (GOI). This is possible by using block E which encode a 3′-tag fused to the GOI which can be used to add a degradation tag.

A key consideration when assembling an MLC is the choice of the secondary, level 2 (L2) regulator used. Our toolkit includes a THS, DC, and two STAR L2 regulators, with the possibility of using computational tools to design new THS and STAR sequences as needed [19, 22]. Previous testing of these different L2 regulator types showed that all reduced relative basal expressions, with both THS and STAR regulators, led to more rapid digital-like transitions between "off" and "on" states when increasing the concentration of an input inducer [18]. Overall, THS L2 regulators displayed the best performance, although STARs were found to work more consistently when used across in vivo and in vitro cell-free protein expression (CFPE) environments, making them a good choice if a user is rapidly prototyping a system using CFPE for ultimate deployment in living cells.

3.1 Part Creation by DNA Synthesis

The simplest way to generate a new part for use with the toolkit is through direct DNA synthesis supplied as a sequence-verified plasmid.

1. Order plasmid containing the desired part sequence such that digestion by BsaI will lead to appropriate 4 bp 5′-overhangs needed for the correct ligation of the multipart construct. These flanking sequences are provided in Table 1 (*see* **Note 1**). Ensure that the synthesized part sequence is provided in a plasmid containing an ampicillin resistance marker for antibiotic selection (*see* **Note 2**).
2. Amplify the synthesized part plasmids by transforming *E. coli* cells with the plasmid.
3. Perform an overnight culture in LB liquid media with appropriate antibiotics.
4. Carry out plasmid purification using standard methods. It is recommended that purified part plasmids are kept at a concentration of 20–50 ng/μL for all assembly steps and are stable when stored at −20 °C for at least 1 year.

3.2 Part Creation Through PCR Amplification

New compatible parts can also be generated through PCR amplification using an existing DNA sequence as a template and a sequence-specific primer pair (*see* **Notes 3** and **4**). This is useful for adapting existing building blocks and creating new multipart assemblies (e.g., placing a synthesized part in a different location of a new multipart design) by simply changing the flanking 4 bp 5′-overhangs (Table 1).

1. On ice and in a 0.2 μL PCR tube, prepare the following reaction: 2.5 μL of 10 μM forward primer, 10 μL of 5× Phusion HF buffer, 1 μL of 10 mM dNTPs, 1 pg–10 ng of DNA template, 0.5 μL of Phusion polymerase (1 unit per 50 μL final concentration), and nuclease-free water to a total reaction volume of 50 μL (*see* **Note 5**).
2. Place the reaction in a thermocycler and execute the following program: 1 cycle at 98 °C for 30 s (initial denaturation), 30 cycles of denaturing the dsDNA for 10 s at 98 °C, anneal the primers for 25 s at the primers' annealing temperature, and elongate for the appropriate time, depending on the length of the final amplicons (30 s per kilobase), at 72 °C. Finally, perform a final extension for 5 min at 72 °C and hold at 4 °C. The previous two steps are written if using a Phusion polymerase and should be adjusted to the manufacturer's recommendations if using a different polymerase.
3. Verify that the amplicons have the correct molecular weight by DNA electrophoresis. We recommend loading ~3 μL of the PCR product on a 1% agarose gel containing an appropriate DNA dye. This should be run at a constant 80 V for 30 min, or until good band separation is achieved, and the length of the PCR product is compared to an appropriate DNA ladder.

4. Purify the DNA amplicons using a PCR cleanup kit to remove residual primers, enzymes, and general reaction products that could interfere with the assembly, and elute in >6 μL of nuclease-free water such that a final DNA concentration of >50 ng/μL is achieved.
5. Amplicons are now ready for Golden Gate assembly (*see* Subheading 3.4).

3.3 Part Creation Using Annealed DNA Oligos

For short parts like inert spacers or RBSs, it is possible to synthesize complementary ssDNA oligos (*see* **Note 6**) that anneal to create a predigested part (i.e., they contain appropriate 4 bp 5′-overhangs needed for the Golden Gate assembly).

1. Design two complementary ssDNA oligos such that the first ssDNA oligo consists of the sequence of the part read from the 5′- to 3′-end plus the appropriate 4 bp overhang at its 5′-end. The second ssDNA oligo sequence should consist of the reverse complement of the part and the appropriate 4 bp overhang at its 5′-end. Benchling can be used to ensure that the overhangs are exposed after annealing and that they are in the right orientation for the Golden Gate assembly, as shown in Fig. 2a.
2. Upon receiving the oligos, resuspend each oligo to a working concentration of 100 μM in annealing buffer. It is advised that care is taken to ensure the DNA is at the bottom of any tubes before resuspending by brief centrifugation.
3. In a sterile 1.5 mL microcentrifuge tube, add 2 μg of each oligo and annealing buffer to a final volume of 50 μL.
4. Place the tube in a heat block at 95 °C for 5 min.
5. Turn off the heating block, and let the block and content of the tube cool down slowly to room temperature.
6. To enable ligation during the Golden Gate assembly, it is necessary to phosphorylate the 5′-end of the overhangs by adding to a sterile 0.2 μL PCR tube: 100 pmol of the annealed oligo mix from the previous step, 5 μL of T4 ligase buffer, and 1 μL of phosphokinase enzyme and nuclease-free water to a 50 μL total reaction volume.
7. Incubate this tube in a thermocycler at 37 °C for 30 min and 65 °C for 20 min, and then remove from the thermocycler and allow to cool slowly to room temperature.
8. The oligos are now annealed and ready for Golden Gate assembly (*see* Subheading 3.4). Place on ice for immediate use, otherwise store at −20 °C for up to 2 weeks.

3.4 MLC Assembly

1. Calculate the amount of the required parts and backbone and mix in a 20 μL Golden Gate reaction. In assemblies where the

inserts are in a pre-cloned form (Table 1), use 75 ng of each plasmid. In assemblies where the inserts are amplicons, use a 2: 1 molar ratio (insert: vector). In assemblies where oligo parts are used, we recommend 3.2 ng (88 fmoles) of each annealed oligo part.

2. Mix thoroughly and incubate in a thermocycler at 37 °C for 5 min and at 16 °C for 5 min. Repeat this for 30 cycles followed by 5 min at 60 °C. Finally, hold at 4 °C.
3. The product should be kept at 4 °C for immediate use or stored at −20 °C for up to 2 weeks.
4. Transform competent *E. coli* cells with 2 μL of the reaction mix, and plate out all the transformation mix in serial dilutions on LB agar plates containing 50 μg/mL kanamycin to select for uptake of the destination vector.
5. Incubate plates overnight at 37 °C.
6. Place the plates on a blue light transilluminator to visualize possible fluorescent (incorrectly assembled) colonies. Only white (nonfluorescent) colonies contain an insert in the destination vector.
7. Select only white colonies and verify correct MLC assembly by Sanger sequencing (*see* **Note** 7).

4 Notes

1. Care should be taken that part sequence has the correct orientation, as shown in Fig. 2a, and for parts containing open reading frames that the assembly will result in an in-frame translation of the protein product. This can be checked by using the Golden Gate assembly tool that is available in Benchling.
2. It should be ensured that part plasmids use an ampicillin resistance marker so that no issues with plasmid carried through during assembly occur. This is because kanamycin is used for antibiotic selection of the destination vector, and thus transformants containing only part plasmids will not grow.
3. The sequence-specific primer pair must have a suitable GC content (40–60%) and length (18–22 bp) for PCR amplification, and the 4 bp 5′-overhang sequences must be flanked by BsaI recognition sites in the right orientation (i.e., such that the recognition site is removed after correct assembly). In addition, BsaI restriction sites need to be at least six nucleotides away from the end of a fragment to be efficiently digested by BsaI.

4. Be aware of the possibility for primers to generate unwanted secondary PCR products or to create secondary structures that could interfere with the generation of new parts. We recommend making sure the GC content is between 40% and 60% and checking for secondary structures and off-target binding sites using the primer analysis tool available in Benchling. Furthermore, the final orientation of the parts and the translation frame of the final translation unit can be checked in Benchling, as suggested in Subheading 3.1, **step 1**.
5. To ensure a high-quality PCR product and to keep the mutation rate to a minimum, it is important to choose a high-fidelity DNA polymerase (we recommend Phusion high-fidelity DNA polymerase, M0530S, New England Biolabs, or Q5 high-fidelity DNA polymerase, M0491S, New England Biolabs).
6. We recommend that when oligos are used directly that they are <60 bp long, and polyacrylamide gel electrophoresis (PAGE) purification is requested to ensure high-quality starting material. However, we have been successful in assembling MLCs with standard desalted oligos without further purification, although with a lower perfect assembly efficiency.
7. If low transformation efficiency is obtained, repeat ensuring that the correct concentrations of parts are present in the Golden Gate reaction. If problems persist, it may help to clone all parts into plasmids for the assembly.

Acknowledgments

F.V.G. was supported by a Royal Society PhD studentship. T.E.G. and C.S.G. were supported by BrisSynBio, a BBSRC/EPSRC Synthetic Biology Research Centre (grant BB/L01386X/1). T.E.G. was further supported by a Royal Society University Research Fellowship (grant UF160357).

References

1. Greco FV, Tarnowski MJ, Gorochowski TE (2019) Living computers powered by biochemistry. Biochemist 41(3):14–18. https://doi.org/10.1042/BIO04103014
2. Brophy JAN, Voigt CA (2014) Principles of genetic circuit design. Nat Methods 11:508
3. Chen Y-J et al (2013) Characterization of 582 natural and synthetic terminators and quantification of their design constraints. Nat Methods 10:659
4. Salis HM, Mirsky EA, Voigt CA (2009) Automated design of synthetic ribosome binding sites to control protein expression. Nat Biotechnol 27(10):946–950. https://doi.org/10.1038/nbt.1568
5. Stanton BC, Nielsen AAK, Tamsir A, Clancy K, Peterson T, Voigt CA (2014) Genomic mining of prokaryotic repressors for orthogonal logic gates. Nat Chem Biol 10(2):99–105. https://doi.org/10.1038/nchembio.1411
6. Gallivan JP (2007) Toward reprogramming bacteria with small molecules and RNA. Curr Opin Chem Biol 11(6):612–619. https://doi.org/10.1016/j.cbpa.2007.10.004
7. Sen S, Apurva D, Satija R, Siegal D, Murray RM (2017) Design of a toolbox of RNA

thermometers. ACS Synth Biol 6(8): 1461–1470. https://doi.org/10.1021/acssynbio.6b00301
8. Baumschlager A, Aoki SK, Khammash M (2017) Dynamic blue light-inducible T7 RNA polymerases (Opto-T7RNAPs) for precise spatiotemporal gene expression control. ACS Synth Biol 6(11):2157–2167. https://doi.org/10.1021/acssynbio.7b00169
9. Castillo-Hair SM, Baerman EA, Fujita M, Igoshin OA, Tabor JJ (2019) Optogenetic control of Bacillus subtilis gene expression. Nat Commun 10(1):3099. https://doi.org/10.1038/s41467-019-10906-6
10. de Boer HA, Comstock LJ, Vasser M (1983) The tac promoter: a functional hybrid derived from the trp and lac promoters. Proc Natl Acad Sci 80(1):21. https://doi.org/10.1073/pnas.80.1.21
11. Bartoli V, Meaker GA, di Bernardo M, Gorochowski TE (2020) Tunable genetic devices through simultaneous control of transcription and translation. Nat Commun. https://doi.org/10.1038/s41467-020-15653-7
12. Elowitz MB, Levine AJ, Siggia ED, Swain PS (2002) Stochastic gene expression in a single cell. Science 297(5584):1183. https://doi.org/10.1126/science.1070919
13. Raj A, van Oudenaarden A (2008) Nature, nurture, or chance: stochastic gene expression and its consequences. Cell 135(2):216–226. https://doi.org/10.1016/j.cell.2008.09.050
14. Süel GM, Garcia-Ojalvo J, Liberman LM, Elowitz MB (2006) An excitable gene regulatory circuit induces transient cellular differentiation. Nature 440(7083):545–550. https://doi.org/10.1038/nature04588
15. Gorochowski TE et al (2017) Genetic circuit characterization and debugging using RNA-seq. Mol Syst Biol 13(11):952. https://doi.org/10.15252/msb.20167461
16. Ceroni F et al (2018) Burden-driven feedback control of gene expression. Nat Methods 15(5):387–393. https://doi.org/10.1038/nmeth.4635
17. Gorochowski TE, Chelysheva I, Eriksen M, Nair P, Pedersen S, Ignatova Z (2019) Absolute quantification of translational regulation and burden using combined sequencing approaches. Mol Syst Biol 15(5):e8719. https://doi.org/10.15252/msb.20188719
18. Greco FV, Pandi A, Erb TJ, Grierson CS, Gorochowski TE (2021) Harnessing the central dogma for stringent multi-level control of gene expression. Nat Commun 12(1):1738. https://doi.org/10.1038/s41467-021-21995-7
19. Green AA, Silver PA, Collins JJ, Yin P (2014) Toehold switches: de-novo-designed regulators of gene expression. Cell 159(4):925–939. https://doi.org/10.1016/j.cell.2014.10.002
20. Green AA, Kim J, Ma D, Silver PA, Collins JJ, Yin P (2017) Complex cellular logic computation using ribocomputing devices. Nature 548(7665):117–121. https://doi.org/10.1038/nature23271
21. Chappell J, Takahashi MK, Lucks JB (2015) Creating small transcription activating RNAs. Nat Chem Biol 11(3):214–220. https://doi.org/10.1038/nchembio.1737
22. Chappell J, Westbrook A, Verosloff M, Lucks JB (2017) Computational design of small transcription activating RNAs for versatile and dynamic gene regulation. Nat Commun 8(1):1051. https://doi.org/10.1038/s41467-017-01082-6
23. Westbrook AM, Lucks JB (2017) Achieving large dynamic range control of gene expression with a compact RNA transcription–translation regulator. Nucleic Acids Res 45(9):5614–5624. https://doi.org/10.1093/nar/gkx215
24. Woodruff LBA et al (2016) Registry in a tube: multiplexed pools of retrievable parts for genetic design space exploration. Nucleic Acids Res 45(3):1553–1565. https://doi.org/10.1093/nar/gkw1226
25. Pryor JM, Potapov V, Kucera RB, Bilotti K, Cantor EJ, Lohman GJS (2020) Enabling one-pot Golden Gate assemblies of unprecedented complexity using data-optimized assembly design. PLoS One 15(9):e0238592. https://doi.org/10.1371/journal.pone.0238592
26. Lee ME, DeLoache WC, Cervantes B, Dueber JE (2015) A highly characterized yeast toolkit for modular, multipart assembly. ACS Synth Biol 4(9):975–986. https://doi.org/10.1021/sb500366v
27. Park Y, Espah Borujeni A, Gorochowski TE, Shin J, Voigt CA (2020) Precision design of stable genetic circuits carried in highly-insulated E. coli genomic landing pads. Mol Syst Biol 16(8):e9584. https://doi.org/10.15252/msb.20209584
28. Engler C, Gruetzner R, Kandzia R, Marillonnet S (2009) Golden Gate shuffling: a one-pot DNA shuffling method based on type IIs restriction enzymes. PLoS One 4(5):e5553. https://doi.org/10.1371/journal.pone.0005553
29. Meyer AJ, Segall-Shapiro TH, Glassey E, Zhang J, Voigt CA (2019) Escherichia coli "marionette" strains with 12 highly optimized

small-molecule sensors. Nat Chem Biol 15(2): 196–204. https://doi.org/10.1038/s41589-018-0168-3

30. Baig H et al (2020) Synthetic biology open language visual (SBOL visual) version 2.2. J Integr Bioinform 17(2–3). https://doi.org/10.1515/jib-2020-0014

31. Beal J et al (2019) Communicating structure and function in synthetic biology diagrams. ACS Synth Biol 8(8):1818–1825. https://doi.org/10.1021/acssynbio.9b00139

32. Der BS et al (2017) DNAplotlib: programmable visualization of genetic designs and associated data. ACS Synth Biol 6(7):1115–1119. https://doi.org/10.1021/acssynbio.6b00252

33. Bartoli V, Dixon DOR, Gorochowski TE (2018) Automated visualization of genetic designs using DNAplotlib. In: Braman JC (ed) Synthetic biology: methods and protocols. Springer New York, New York, pp 399–409. https://doi.org/10.1007/978-1-4939-7795-6_22

Chapter 7

Model-Based Design of Synthetic Antisense RNA for Predictable Gene Repression

Tae Seok Moon

Abstract

Our enhanced understanding of RNA folding and function has increased the use of small RNA regulators. Among these RNA regulators, synthetic antisense RNA (asRNA) is designed to contain an RNA sequence complementary to the target mRNA sequence, and the formation of double-stranded RNA (dsRNA) facilitates gene repression due to dsRNA degradation or prevention of ribosome access to the mRNA. Despite the simple complementarity rule, however, predictably tunable repression has been challenging when synthetic asRNAs are used. Here, the protocol for model-based asRNA design is described. This model can predict synthetic asRNA-mediated repression efficiency using two parameters: the change in free energy of complex formation (ΔG_{CF}) and percent mismatch of the target binding region (TBR). The model has been experimentally validated in both Gram-positive and Gram-negative bacteria as well as for target genes in both plasmids and chromosomes. These asRNAs can be created by simply replacing the TBR sequence with one that is complementary to the target mRNA sequence of interest. In principle, this protocol can be applied to design and build asRNAs for predictable gene repression in various contexts, including multiple target genes and organisms, making asRNAs predictably tunable regulators for broad applications.

Key words Synthetic biology, Predictive model, RNA regulator, Antisense RNA, Gene repression

1 Introduction

In many biological studies and biotechnological applications, gene regulation has been achieved using regulatory proteins [1], including transcription factors and sigma factors [2–6]. These proteins are repurposed to perform their native function, and their use generally leads to large dynamic ranges in synthetic circuits. Despite this advantage, protein regulators have several drawbacks. First, it is challenging to design new protein regulators with novel functions due to the difficulty in predicting protein structure and function. Second, protein regulators must be transcribed and translated, costing cellular resources. Third, many protein regulators interact with a target DNA sequence, meaning that the target sequence

James Chappell and Melissa K. Takahashi (eds.), *Riboregulator Design and Analysis*,
Methods in Molecular Biology, vol. 2518, https://doi.org/10.1007/978-1-0716-2421-0_7,

should be introduced into the chromosome to regulate chromosomal genes. To simultaneously regulate multiple chromosomal genes, genome engineering often requires cumbersome experimental procedures, especially in non-model organisms, and can sometimes cause undesirable context effects on neighboring genes [7, 8].

As an alternative strategy, RNA regulators have been developed, including small transcription activating RNA [9], attenuator [10], CRISPR interference [11], CRISPR activation [12], toehold switch [13], synthetic transacting antisense RNA (asRNA) [14, 15], riboswitch [16, 17], RNA thermosensor [18], and aptazyme [19, 20]. Among these RNA regulators, asRNA contains a sequence complementary to the target mRNA sequence, and its binding to the target mRNA leads to gene repression through the prevention of ribosome access to the mRNA or RNA degradation [21–24]. Additionally, asRNA has been repurposed to control other RNA regulators [25, 26] and combined with protein regulators to build complex genetic circuits [27, 28].

As a regulator, asRNA has many advantages over proteins. First, asRNA is easy to design. By simply transcribing the reverse complement of the target mRNA, a new regulator can be generated. Second, because of this simple base-pairing mechanism, asRNAs are generally orthogonal [27], such that a given regulator usually represses its target gene only. Third, unlike many RNA regulators that require specifically designed targets [9, 13, 16–20] and thus modification of the chromosome, the asRNA sequence is tailored specifically to the existing DNA sequence. This design strategy for asRNA does not require modification of the chromosome when chromosomal genes are the targets, making asRNA ideal for regulating multiple chromosomal genes at the same time. Fourth, like other RNA regulators, asRNA is not translated into a protein, potentially reducing cellular costs. Finally, asRNAs have been found in many microbes [29–42], allowing them to be developed as regulators with broad applicability in a diverse set of hosts.

Despite these advantages, asRNA had been used in a relatively limited number of species as a synthetic tool to regulate gene expression. This scarcity was partly due to the difficulty in achieving predictably tunable repression of target genes. Recently, a mathematical model has been developed and experimentally validated for predictable asRNA-mediated repression in diverse organisms, including *Escherichia coli* DH10B, *E. coli* Nissle 1917, and *Bacillus subtilis* 168 [15]. This prediction model is based on two design parameters that have been identified by Hoynes-O'Connor and Moon [27]. One parameter is a change in free energy of complex formation (ΔG_{CF}), and the other is a percent mismatch. ΔG_{CF} is the change in ΔG when the target binding region (TBR) of an asRNA binds to the mRNA. It is calculated using the equation $\Delta G_{\text{asRNA:mRNA}} - \Delta G_{\text{asRNA}} - \Delta G_{\text{mRNA}}$, where each ΔG is estimated using

NUPACK [43] as described in the Methods section in detail. The percent mismatch is calculated by dividing the number of mismatched nucleotides by the total number of TBR nucleotides. Notably, ΔG_{CF} and percent mismatch are not correlated, allowing these two predictors to be used for multiple linear regression analyses. Using 434 different strain-asRNA combinations, this multivariate model has been built and experimentally validated, enabling predictable tunability of asRNA-mediated repression [15].

In this protocol, the detailed procedures for the design and construction of asRNAs are described. Because various gene cloning methods can be used to construct plasmids that contain synthetic asRNA expression cassettes, this protocol focuses on the detailed steps for the model-based asRNA design instead of detailed cloning steps. An asRNA sequence consists of TBR, the Hfq binding site (MicF M7.4), and a transcription terminator. MicF M7.4 was shown to facilitate high asRNA-mediated repression with minimal off-target effect [15, 27]. As long as the described design rules are correctly followed, predictable asRNA-mediated repression can be achieved. The construction of a plasmid itself is relatively simple, only requiring the replacement of an existing TBR with one that is complementary to the target mRNA sequence while retaining the rest sequence of the platform plasmid (*see* the Materials and Methods sections) [15].

2 Materials

2.1 asRNA Design

1. The target mRNA (or the corresponding DNA) sequence (e.g., 5′TAGCGAATTCACTTATTAAAGAACAGGAGTAAGTAATGAGTAAAGGAGAAGAACTTTTCACTGGAGTTGTCCCAA3′).
2. The TBR (or the corresponding DNA) sequence (e.g., 5′CTCCTTTTCTCATTTCTTACTCCTCTTCTTTAATAAGTGA3′).
3. The asRNA (or the corresponding DNA) sequence (e.g., 5′CTCCTTTTCTCATTTCTTACTCCTCTTCTTTAATAAGTGACGTCCCGCAAGGATGCGGGTCTGTTTACCCCTATTTCAACCGGCCGCCTCGCGGCCGGTTTTTTTTTGCTTAATTAGCTGAGCTTGGACTCCTGTTGATAGATCCAGTAATGACCTCAGAACTCCATCTGGATTTGTTCAGAACGCTCGGTTGCCGCCGGGCGTTTTTTATTGGTGAGAATCCA3′).
4. NUPACK [43], an RNA secondary structure prediction software.

2.2 asRNA Plasmid Construction

1. pG16, the platform plasmid (ColE1 origin and ampicillin resistance; 3199 base pairs) [15] that transcribes asRNA using the aTc-inducible P_{Tet} promoter (aTc, anhydrotetracycline). This plasmid contains the asRNA sequence that includes the original TBR (5′ATAAGTGAATTCGCTA3′), the MicF M7.4 Hfq binding site, and the transcription terminator.
2. Forward asRNA primer (e.g., 5′TCCTCTTCTTTAATAAGTGACGTCCCGCAAGGATGC3′).
3. Reverse asRNA primer (e.g., 5′GTAAGAAATGAGAAAAGGAGAGATGTGCTCAGTATCTCTATCACTGATAG3′).
4. Phusion DNA polymerase and 5× Phusion HF Buffer.
5. 10 mM dNTPs.
6. Dimethyl sulfoxide (DMSO).
7. DpnI, T4 polynucleotide kinase (T4 PNK), T4 DNA ligase, and 10× ligation buffer.
8. Gel extraction kit, PCR purification kit, and plasmid miniprep kit.
9. *E. coli* DH10B competent cells.
10. LB agar plates with 100 μg/mL ampicillin and LB liquid media with 100 μg/mL ampicillin.
11. Purified water (DNase and RNase free, ddH_2O).
12. Sequencing primer (5′CGACCTCATTAAGCAGCTCTAATG3′).
13. SYBR Safe DNA Gel Stain, 1 kb DNA ladder, and gel loading dye.
14. Agarose.

2.3 GFP Repression Assays

1. Test strain with the target gene (e.g., *E. coli* DH10B variant strain containing *gfpmut3* in the chromosome, called *E. coli* DH10B-GFP [15]).
2. *E. coli* DH10B (no-GFP control).
3. LB agar plates without and with 100 μg/mL ampicillin and LB liquid media without and with 100 μg/mL ampicillin.
4. Anhydrotetracycline (aTc).
5. Phosphate-buffered saline, pH 8.0.
6. 2 mg/mL kanamycin in phosphate-buffered saline, pH 8.0
7. Deep 96-well plate and breathable sealing membrane.

3 Methods

In this section, the *gfpmut3* gene integrated into the *E. coli* chromosome is used as an example target gene. The same design procedure can be applied to repress other genes in other species (both chromosomal genes and plasmid-encoded genes), as demonstrated in the previous report [15].

3.1 Model-Based asRNA Design

Various asRNAs with different predicted repression efficiencies can be designed. Here, one asRNA with 56.6% predicted repression efficiency is designed as an example (*see* Fig. 1).

1. Select the target mRNA sequence in the translation initiation region (TIR) of *gfpmut3* (*see* **Note 1**).
2. Decide the TBR sequence that is complementary to the target mRNA sequence, and paste it upstream of the MicF M7.4 Hfq binding site, as shown in Fig. 1 (*see* **Note 2**).
3. Calculate ΔG using the NUPACK's analysis tool (setting the nucleic acid type to RNA and the temperature to 37 °C) [43]. $\Delta G_{\mathrm{asRNA}}$ (−65.5 kcal/mol) is estimated by entering the entire asRNA sequence (224 nucleotides in Fig. 1; *see* **Note 3**). ΔG_{mRNA} (−5.5 kcal/mol) is estimated by entering the target mRNA sequence and one extra nucleotide at the 3′-end (41 nucleotides in Fig. 1; *see* **Note 4**). The extra nucleotide is added to consider the stacking contribution of neighboring base pairs [44]. Similarly, $\Delta G_{\mathrm{asRNA:mRNA}}$ (−111.38 kcal/mol, change in free energy of asRNA-mRNA complex) is estimated by entering the two sequences used above (1:1 concentration ratio). Finally, ΔG_{CF} (−40.38 kcal/mol) is calculated by using the equation: $\Delta G_{CF} = \Delta G_{\mathrm{asRNA:mRNA}} - \Delta G_{\mathrm{asRNA}} - \Delta G_{\mathrm{mRNA}}$.
4. Determine the percent mismatch (7.5%), as shown in Fig. 1. The percent mismatch is calculated by dividing the number of mismatched nucleotides (3 nucleotides) by the total number of TBR nucleotides (40 nucleotides).
5. Calculate the predicted repression efficiency ($F = 0.566$) using the experimentally validated model (*see* **Note 5**): $F(X_1, X_2) = [0.3848–0.0068X_1\_0.0125X_2 + \varepsilon]$ ($R^2 = 0.685$, p-value < 0.001) where X_1 is ΔG_{CF} (in kcal/mol), X_2 is percent mismatch (in %), and ε is the standard error ($\varepsilon = 0.123$). For reliable prediction, ΔG_{CF} should be from −59.88 to −6.58 kcal/mol, and percent mismatch should be from 0% to 32.1% (*see* **Note 6**).

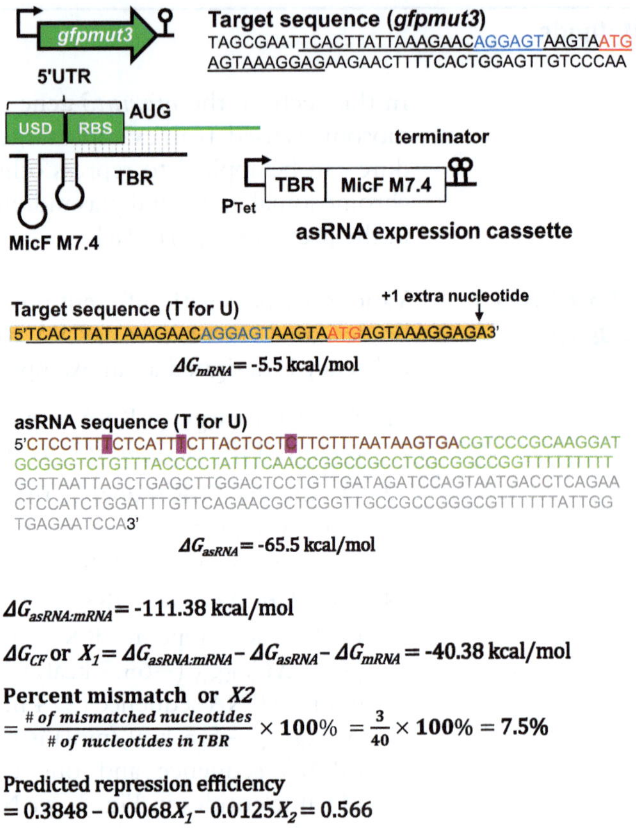

Fig. 1 Model-based asRNA design. Different DNA sequence elements are shown in different colors: Shine-Dalgarno sequence (blue), start codon (red), the target sequence (underlined) plus one extra nucleotide on the 3′-end (highlighted in yellow), TBR (brown), MicF M7.4 Hfq binding site (green), transcription terminator (gray), and mismatched nucleotide (highlighted in magenta). Abbreviation: upstream of Shine-Dalgarno sequence (USD), ribosome binding site (RBS), 5′ untranslated region (5′ UTR)

3.2 Construction of the asRNA Plasmid Using Inverse PCR

1. Design primers for inverse PCR (*see* Subheading 2.2 for their sequences). The forward primer contains the 3′ half of the TBR (20 nucleotides) and anneals to the part of pG16 (i.e., the part of the MicF M7.4 Hfq binding site, 16 nucleotides). The reverse primer contains the 5′ half of the TBR (reverse complement, 20 nucleotides) and anneals to the part of pG16 (i.e., the part of the P_{Tet} promoter (reverse complement), 30 nucleotides).
2. Perform inverse PCR using the following conditions (*see* **Note 7**).

Water (ddH_2O)	37 μL
5× HF Phusion buffer	10 μL
10 mM dNTPs	1 μL
25 μM forward primer	0.5 μL
25 μM reverse primer	0.5 μL
Template DNA (pG16)	0.5 μL
Phusion DNA polymerase	0.5 μL

Thermal cycle temperature	Duration	Repeat
98 °C	30 s	1 cycle
98 °C	10 s	30 cycles
60 °C	20 s	
72 °C	80 s	
72 °C	5 min	1 cycle

3. Purify the size-confirmed, 3223 base pair PCR products (or DpnI-treated PCR products; *see* **Note 8**) using a gel extraction kit or a PCR purification kit (following the manufacturer's instruction).
4. Perform simultaneously phosphorylation and blunt-end ligation at room temperature for 1 h (*see* **Note 9**).

T4 DNA ligase	1 μL
10× ligation buffer	1 μL
Purified DNA	7.8 μL
T4 PNK	0.2 μL

5. Transform competent *E. coli* cells, select ampicillin-resistant colonies using LB agar plates with 100 μg/mL ampicillin, and prepare the constructed asRNA plasmid (called pG43B2; 3223 base pairs) using a plasmid miniprep kit according to the manufacturer's instruction (*see* **Note 10**). The DNA sequence should be confirmed using the sequencing primer (*see* Subheading 2.2 for the primer sequence), which can cover the entire P_{Tet} promoter, the entire asRNA region (the replaced TBR sequence, the MicF M7.4 Hfq binding site, and the lambda T0 transcription terminator), and the extra terminator (*rrnB* T1 terminator).

3.3 GFP Repression Assays

1. Transform *E. coli* DH10B-GFP (*see* **Note 11**) using the sequence-confirmed asRNA plasmid (e.g., pG43B2).
2. Incubate the transformed cells on LB agar plates containing 100 μg/mL ampicillin overnight at 37 °C. Incubate *E. coli* DH10B (no-GFP control) on LB agar plates without any antibiotic overnight at 37 °C.
3. Pick colonies and grow them in 1 mL LB liquid media with or without 100 μg/mL ampicillin overnight at 37 °C and 250 rpm. For all liquid cultures, use deep 96-well plates with breathable sealing membranes.
4. Transfer the overnight cultures (1% v/v) into 1 mL fresh LB media with or without 100 μg/mL ampicillin, and grow them for 2 h at 37 °C and 250 rpm.
5. Transfer the subcultures (1.67% v/v) into 0.6 mL fresh LB media with or without 100 μg/mL ampicillin and 250 ng/mL aTc (*see* **Note 12**). Grow them for 8 h at 37 °C and 250 rpm.
6. Pellet the cells by centrifugation (for 10 min at 2200 *g*), and resuspend them in 0.2 mL phosphate-buffered saline containing 2 mg/mL kanamycin (*see* **Note 13**).
7. Measure the population fluorescence ($F_{\text{experimental}}$) and the absorbance at 600 nm ($Abs_{\text{experimental}}$) using a microplate reader. The fluorescence is measured with excitation at 483 nm and emission at 530 nm.
8. Normalize the measured fluorescence value by the equation: $F_{\text{norm}} = (F_{\text{experimental}}/Abs_{\text{experimental}}) - (F_{\text{no-GFP control}}/Abs_{\text{no-GFP control}})$ where the no-GFP control is *E. coli* DH10B without a plasmid (i.e., the parent DH10B strain without *gfpmut3* integrated into the genome). Calculate the repression efficiency by the equation: $1 - (F_{\text{aTc+}}/F_{\text{aTc-}})$ where $F_{\text{aTc-}}$ and $F_{\text{aTc+}}$ are the normalized fluorescence values without and with aTc, respectively (*see* **Note 14**).

4 Notes

1. In this protocol, the target region is restricted to 75 nucleotides that cover the translation initiation region (i.e., the sequence from −35 to +40 with the ATG start codon's A as +1). The TIR includes the upstream of Shine-Dalgarno sequence (USD), the ribosome binding site (RBS or the SD sequence), and the start codon of the target gene. Targeting the TIR is essential to generate asRNAs with high and reliable repression efficiencies [15].
2. Mismatches can be introduced in a TBR sequence through base substitutions (e.g., T to A).

3. While both bacteriophage lambda T0 (the sequence shown in Fig. 1) and *rrnB* T1 (not shown in Fig. 1) terminators are used to ensure transcription termination of the asRNA, the actual asRNA is likely to contain only the upstream terminator (lambda T0), not both. Thus, $\Delta G_{\mathrm{asRNA}}$ is estimated by entering the sequence of the TBR, the MicF M7.4 Hfq binding site, and the lambda T0 terminator, which excludes the *rrnB* T1 terminator sequence.
4. It is assumed that the coupling of transcription and translation in bacteria leads to local folding [45]. Thus, only the target mRNA region plus one extra nucleotide at the 3′-end (41 nucleotides) is considered in this model, as opposed to the entire mRNA sequence.
5. These two parameters had been selected as the main contributors for asRNA-mediated gene repression from the 12 initially considered parameters, including target location, double-stranded RNA length, Hfq binding site, ribosome interaction, and YUNR motif [15, 27]. To prevent the target location effect on repression, the target region should be restricted to the TIR, as mentioned above. The double-stranded RNA length and ΔG_{CF} show strong multicollinearity, requiring the elimination of one of the two parameters in the model. As discussed below, this predictive equation should be used only for asRNAs containing MicF M7.4 in *E. coli* and asRNAs lacking an Hfq binding site sequence in *B. subtilis*. In *E. coli*, this model should not be used for asRNAs lacking an Hfq binding site sequence. The rest initially considered parameters, as well as the newly visited parameters (e.g., GC contents of the paired asRNA: mRNA sequence and target accessibility), did not affect the model predictability.
6. Considering the desired target repression efficiency and off-target repression levels, the users can design TBR sequences with different lengths/target locations ($\Delta G_{CF} = -59.88$ to -6.58 kcal/mol) and percent mismatch (0–32.1%). However, this data-driven model must be used only within the tested parameter ranges ($\Delta G_{CF} = -59.88$ to -6.58 kcal/mol; 0–32.1% mismatch) to obtain a reliable prediction. As mentioned above, this model must also be used only for TIR targeting asRNAs. Thus, it is unlikely to accurately predict potential off-target repression levels in which the parameter values are most likely to be outside of the acceptable ranges. Notably, the users can run BLAST searches against the entire genome of interest using the target sequence or the TBR sequence to identify potential off-target sites. Predicting these sites or their off-target repression levels is an interesting future work and beyond the scope of this protocol.

7. Different labs use different reagents, and the users can follow the manufacturer's instructions to determine their PCR conditions. DMSO (up to 8%) can be added when using DNA templates with GC-rich sequences or secondary structures. When DMSO is used, the annealing temperature should be lowered (up to ~5 °C).
8. If no DMSO (or enzyme-denaturing chemical) is used for the PCR step, DpnI can be added directly into the PCR reaction mixture after the PCR step (1 μL per 50 μL PCR mixture). DpnI treatment at 37 °C for 1 h would be sufficient to degrade the template plasmid. If enzyme-denaturing chemicals are used, the PCR products should be purified first, followed by DpnI treatment and additional purification according to the manufacturer's instruction.
9. Because primers are usually synthesized without 5-′-phosphorylation, T4 PNK treatment is necessary for the ligation reaction. While T4 PNK's optimum temperature is 37 °C, this simultaneous phosphorylation/ligation reaction is usually efficient at room temperature for blunt-end ligation of inverse PCR products.
10. These steps can be modified according to each lab's practice and protocol as long as the DNA sequence of the constructed plasmid is verified. The confirmed sequence should cover the entire P_{Tet} promoter, the entire asRNA region, and the extra terminator.
11. The prediction model has been experimentally validated for asRNAs containing MicF M7.4 in *E. coli* (DH10B and Nissle 1917) and asRNAs lacking an Hfq binding site sequence in *B. subtilis* 168 [15]. The repression targets can be either chromosomal genes or plasmid-encoded genes if the expressed asRNA level is in excess of the target mRNA level. The repression targets can also be native genes other than fluorescent reporter genes, as the model was experimentally validated in the previous report [15]. Similar models can be applied to other strains than the abovementioned strains, but the constants in $F(X_1, X_2)$ would need to be determined for accurate prediction, as described in the previous report (i.e., a, b, and c should be determined in $F = a - bX_1 - cX_2$) [15]. For the abovementioned strains, the experimentally measured repression efficiencies of 70% of the tested asRNAs (117 out of 168) were within a range of one standard error (1ε) of the predicted repression efficiencies (95% within a range of 2ε and 100% within 3ε) [15]. For non-TIR targeting asRNAs, this model cannot be used.
12. As shown in the previous report [15], the target gene repression level increases with the asRNA transcript level. The

prediction model was built on the premise that the asRNA would be in excess of the target mRNA such that asRNA abundance would not be a determining factor for the repression efficiency. To ensure that asRNA levels are in excess of target mRNA levels, a high copy number plasmid (e.g., pG43B2 with ColE1 origin) should be used, and the P_{Tet} promoter should be induced maximally.

13. A 2 mg/mL kanamycin is used to stop bacterial translation (e.g., GFP synthesis) after sampling. Cells resuspended in phosphate-buffered saline containing kanamycin can be stored at 4 °C before fluorescence measurement. Some fluorescent proteins require a long maturation time, especially when aeration is inadequate. Pipetting cells multiple times during the resuspension step and storing them at 4 °C before analysis would help fluorescent proteins' maturation without further bacterial translation.

14. The repression efficiency (RE) should be defined as 1 – (F_{aTc+}/F_{aTc-}) for accurate comparison with the model's predicted value. This definition gives conservative RE values in that a basal asRNA level (without aTc) due to promoter leakiness reduces F_{aTc-} and thus RE. If the RE is alternatively defined as 1 – (F_{asRNA+}/F_{asRNA-}), where F_{asRNA-} (e.g., DH10B-GFP without any "leakiness" effect) and F_{asRNA+} (e.g., DH10B-GFP + an asRNA plasmid) are the normalized fluorescence values without and with an asRNA plasmid containing a constitutively and maximally expressed asRNA cassette, the RE value would be higher than that of the original definition (1 – (F_{aTc+}/F_{aTc-})). This alternative definition allows for reporting higher RE values, but the fluorescence ratio (F_{asRNA+}/F_{asRNA-}) is obtained by comparing genetically different strains (DH10B-GFP vs. DH10B-GFP + an asRNA plasmid), as opposed to the genetically identical strain (DH10B-GFP + pG43B2; without vs. with aTc). Additionally, inducible asRNA expression, instead of constitutive asRNA expression, would make a better tool for gene expression control. Thus, considering the fair comparison (conservative reporting) and the RNA tool's utility, this model has been developed based on the experimental data obtained by using the original definition. The same point was discussed in a previous report covering a topic of engineering toehold switches in which RNA regulators are also used [46].

Acknowledgments

This work was supported by the National Science Foundation (MCB-1714352 and MCB-2001743). The author declares no conflict of interest.

References

1. Chappell J, Watters KE, Takahashi MK, Lucks JB (2015) A renaissance in RNA synthetic biology: new mechanisms, applications and tools for the future. Curr Opin Chem Biol 28: 47–56. https://doi.org/10.1016/j.cbpa.2015.05.018
2. Rhodius VA, Segall-Shapiro TH, Sharon BD, Ghodasara A, Orlova E, Tabakh H, Burkhardt DH, Clancy K, Peterson TC, Gross CA, Voigt CA (2013) Design of orthogonal genetic switches based on a crosstalk map of sigmas, anti-sigmas, and promoters. Mol Syst Biol 9: 702. https://doi.org/10.1038/msb.2013.58
3. Shopera T, Henson WR, Moon TS (2017) Dynamics of sequestration-based gene regulatory cascades. Nucleic Acids Res 45: 7515–7526. https://doi.org/10.1093/nar/gkx465
4. Nielsen AA, Der BS, Shin J, Vaidyanathan P, Paralanov V, Strychalski EA, Ross D, Densmore D, Voigt CA (2016) Genetic circuit design automation. Science (New York, NY) 352(6281):aac7341. https://doi.org/10.1126/science.aac7341
5. Gardner TS, Cantor CR, Collins JJ (2000) Construction of a genetic toggle switch in Escherichia coli. Nature 403(6767):339–342
6. Elowitz MB, Leibler S (2000) A synthetic oscillatory network of transcriptional regulators. Nature 403(6767):335–338
7. Venturelli OS, Egbert RG, Arkin AP (2016) Towards engineering biological systems in a broader context. J Mol Biol 428(5, Part B):928–944. https://doi.org/10.1016/j.jmb.2015.10.025
8. Lou C, Stanton B, Chen YJ, Munsky B, Voigt CA (2012) Ribozyme-based insulator parts buffer synthetic circuits from genetic context. Nat Biotechnol 30(11):1137–1142. https://doi.org/10.1038/nbt.2401
9. Chappell J, Takahashi MK, Lucks JB (2015) Creating small transcription activating RNAs. Nat Chem Biol 11:214–220. https://doi.org/10.1038/nchembio.1737
10. Lucks JB, Qi L, Mutalik VK, Wang D, Arkin AP (2011) Versatile RNA-sensing transcriptional regulators for engineering genetic networks. Proc Natl Acad Sci U S A 108: 8617–8622. https://doi.org/10.1073/pnas.1015741108
11. Qi LS, Larson MH, Gilbert LA, Doudna JA, Weissman JS, Arkin AP, Lim WA (2013) Repurposing CRISPR as an RNA-guided platform for sequence-specific control of gene expression. Cell 152:1173–1183. https://doi.org/10.1016/j.cell.2013.02.022
12. Bikard D, Jiang WY, Samai P, Hochschild A, Zhang F, Marraffini LA (2013) Programmable repression and activation of bacterial gene expression using an engineered CRISPR-Cas system. Nucleic Acids Res 41:7429–7437. https://doi.org/10.1093/nar/gkt520
13. Green AA, Silver PA, Collins JJ, Yin P (2014) Toehold switches: de-novo-designed regulators of gene expression. Cell 159:925–939. https://doi.org/10.1016/j.cell.2014.10.002
14. Na D, Yoo SM, Chung H, Park H, Park JH, Lee SY (2013) Metabolic engineering of Escherichia coli using synthetic small regulatory RNAs. Nat Biotechnol 31:170–174. https://doi.org/10.1038/nbt.2461
15. Lee YJ, Kim SJ, Amrofell MB, Moon TS (2019) Establishing a multivariate model for predictable antisense RNA-mediated repression. ACS Synth Biol 8(1):45–56. https://doi.org/10.1021/acssynbio.8b00227
16. Winkler W, Nahvi A, Breaker RR (2002) Thiamine derivatives bind messenger RNAs directly to regulate bacterial gene expression. Nature 419:952–956. https://doi.org/10.1038/nature01145
17. Jang S, Jang S, Yang J, Seo SW, Jung GY (2018) RNA-based dynamic genetic controllers: development strategies and applications. Curr Opin Biotechnol 53(Supplement C):1–11. https://doi.org/10.1016/j.copbio.2017.10.005
18. Hoynes-O'Connor A, Hinman K, Kirchner L, Moon TS (2015) De novo design of heat-repressible RNA thermosensors in E. coli. Nucleic Acids Res 43(12):6166–6179. https://doi.org/10.1093/nar/gkv499
19. Win MN, Smolke CD (2007) A modular and extensible RNA-based gene-regulatory

platform for engineering cellular function. Proc Natl Acad Sci U S A 104:14283–14288. https://doi.org/10.1073/pnas.0703961104

20. Carothers JM, Goler JA, Juminaga D, Keasling JD (2011) Model-driven engineering of RNA devices to quantitatively program gene expression. Science (New York, NY) 334(6063): 1716–1719. https://doi.org/10.1126/science.1212209
21. Prevost K, Desnoyers G, Jacques JF, Lavoie F, Masse E (2011) Small RNA-induced mRNA degradation achieved through both translation block and activated cleavage. Genes Dev 25(4): 385–396. https://doi.org/10.1101/gad.2001711
22. Mizuno T, Chou MY, Inouye M (1984) A unique mechanism regulating gene expression: translational inhibition by a complementary RNA transcript (micRNA). Proc Natl Acad Sci U S A 81(7):1966–1970
23. Bouvier M, Sharma CM, Mika F, Nierhaus KH, Vogel J (2008) Small RNA binding to 5′ mRNA coding region inhibits translational initiation. Mol Cell 32(6):827–837. https://doi.org/10.1016/j.molcel.2008.10.027
24. Lee YJ, Moon TS (2018) Design rules of synthetic non-coding RNAs in bacteria. Methods 143:58–69. https://doi.org/10.1016/j.ymeth.2018.01.001
25. Lee YJ, Kim S-J, Moon TS (2018) Multilevel regulation of bacterial gene expression with the combined STAR and antisense RNA system. ACS Synth Biol 7(3):853–865. https://doi.org/10.1021/acssynbio.7b00322
26. Lee Y, Hoynes-O'Connor A, Leong MC, Moon TS (2016) Programmable control of bacterial gene expression with the combined CRISPR and antisense RNA system. Nucleic Acids Res 44:2462–2473
27. Hoynes-O'Connor A, Moon TS (2016) Development of design rules for reliable antisense RNA behavior in E. coli. ACS Synth Biol 5(12):1441–1454. https://doi.org/10.1021/acssynbio.6b00036
28. Hoynes-O'Connor A, Shopera T, Hinman K, Creamer JP, Moon TS (2017) Enabling complex genetic circuits to respond to extrinsic environmental signals. Biotechnol Bioeng 114:1626–1631. https://doi.org/10.1002/bit.26279
29. Dornenburg JE, Devita AM, Palumbo MJ, Wade JT (2010) Widespread antisense transcription in Escherichia coli. MBio 1(1). https://doi.org/10.1128/mBio.00024-10
30. Lee JM, Zhang S, Saha S, Santa Anna S, Jiang C, Perkins J (2001) RNA expression analysis using an antisense Bacillus subtilis genome array. J Bacteriol 183(24):7371–7380. https://doi.org/10.1128/jb.183.24.7371-7380.2001
31. Irnov I, Sharma CM, Vogel J, Winkler WC (2010) Identification of regulatory RNAs in Bacillus subtilis. Nucleic Acids Res 38(19): 6637–6651. https://doi.org/10.1093/nar/gkq454
32. Rasmussen S, Nielsen HB, Jarmer H (2009) The transcriptionally active regions in the genome of Bacillus subtilis. Mol Microbiol 73(6):1043–1057. https://doi.org/10.1111/j.1365-2958.2009.06830.x
33. Csiszar K, Houmard J, Damerval T, Demarsac NT (1987) Transcriptional analysis of the cyanobacterial Gvpabc operon in differentiated cells – occurrence of an antisense RNA complementary to 3 overlapping transcripts. Gene 60(1):29–37. https://doi.org/10.1016/0378-1119(87)90210-1
34. Duhring U, Axmann IM, Hess WR, Wilde A (2006) An internal antisense RNA regulates expression of the photosynthesis gene isiA. Proc Natl Acad Sci U S A 103(18): 7054–7058. https://doi.org/10.1073/pnas.0600927103
35. Eisenhut M, Georg J, Klaehn S, Sakurai I, Mustila H, Zhang P, Hess WR, Aro E-M (2012) The antisense RNA As1_flv4 in the cyanobacterium Synechocystis sp PCC 6803 prevents premature expression of the flv4-2-operon upon shift in inorganic carbon supply. J Biol Chem 287(40):33153–33162. https://doi.org/10.1074/jbc.M112.391755
36. Flaherty BL, Van Nieuwerburgh F, Head SR, Golden JW (2011) Directional RNA deep sequencing sheds new light on the transcriptional response of Anabaena sp strain PCC 7120 to combined-nitrogen deprivation. BMC Genomics 12. https://doi.org/10.1186/1471-2164-12-332
37. Gong Y, Xu X (2012) A small internal antisense RNA (aftsH) of all3642 (ftsH) in Anabaena sp PCC 7120. Chin Sci Bull 57(7):756–761. https://doi.org/10.1007/s11434-011-4929-6
38. Hernandez JA, Alonso I, Pellicer S, Luisa Peleato M, Cases R, Strasser RJ, Barja F, Fillat MF (2010) Mutants of Anabaena sp PCC 7120 lacking alr1690 and alpha-furA antisense RNA show a pleiotropic phenotype and altered photosynthetic machinery. J Plant Physiol 167(6): 430–437. https://doi.org/10.1016/j.jplph.2009.10.009
39. Hernandez JA, Muro-Pastor AM, Flores E, Bes MT, Peleato ML, Fillat MF (2006) Identification of a furA cis antisense RNA in the cyanobacterium Anabaena sp PCC 7120. J Mol Biol

355(3):325–334. https://doi.org/10.1016/j.jmb.2005.10.079

40. Merino-Puerto V, Herrero A, Flores E (2013) Cluster of genes that encode positive and negative elements influencing filament length in a heterocyst-forming cyanobacterium. J Bacteriol 195(17):3957–3966. https://doi.org/10.1128/jb.00181-13
41. Sakurai I, Stazic D, Eisenhut M, Vuorio E, Steglich C, Hess WR, Aro E-M (2012) Positive regulation of psbA gene expression by cis-encoded antisense RNAs in Synechocystis sp PCC 6803. Plant Physiol 160(2):1000–1010. https://doi.org/10.1104/pp.112.202127
42. Georg J, Hess WR (2011) cis-antisense RNA, another level of gene regulation in bacteria. Microbiol Mol Biol Rev 75(2):286–300. https://doi.org/10.1128/MMBR.00032-10
43. Zadeh JN, Steenberg CD, Bois JS, Wolfe BR, Pierce MB, Khan AR, Dirks RM, Pierce NA (2011) NUPACK: analysis and design of nucleic acid systems. J Comput Chem 32(1):170–173. https://doi.org/10.1002/jcc.21596
44. Vazquez-Anderson J, Mihailovic MK, Baldridge KC, Reyes KG, Haning K, Cho SH, Amador P, Powell WB, Contreras LM (2017) Optimization of a novel biophysical model using large scale in vivo antisense hybridization data displays improved prediction capabilities of structurally accessible RNA regions. Nucleic Acids Res 45(9):5523–5538. https://doi.org/10.1093/nar/gkx115
45. Shao Y, Wu Y, Chan CY, McDonough K, Ding Y (2006) Rational design and rapid screening of antisense oligonucleotides for prokaryotic gene modulation. Nucleic Acids Res 34(19):5660–5669. https://doi.org/10.1093/nar/gkl715
46. Kim S-J, Leong M, Amrofell MB, Lee YJ, Moon TS (2019) Modulating responses of toehold switches by an inhibitory hairpin. ACS Synth Biol 8(3):601–605. https://doi.org/10.1021/acssynbio.8b00488

Chapter 8

Design of a Toolbox of RNA Thermometers

Shaunak Sen, Abhilash Patel, and Krishan Kumar Gola

Abstract

RNA thermometers are RNA regulatory elements that convert temperature into a functional biological response through a temperature-induced conformational change. These regulatory elements have been investigated in numerous natural contexts and have been designed for synthetic biology as well. A basic challenge has been the design of an RNA thermometer whose final activity in response to temperature matches a prespecified response, in terms of its sensitivity, threshold, and leakiness. This chapter provides a methodology for the design of a toolbox of RNA thermometers. We describe considerations for the conceptual design, a computational assessment, and strategies for experimental synthesis and measurement.

Key words RNA thermometers, Hairpin loop, Minimum free energy (MFE), Melt profile, Untranslated region (UTR), Green fluorescent protein (GFP)

1 Introduction

The temperature dependence of an RNA structure makes it possible to convert environmental temperature into biological activity, also referred to as an RNA thermometer [1–6]. A simple scheme for an RNA thermometer has a hairpin loop in the 5′-UTR of a gene (*see* Fig. 1). The stem of the hairpin contains the ribosome binding site (RBS). The hairpin structure unfolds as the temperature is increased, allowing the ribosome to access the RBS and initiate the translation of the protein. There are variations to this basic scheme for temperature-dependent gene expression, including an RNA thermometer that works in conjunction with an enzyme [7] and using the temperature dependence of the terminator structure to modulate gene expression [8]. Most synthetic RNA thermometers have been designed based on this basic scheme [9–13], whereas naturally occurring RNA thermometers have more complex structures including multiple bulges and multiple hairpins, and

This article is dedicated to the memory of Prof. Mashuq un Nabi.

James Chappell and Melissa K. Takahashi (eds.), *Riboregulator Design and Analysis*,
Methods in Molecular Biology, vol. 2518, https://doi.org/10.1007/978-1-0716-2421-0_8,

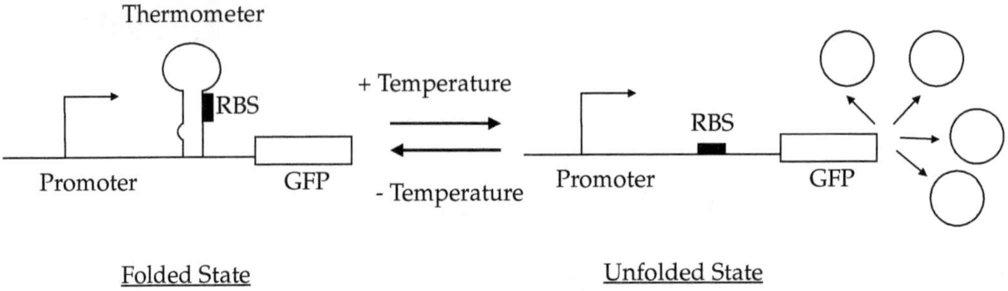

Fig. 1 A simple scheme for the working of an RNA thermometer

may also have pseudoknots [3]. Naturally occurring RNA thermometers are found in cellular contexts where sensing the environmental temperature is functionally important, such as in the cases of heat shock response [14–16], cold shock response [17, 18], and virulence [19]. Analogous to this, synthetically designed RNA thermometers are likely to find applications in synthetic biology [20], including in the spatiotemporal imaging and control of gene expression [21] and in designing temperature compensation for biomolecular circuits [22, 23].

The generation of a library of RNA thermometers based on the above scheme has been nicely summarized in the methods paper [24]. The use of a longer stem region increases the stability of the structure and shifts the response toward higher temperatures. Conversely, an increase in the number of mismatches, bulges, and loops decreases the stability of the structure and shifts the response toward lower temperatures. These principles underlie most of the designs based on this scheme [9, 10, 12, 13]. While the computational expectations do not always match the experimental measurements, these studies have been invaluable in understanding RNA thermometer design. More recently, finer resolution structural mapping of naturally occurring RNA thermometers has been performed [25], which might inform the future design of RNA thermometers.

An ideal design methodology would provide the steps to proceed from a prespecified temperature response to a sequence that exhibits the desired response. Challenges lie on both the theoretical and experimental ends of the spectrum. Theoretically predicting the structure from a sequence, or inversely the sequence from a structure, involves assumptions about the parameters and their temperature dependence as well as about the structures that are possible. The experimental challenge lies in the global role of temperature that makes it difficult to disentangle the effects of the thermometer response from the other aspects of the measurements that could be temperature dependent, especially in realistic systems such as a cell. It is likely that efforts to design RNA thermometers can provide the means to address both these challenges.

The following three broad stages may be considered in the design of a toolbox of RNA thermometers. The first stage is the choice of the conceptual scheme. The simplest choice is the basic scheme outlined above. Alternative starting points include the available naturally occurring or synthetically designed RNA thermometer sequences. The second stage is a computational assessment of the secondary structures and their temperature dependence. While computational methods may not be exact, such assessments provide support to the design scheme. The third stage is the experimental synthesis and measurement. Experimental synthesis may proceed in any from a multiple set of ways to construct the RNA thermometer sequences and integrate them into a suitable protein expression system, with the green fluorescent protein (GFP)-based systems being the most widely used. Similarly, experimental measurements can be performed in multiple contexts, including in cells and in cell-free systems. A particular variation of these stages is outlined below. A possible fourth stage is the comparison of the experimental measurements with the theoretical predictions which could inform the overall process of thermometer design.

2 Materials

1. The construct containing the RNA thermometer is encoded on a plasmid. In addition to the RNA thermometer, the construct should contain a reporter gene, a constitutive promoter, and terminators in the beginning and at the end. A widely used reporter gene is GFP. The RNA thermometer sequence forms a part of the 5'-UTR of the reporter gene. Apart from this construct, the plasmid should have an origin of replication and an antibiotic selection marker. A typical configuration is shown in Fig. 2.
2. *E. coli* JM109.
3. Lysogeny broth (LB).
4. Minimal media (33.9 g/L Na_2HPO_4, 15 g/L KH_2PO_4, 2.5 g/L NaCl, 5 g/L NH_4Cl) supplemented with 0.2% casamino acid, 0.4% glucose, 100 mM thiamine, 1 M $MgSO_4$, and 1 M $CaCl_2$.
5. Cell-free expression system.
6. 96-well, black transparent-bottomed plates for *E. coli* measurements.
7. 384-well, black transparent-bottomed plates for cell extract measurements.
8. A standard plate reader with options to measure the optical density of cells (absorbance at 600 nm) and the GFP fluorescence (filter cube-based or monochromator-based). The typical excitation and emission wavelengths that can be used are 485 nm and 525 nm, respectively.

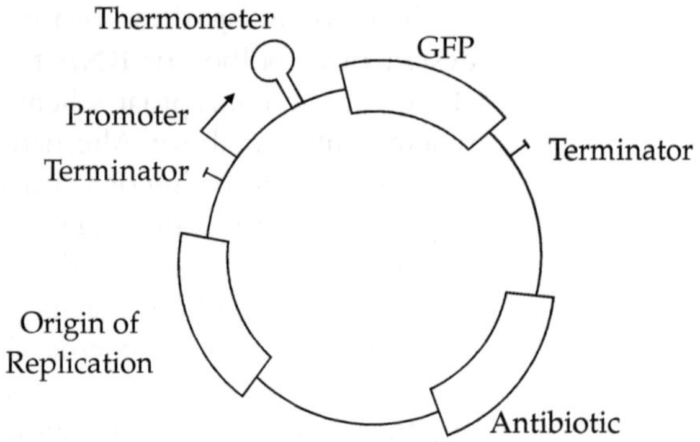

Fig. 2 Schematic of a plasmid containing an RNA thermometer and other essential elements

3 Methods

3.1 Design Concept

1. Identify the sequence of the RNA thermometer that is the starting point of the library. This could be based on a naturally occurring RNA thermometer or on a synthetically designed one.
2. The library can be designed through point mutations in the sequence of the starting RNA thermometer. Alternative possibilities include reconfiguring the various structural elements using different sequences as well as changing the number and the size of the various structural elements.

3.2 Computational Assessment

1. A computational assessment of the library is strongly recommended. This can give an estimate of the diversity in the library as well as a benchmark to compare the experimental measurements with (*see* **Note 1**). Among the options for secondary structure prediction, NUPACK provides a simple, elegant, and feature-rich package [26]. There are multiple computational packages available to perform such an assessment (*see* **Note 2**).
2. The free energy of the minimum free energy (MFE) structure is the most widely used metric that provides an estimate of the temperature response of the RNA thermometer. The more negative the free energy, the more stable is the structure. NUPACK provides the MFE structure as well as this energy. Both the structure and the free energy can be computed for a range of temperatures.
3. The MFE structure is one among an ensemble of structures that the RNA molecule can fold into. These structures can have different degrees to which the RBS is bound. Therefore, checking the fraction of the MFE structure in the ensemble as well as the ensemble free energy, both available in NUPACK, provides a more realistic picture of the temperature response. A large

fraction of RNA molecules in the MFE structure indicates that the design considerations based primarily on the MFE structure work. The computation of the ensemble free energy provides an estimate of the overall stability of the RNA molecule.

4. A melt profile tracks the fraction of bases that are unpaired as a function of the temperature. This is also an important metric of the RNA thermometer activity. The basic idea closely resembles the melt curve that is used in various biological investigations. The slope of the melt profile provides useful information on the sensitivity of the temperature response as well as the order in which the possibly multiple different structural elements in the RNA unfold. The melt curve can also be computed at the resolution of a single base, in which case it signifies the probability that the chosen base is unpaired over a range of temperatures.

3.3 Experimental Synthesis

1. The library of RNA thermometers can be synthesized directly. It can also be synthesized through a variety of schemes ranging from the standard restriction enzyme cloning and more recent scarless synthesis followed by mutation.
2. The essential elements that are recommended in addition to the RNA thermometer are the reporter gene, which is typically a GFP variant, a promoter, and terminators enclosing the construct (*see* Fig. 2).
3. It should be ensured that the spacing between the RBS and the start codon of the gene is 7–9 bases [27, 28].
4. The final sequence in the synthesized plasmids should be verified by sequencing (*see* **Note 3**).

3.4 Experimental Measurement in a Cell-Free Expression System

1. The measurement protocol uses reactions consisting of cell extract, buffer, plasmid DNA, and water in a total volume of 10 μL. *See* **Notes 4** and **5**.
2. To determine the plasmid concentration to be used, the fluorescence of cell-free reactions with different concentrations of the starting RNA thermometer plasmid is measured. The plasmid concentration around which the fluorescence of the reaction varies linearly with the plasmid concentration is in a linear regime and is chosen as the final plasmid concentration to be used in further measurements.
3. The reactions are pipetted into the wells of a black transparent-bottomed 384-well plate. It is important to avoid bubbles in the pipetting process as they can corrupt the fluorescence measurements.
4. The plate is covered with a membrane and then spun down so that the well contents settle. Next, it is placed in a plate reader for measurement.

5. The plate reader is set to the desired temperature, and measurements can be taken every 3 min for 8 h to obtain a temporal profile of each thermometer. The GFP fluorescence is measured with the excitation/emission setting suitable for the GFP variant.
6. **Step** 3.4.5 can be repeated at different temperatures to obtain the temporal profiles at those temperatures.
7. Alternatively, instead of **steps** 3.4.5 and 3.4.6, the reactions can be incubated in PCR tubes at different temperatures simultaneously for 150 min (*see* **Note 6**). Subsequently, the reactions can be transferred to a plate as in **step** 3.4.5, and a single timepoint measurement can be made.
8. A reaction without the plasmid DNA is used as the background fluorescence reading. This is subtracted from the fluorescence readings of the different thermometers. *See* **Note 8**.

3.5 Experimental Measurement in Cells

1. The plasmid is transformed into an *E. coli* strain for measurement.
2. A colony is picked from an agar plate and inoculated overnight in LB media with suitable antibiotic. This culture is grown overnight at 37 °C in a shaking incubator. *See* **Note** 7.
3. The overnight culture is subsequently diluted 1:200 in fresh minimal media (with antibiotic) and grown for 2 h in a shaking incubator at the desired measurement temperature.
4. 200 μL of this culture is pipetted into five wells of a 96 well black transparent-bottomed plate.
5. Wells with 200 μL of media without cells are used for estimating the background absorbance and fluorescence.
6. The plate is placed in the plate reader for measurements, which are taken every 5 min for 8 h. The temperature of the plate reader is set to the desired temperature. The GFP fluorescence measurement is used to measure the thermometer activity while the optical density measurement is used to estimate the growth of the cell.
7. The activity of the RNA thermometer at this temperature is computed as the fluorescence of the well divided by the optical density of the well, where the backgrounds are subtracted from the values of the fluorescence and the optical density. *See* **Note 8**.
8. The promoter activity can also be quantified as follows: Let N be the number of bacterial cells and x be the amount of GFP per cell so that $Nx = F$ is the total fluorescence. N and F can be estimated from the optical density and the fluorescence

measurements, respectively. The promoter activity is $\frac{1}{N}\frac{dF}{dt}$, where t is time. The derivative may be approximated by a finite difference or applied on a filtered version of the measured data, for example, using a moving average filter.

4 Notes

1. It has often been noted that the computations do not explicitly predict the thermometer performance. This could be due to a variety of reasons including the simulation parameters, which are not available at all temperatures, as well as various modeling assumptions about the possible structures and the ignored dynamics. Nevertheless, computations, even if not exact, can provide good estimates of the general properties.
2. There are other secondary structure prediction algorithms available that can also be used, for example [29, 30].
3. Sequencing the plasmid with thermometers is important because it is possible that thermometers having high expression of GFP may accumulate mutations in the promoter region. In such cases, the plates can be grown at a lower temperature after transformation so that the expression levels are not high.
4. The preparation of an *E. coli* cell extract and the related buffers are detailed in [31]. Other cell-free expression systems have also been used for measurements on RNA thermometers [12, 13].
5. An important positive control for the cell-free measurements is a constitutively active promoter that tests for the working of the cell-free reaction. A reaction without any plasmid DNA may be used as a negative control.
6. For measurements in the cell-free system, the initial measurement can be used to further fine-tune the protocol, for example, in the concentrations of the plasmid DNA to be used and in the incubation time for the single timepoint measurement.
7. For measurements in cells, a qualitative check for the temperature dependence can be made by spotting the cells onto LB agar plates with suitable antibiotic and incubating them overnight at different temperatures. The fluorescence can be imaged in UV light in a gel imager.
8. Temperature is a global environmental variable and can affect multiple aspects of measurements ranging from GFP properties to cell growth rates. Ideally, independent calibration of these properties is needed. However, an exhaustive characterization may not be possible. The following three points list some measures that can be used to disentangle these global effects of temperature:

8.1. Purified GFP can be used, and its brightness at different temperatures can be measured as a calibration. While this does not account for all the GFP properties that can change with temperature, it provides a first step toward normalizing for the effect of temperature on GFP brightness.

8.2. With measurements of a thermometer library, the relative difference between the different thermometers should be compared rather than viewing the thermometer response in isolation. As the constructs are identical, except for the thermometer sequence, the relative differences can directly be attributed to the difference in the thermometer sequences.

8.3. A thermometer control whose secondary structure does not change with the temperature, if available, can be used as a control. There are some thermometers that have been used for this purpose [9, 25], and further development of these would be helpful.

Acknowledgments

Support from DBT (BT/PR28513/med/32/672/2019) and the Discover and Learn Program, IRD, IIT Delhi is gratefully acknowledged.

References

1. Narberhaus F, Waldminghaus T, Chowdhury S (2005) RNA thermometers. FEMS Microbiol Rev 30:3–16
2. Klinker B, Narberhaus F (2009) Microbial thermosensors. Cell Mol Life Sci 66:2661–2676
3. Kortmann J, Narberhaus F (2012) Bacterial RNA thermometers: molecular zippers and switches. Nat Rev Microbiol 10:255–265
4. De la Fuente M, Valera S, Martínez-Guitarte (2012) ncRNAs and thermoregulation: a view in prokaryotes and eukaryotes. FEBS Lett 586: 4061–4069
5. Krajewski SS, Narberhaus F (2014) Temperature-driven differential gene expression by RNA thermosensors. Biochim Biophys Acta 1839:978–988
6. Mandin P, Johansson J (2020) Feeling the heat at the millennium: thermosensors playing with fire. Mol Microbiol 113:588–592
7. Hoynes-O'Connor A, Hinman K, Kirchner L et al (2015) *De novo* design of heat-repressible RNA thermosensors in *E. coli*. Nucleic Acids Res 43:6166–6179
8. Roßmanith J, Weskamp M, Narberhaus F (2018) Design of a temperature-responsive transcription terminator. ACS Synth Biol 7: 613–621
9. Neupert J, Karcher D, Bock R (2008) Design of simple synthetic RNA thermometers for temperature-controlled gene expression in *Escherichia coli*. Nucleic Acids Res 36:e124
10. Jia H, Sun X, Sun H et al (2016) Intelligent microbial heat-regulating engine (IMHeRE) for improved thermo-robustness and efficiency of bioconversion. ACS Synth Biol 5:312–320
11. Sen S, Apurva D, Satija R et al (2017) Design of a toolbox of RNA thermometers. ACS Synth Biol 6:1461–1470
12. Sadler FW, Dodevski I, Sarkar CA (2018) RNA thermometers for the PURExpress system. ACS Synth Biol 7:292–296

13. Jia H, Heymann M, Härtel T et al (2019) Temperature-sensitive protein expression in protocells. Chem Commun 55:6421–6427
14. Morita M, Kanemori M, Yanagi H et al (1999) Heat-induced synthesis of σ^{32} in *Escherichia coli*: structural and functional dissection of *rpoH* mRNA secondary structure. J Bacteriol 181:401–410
15. Morita MT, Tanaka Y, Kodama TS et al (1999) Translational induction of heat shock transcription factor σ^{32}: evidence for a built-in thermosensor. Genes Dev 13:655–665
16. Waldminghaus T, Heidrich N, Brantl S et al (2007) FourU: a novel type of RNA thermometer in *Salmonella*. Mol Microbiol 65:413–424
17. Eriksson S, Hurme R, Rhen M (2002) Low-temperature sensors in bacteria. Philos Trans R Soc Lond B 357:887–893
18. Giuliodori AM, Di Pietro F, Marzi S et al (2010) The *cspA* mRNA is a thermosensor that modulates translation of the cold-shock protein CspA. Mol Cell 37:21–33
19. Shapiro RS, Cowen LE (2012) Thermal control of microbial development and virulence: molecular mechanisms of microbial temperature sensing. MBio 3:e00238–e00212
20. Wang Y-H, Wei KY, Smolke CD (2013) Synthetic biology: advancing the design of diverse genetic systems. Annu Rev Chem Biomol Eng 4:69–102
21. Piraner DI, Abedi MH, Moser BA et al (2017) Tunable thermal bioswitches for *in vivo* control of microbial therapeutics. Nat Chem Biol 13: 75–80
22. Hussain F, Gupta C, Hirnin AJ et al (2014) Engineered temperature compensation in a synthetic genetic clock. Proc Natl Acad Sci U S A 111:972–977
23. Sen S, Kim J, Murray RM (2014) Designing robustness to temperature in a feedforward loop circuit. In: Proceedings of 53rd IEEE Conference on Decision and Control, pp 4629–4634
24. Neupert J, Bock R (2009) Designing and using synthetic RNA thermometers for temperature-controlled gene expression in bacteria. Nat Protoc 4:1262–1273
25. Meyer S, Carlson PD, Lucks JB (2017) Characterizing the structure-function relationship of a naturally occurring RNA thermometer. Biochemistry 56:6629–6638
26. Zadeh JN, Steenberg CD, Bois JS et al (2011) NUPACK: analysis and design of nucleic acid systems. J Comput Chem 32:170–173
27. Berwal SK, Sreejith RK, Pal JK (2010) Distance between RBS and AUG plays an important role in overexpression of recombinant proteins. Anal Biochem 405:275–277
28. Vellanoweth RL, Rabinowitz JC (1992) The influence of ribosome-binding-site elements on translational efficiency in *Bacillus subtilis* and *Escherichia coli in vivo*. Mol Microbiol 6: 1105–1114
29. Lorenz R, Bernhart SH, zu Siederdissen H et al (2011) ViennaRNA Package 2.0. Algorithms Mol Bio 6:26
30. Garcia-Martin JA, Dotu I, Fernandez-Chamorro J (2016) RNAiFold2T: constraint programming design of thermo-IRES switches. Bioinformatics 32:i360–i368
31. Sun ZZ, Hayes CA, Shin J (2013) Protocols for implementing and *Escherichia coli* based TX-TL cell-free expression system for synthetic biology. J Vis Exp:e50762

Chapter 9

Development of Synthetic Riboswitches to Guide the Evolution of Metabolite Production in Microorganisms

Minsun Kim, Sungho Jang, and Gyoo Yeol Jung

Abstract

The untranslated region (UTR) of prokaryotic mRNA contains riboswitches, which are gene regulating modules. Riboswitches can be used as biosensors to regulate the expression of a gene or an operon depending on the intracellular level of a target molecule and consequently modulate the cellular responses. In evolutionary engineering, riboswitch-based biosensors have been widely applied for high-throughput screening or selection of target phenotypes. Evolutionary approaches can overcome the limitations of rational approaches in metabolic engineering. Previous studies have reported synthetic riboswitches equipped with novel aptamers and marker genes based on a deep understanding of the operation mechanism of the riboswitch. Here, we introduce the development process of novel synthetic riboswitches for applications in metabolic engineering.

Key words SELEX, Aptamer, Riboswitch, Dual selection, High-throughput screening, Metabolic engineering, Synthetic biology

1 Introduction

A riboswitch is a well-conserved genetic regulatory machinery in bacterial and archaeal species. In response to environmental cues, the riboswitch allocates limited resources to promote the survival of the host cells. Riboswitches are involved in the anabolic operon of key metabolites (amino acids and nucleotides) [1–4] and cofactors (thiamine pyrophosphate, coenzyme B12, tetrahydrofolate, and flavin mononucleotide (FMN)) in the bacterial genome [5–8].

Riboswitches are *cis*-regulatory modules located at the 5′-untranslated region (UTR) or 3′-UTR of an mRNA and ligand-dependently regulate the rate of transcription or translation of a downstream gene [9] (*see* Fig. 1). Thus, the gene expression can be modulated using mRNA without the involvement of protein components. Moreover, riboswitches enable the designing of circuits with various combinations of input and output by rearranging the

James Chappell and Melissa K. Takahashi (eds.), *Riboregulator Design and Analysis*,
Methods in Molecular Biology, vol. 2518, https://doi.org/10.1007/978-1-0716-2421-0_9,

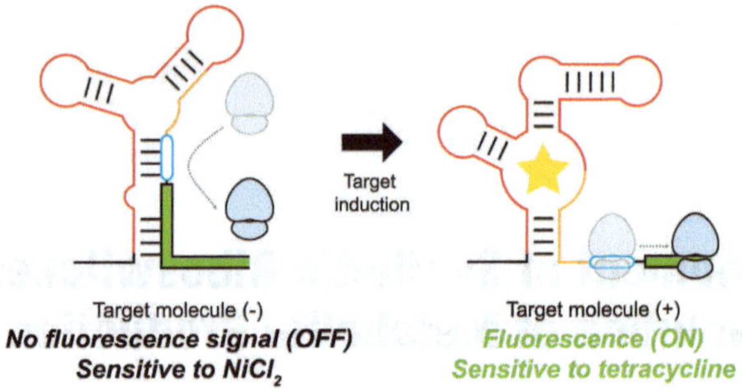

Fig. 1 Scheme of library design and operation mechanism of the synthetic translation regulating riboswitch. Mechanism of ligand-dependent gene expression regulation of the translation regulating riboswitch. Aptamer is indicated in red, linker in yellow, ribosome binding site (RBS) in light blue, and *tetAsgfp* in green. Star indicates the target ligand. The ribosome becomes accessible when the RBS is exposed and consequently the translation is initiated

genetic regulatory modules, which will result in the signal-dependent cascade of gene expression in vivo [10].

Conventional metabolic engineering has heavily relied on rational and static engineering strategies to achieve optimal combinations of several factors that affect the expression of metabolic genes. The number of variants to be screened increases logarithmically with the increase in the number of factors involved in gene expression. However, the recent advances in metabolite-sensing gene regulators have enabled the development of novel strategies for metabolic engineering, including combinatorial and evolutionary engineering [11, 12].

The amount of metabolites produced through microbial fermentation can be calculated by multiplying the specific production and the total cell mass. Therefore, the aim of microbial production is to maximize both specific production and cell mass. However, the cellular resources are often shared between metabolite production and cell growth. Therefore, the metabolic flux between growth and production must be optimally balanced at a key node [13, 14]. The precise balancing of metabolic flux requires the identification of the key node that distributes essential metabolites and cofactors. After the key node is identified, the major hurdle for metabolic engineering remains: building and testing a large number of metabolic pathways to find the optimally balanced combination. Moreover, gene overexpression does not maximize metabolic flux in some cases. Thus, enzyme engineering must be performed, which involves testing numerous variants [15].

Biosensors can report the intracellular levels of target metabolites and consequently redirect the cell physiology by regulating the expression of functional genes. As biosensors can be genetically encoded, they can mediate various processes in individual cells.

The applications of biosensors in metabolic engineering include evolutionary engineering utilizing biosensors with selection markers and high-throughput screening using cell sorters and biosensors with fluorescent protein-encoding genes [16, 17]. RNA aptamers and synthetic riboswitches responsive to new target molecules have been developed to generate riboswitch-based biosensors [18–21]. The increasing demand for the development and engineering of riboswitches indicates the potential for high-throughput analysis of the intracellular concentration of small molecules.

The development of a riboswitch involves screening a large mutational landscape to select the ligand-responsive gene regulator from a variant library. The following two states of the biosensor should be considered for designing a functional translation regulating riboswitch: a ligand binding state that induces the exposure of ribosome binding site (RBS) and the ligand-free conformation in which the RBS is buried to suppress the expression of a downstream gene. Furthermore, the flexibility of the conformation transition, which determines the fold activation ratio between two states, is determined by the specific ribonucleotide sequence and affinity of the aptamer to the ligand [22, 23].

Hence, the development of riboswitches is associated with considerable screening load. Here, we introduce systematic evolution of ligands by exponential enrichment (SELEX) [24], a robust strategy based on evolutionary engineering to screen small-molecule binding aptamers, and an in vivo evolution approach for isolating translation regulating riboswitches from the cell library [18, 25]. This chapter comprises the following two parts: (1) aptamer library design, selection bead preparation, and SELEX protocols and (2) translation regulating riboswitch library cloning, screening, and validation protocols.

2 Materials

Prepare all solutions using diethyl pyrocarbonate-treated or RNase-free water to prevent RNase contamination. Mix enzymatic reaction mixtures and DNA samples on ice. Store reagents at −20 °C. Store RNA samples at −80 °C. Use molecular biology grade reagents throughout the protocol.

2.1 Preparation of RNA Aptamer Library

1. Template single-stranded DNA: insert 50 nt random sequences between two primer binding sites. Use 30 pmoles single-stranded DNA/mL PCR mixture in the library-generating PCR reaction (*see* **Note 1**).
2. Reaction buffer: 0.1 M Tris-HCl, 0.5 M KCl, 15 mM $MgCl_2$, and pH 8.3.
3. Taq DNA polymerase, 5 U/μL.

4. Deoxyribonucleotide mix (dNTPs), pH 7.5: equimolar concentrations (2.5 mM) of dATP, dGTP, dCTP, and dTTP.
5. 10 μM forward and reverse primers dissolved in distilled water.
6. Thermocycler.
7. 5× Tris-borate-EDTA (TBE) buffer: 54 g/L Tris base, 27.5 g/L boric acid, and 10 mM EDTA dissolved in distilled water. Adjust pH to 8.3 using HCl.
8. Polyacrylamide gel: 10% polyacrylamide/bisacrylamide [29:1], 1× TBE buffer, 0.3% ammonium persulfate, and 0.3% N,N,N′,N′-tetramethyl-ethylenediamine (TEMED) dissolved in distilled water.
9. 25/100 bp DNA ladder or appropriate range of DNA ladder for the target DNA library.
10. 10× loading buffer: 50% glycerol, 0.9% sodium dodecyl sulfate, and 0.05% bromophenol blue.
11. Mini-PROTEAN tetra cell for polyacrylamide gel electrophoresis (PAGE).
12. 10 mg/mL (20,000× concentration) ethidium bromide (EtBr): Dilute 20,000-fold in 1× TBE buffer for acrylamide gel staining.
13. Gel documentation system.
14. 20 mg/mL glycogen for DNA and RNA ethanol precipitation. Add the glycogen solution at a final concentration of 20 μg/mL to the DNA or RNA solutions to be precipitated. Store at −20 °C.
15. 3 M sodium acetate, pH 5.2 for DNA and RNA ethanol precipitation. Adjust pH with acetic acid; autoclaved.
16. Absolute ethanol and 70% ethanol diluted with distilled water for DNA and RNA precipitation and pellet washing, respectively. Store at −20 °C.
17. Deep freezer.
18. Temperature-controlled microcentrifuge.
19. PCR cleanup kit (*see* **Note 2**).
20. Ultraviolet-visible (UV-Vis) spectrophotometer.
21. 5× T7 RNA polymerase buffer, pH 7.9: 0.2 M Tris-HCl, 30 mM $MgCl_2$, 5 mM dithiothreitol (DTT), and 10 mM spermidine.
22. 50 U/μL T7 RNA polymerase.
23. Ribonucleotide mix (rNTPs), pH 7.5: equimolar concentrations (25 mM) of rATP, rGTP, rCTP, and rUTP.
24. 40 U/μL RNase inhibitor.
25. 1 M $MgCl_2$; autoclaved.

26. 0.1 M molecular biology grade DTT.
27. Denaturing polyacrylamide gel: 10% polyacrylamide/bis-acrylamide (29:1), 1× TBE buffer, 0.3% ammonium persulfate, 0.3% TEMED, and 8 M urea.
28. 10× Tris-EDTA (TE) buffer, pH 8.0: 0.1 M Tris base (pH 8.0 adjusted with HCl), and 10 mM EDTA (pH 8.0 adjusted with NaOH).
29. Thin-layer chromatography (TLC) plate: silica gel coated with the fluorescence indicator F254 (size: 20 × 20 cm).
30. Handheld UV gun for intercalator-free RNA visualization.
31. Rotator mixer for RNA extraction in 1× TE buffer.
32. Centrifugal column with a cellulose acetate membrane (RNase-free and DNase-free).

2.2 Preparation of Ligand-Coupled Agarose Bead for RNA Aptamer Selection

1. Activated agarose beads: The density of the activated group is 19–40 μmol/mL drained medium. Epoxy-activated agarose beads can be used for coupling the hydroxyl groups on the ligand. The mean size of swollen bead is 90 μm. Store powdered beads at −20 °C.
2. 1.1× coupling solution, pH 13.0: 55 mM KCl, 55 mM NaOH (*see* **Note 8**).
3. 1 M ethanolamine, pH 8.0 for blocking remaining active groups on the bead.
4. 0.1 M acetate buffer, pH 4.0 containing 0.5 M NaCl.
5. 0.1 M Tris-HCl buffer, pH 8.0 containing 0.5 M NaCl.
6. Storage buffer: 20% ethanol.

2.3 RNA Aptamer SELEX

1. RNA aptamer library: 8 nmol/mL RNA used for each selection.
2. Selection buffer, pH 7.4: 0.25 M Tris-HCl, pH 8.0, 750 mM NaCl.
3. Washing buffer, pH 7.4: 1× selection buffer, 1 mM $MgCl_2$.
4. Elution buffer, pH 7.4: 25 mM Tris-HCl, pH 8.0, 0.3 M NaCl, 5 mM EDTA, and 5 M urea.
5. 5 mL polypropylene column for packing ligand-coupled bead
6. Rotator mixer for RNA library incubation in selection buffer.
7. Thermocycler for RNA denaturing.
8. 200 U/μL reverse transcriptase
9. Materials Nos. 2–32 in Subheading 2.1.

2.4 Riboswitch Library Preparation

1. Backbone plasmid: pACYC_J23113_tetAsgfp [18].
2. 20 U/μL *Bsa*I restriction enzyme

3. 10× restriction buffer: 0.5 M potassium acetate, 0.2 M Tris-acetate, 0.1 M magnesium acetate, 1 mg/mL recombinant albumin
4. 50× Tris-acetate-EDTA (TAE) buffer: 2 M Tris base, 1 M acetic acid, and 50 mM EDTA
5. 1% agarose gel: 1 g agarose powder dissolved in 100 mL of 1× TAE buffer, supplemented with 0.5 μg/mL EtBr
6. Gel purification kit.
7. 400 U/μL T4 DNA ligase
8. 10× ligation buffer, pH 7.4: 0.5 M Tris-HCl, pH 7.5, 0.1 M $MgCl_2$, 10 mM ATP, 0.1 M DTT
9. Electrocompetent *E. coli* DH10B $T1^R$.
10. S.O.C. medium: 2% tryptone, 0.5% yeast extract, 10 mM NaCl, 2.5 mM KCl, 10 mM $MgCl_2$, 10 mM $MgSO_4$, and 20 mM glucose; pH adjusted to 7.0 with NaOH; autoclaved and stored at 4 °C.
11. LB-agar plate supplemented with appropriate antibiotics.
12. Flame-sterilized spreader.
13. Incubation oven.
14. 1× phosphate-buffered saline (PBS): 8 g/L NaCl, 0.2 g/L KCl, 1.42 g/L Na_2HPO_4, 0.24 g/L KH_2PO_4; filter sterilized before use and stored at 4 °C
15. Centrifuge.
16. Shaking incubator: 37 °C, 220 rpm.

2.5 In Vivo Selection of Riboswitch Library

1. M9 salt: 15 g/L KH_2PO_4, 2.5 g/L NaCl, 33.9 g/L Na_2HPO_4, and 5 g/L NH_4Cl; autoclaved before use.
2. M9 minimal medium: 1× M9 salts, 0.4% glucose, 20 mM $MgSO_4$, and 1 mM $CaCl_2$. All components filter either sterilized or autoclaved before mixing. Stored at 4 °C.
3. Antibiotics for plasmid maintenance.
4. 200 mg/mL tetracycline hydrochloride freshly prepared in 70% ethanol before use
5. 20 mM $NiCl_2$
6. Shaking incubator: 37 °C, 220 rpm.
7. 50% glycerol filter sterilized and stored at 4 °C
8. Cryovial.
9. Deep freezer.

2.6 In Vivo Riboswitch Library Validation

1. M9 minimal medium: 1× M9 salts, 0.4% glucose, 20 mM $MgSO_4$, and 1 mM $CaCl_2$; stored at 4 °C.

Table 1
Oligonucleotides

Name	Sequence (5′ → 3′)
SELEX_Template	**tggtcatgtgatcggcgtatg** $(N)_{50}$ tatcgtgtcatcgtcg**tccctatagtgagtcgtatta**
SELEX_F	taatacgactcactataggga
SELEX_R	tggtcatgtgatcggcgtat
SELEX_T7_F	taatacgactcactatagggacgacgatgacacg

2. Stock solutions of target ligands were prepared in appropriate solvents (*see* **Note 3**).
3. Antibiotics for plasmid maintenance.
4. Deep well plate and breathable sealing cover.
5. Multichannel pipette.
6. LB-agar plate supplemented with appropriate antibiotics.
7. Orbital shaker.
8. Microplate reader.

2.7 Oligonucleotides

See Table 1.

3 Methods

3.1 Preparation of RNA Aptamer Library

Prepare the RNA aptamer library from a single-stranded synthetic DNA oligomer containing a randomized sequence region flanked by primer binding sites at both termini (*see* Fig. 2). Forward and reverse primer binding sites should have similar melting temperatures (55–60 °C) and GC contents (approximately 50%). The primer sets that form homodimers, heterodimers, or any stable secondary structures should be avoided (Table 1). Therefore, the goal is to minimize the probability of enrichment or amplification of nontarget DNAs and RNAs resulting from nonspecific hybridization of the primers to the randomized sequences of the library pool. Moreover, the following general guidelines for RNA experiments, including the use of gloves, face masks, and separate working spaces, should be followed to avoid RNase contamination:

1. Order the single-stranded DNA oligomer containing a randomized sequence region flanked by forward and reverse primer binding sites.
2. Prepare the master PCR mix as shown in Table 2, and transfer a 50 μL aliquot of the PCR mix to a separate PCR tube to perform a pilot PCR as shown in Table 3. Store the remaining PCR mix at 4 °C for the main PCR (*see* **Note 4**).

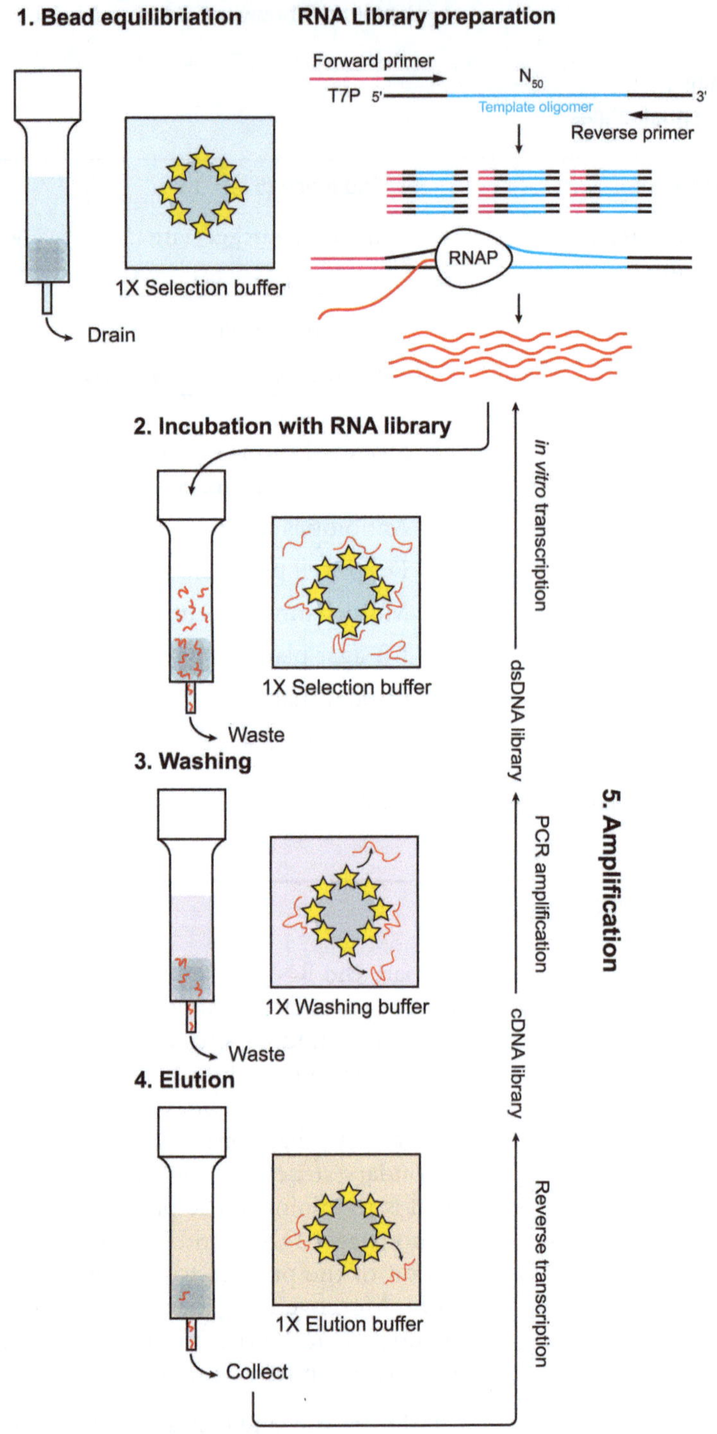

Fig. 2 Scheme of bead-based systematic evolution of ligands by exponential enrichment (SELEX). A bead-packed gravity column and a single bead in a box interacting with RNAs are illustrated for every step of in vitro aptamer selection. Red line, star, and gray circle indicate the RNA library, target molecule, and target-coupled bead, respectively. In the RNA library preparation step, the blue line represents the template random oligomer with primer binding sites. Pink and black on the template oligomer indicate the T7 promoter and primer binding sites, respectively. RNAP, RNA polymerase

Table 2
PCR master mix

Reagent	Final concentration
Reverse transcription mixture	0.1 of total volume
Forward primer (10 μM)	0.5 μM
Reverse primer (10 μM)	0.5 μM
dNTPs (10 mM)	0.4 mM
10× buffer	1×
Taq polymerase (5 U/μL)	2.5 U/50 μL reaction

Table 3
Thermocycling condition for library PCR

Step	Temperature	Time
Initial denaturation	95 °C	30 s
Denaturation	95 °C	30 s
Annealing	55 ~ 65 °C	30 s
Elongation	72 °C	1 kb/min
Final elongation	72 °C	5 min
Number of cycles	Depends on the cDNA concentration in reverse transcription mixture	

3. Subject the mixture to PCR amplification using a thermocycler. Sample 6 μL of the PCR mixture every two cycles (*see* **Note 4**).
4. Mix the sample with 1× gel loading buffer, and resolve alongside a DNA ladder on a 10% polyacrylamide gel in 1× TBE buffer.
5. Determine the optimal amplification cycle using EtBr staining. Submerge the gel in 1× TBE buffer with 0.5 μg/mL EtBr for more than 5 min (*see* **Note 5**).
6. Subject the remaining PCR mixture to PCR amplification based on the number of amplification cycles determined in the preliminary test in **step 5**, under the same conditions as those used in **step 3**.
7. Precipitate double-stranded DNA (dsDNA) using ethanol, sodium acetate, and glycogen.
8. Remove the remaining primers using the PCR purification kit.
9. Determine the concentration and purity of dsDNA library eluent using UV spectrophotometer (*see* **Note 6**).

Table 4
In vitro transcription

Reagent	Final concentration
Template dsDNA	0.2–1 μg/20 μL reaction
10× reaction buffer	1×
rNTPs (100 mM)	2 mM
RNase inhibitor (40 U/μL)	1 U/μL
DTT (0.1 M)	5 mM
T7 RNA polymerase (50 U/μL)	5 U/μL

10. Prepare the in vitro transcription reaction mixture as shown in Table 4, and incubate at 37 °C for 6 h.
11. Mix the reaction mixture obtained in **step 9** with 8 M urea in equal volumes, and resolve the solution using a denaturing polyacrylamide gel (*see* **Note** 7).
12. Identify the target band of RNA on the TLC plate using a UV gun. Excise the gel slice containing the target size of RNA.
13. Crush the gel slices into small pieces, and incubate with 1× TE buffer in a 15 mL conical tube for more than 8 h with rotation to allow the diffusion of RNA into the buffer.
14. Filter the urea-containing gel particles using a centrifugal column with a cellulose acetate membrane.
15. Precipitate the RNA library in TE solution using ethanol, sodium acetate, and glycogen.
16. Determine the concentration and purity of the RNA library (*see* **Note** 7).

3.2 Preparation of Ligand-Coupled Agarose Bead for RNA Aptamer Selection

Ligand-coupled agarose beads provide a high local concentration of the target ligand. This is important for selecting aptamers specific to the hydrophobic molecules as the chance of interaction between hydrophobic molecules in solution, which exhibit poor solubility in the aqueous phase, and RNA is compromised and consequently impairs aptamer enrichment.

The utilization of simple chemical reactions, such as N-hydroxysuccinimide (NHS)/carbodiimide crosslinking, enables the coupling of commercial agarose beads with molecules comprising common reactive groups, such as hydroxyl, thiol, amine, and carboxyl groups. Several vendors provide agarose beads embedded with diverse activated groups at the end of the nonreactive arm stretched from the surface. For example, epoxy-activated beads are recommended for hydroxyl group coupling, while amine-activated beads are recommended for carboxyl group coupling.

The reactive groups to be coupled with the bead should be carefully selected considering the orientation of the target molecule, which determines the specificity of the aptamer. For sufficient spacing between the target molecule and bead, the target molecule can be chemically modified to attach additional linker containing specific reactive groups for coupling chemistry [20]. The composition of the coupling solution differs depending on the chemical moiety of the target ligand and the activated groups on the beads. Abundant ligands must be introduced to the bead, and the conditions must be optimized to maximize the coupling efficiency. Detailed recipes for the reagents are provided in Subheading 2.2:

1. Suspend the bead powder in distilled water. One gram of freeze-dried powder can be used for preparing approximately 3.5 mL of medium.
2. Wash the beads with 200 mL of distilled water per gram of freeze-dried powder on the sintered glass filter.
3. Dissolve the ligand into coupling buffer considering the number of activated groups on the beads. At least 200 μmoles ligand should be suspended per mL medium (*see* **Note 8**).
4. Mix the coupling solution and medium at 1:1 ratio in a stoppered flask, and shake the solution at appropriate temperature for coupling chemistry at 100 rpm, for 16 h (*see* **Note 8**).
5. Wash the medium with ligand-free coupling buffer to remove excess ligand.
6. Transfer the medium in 1 M ethanolamine solution to block remaining active groups, and stand it overnight at 40 °C.
7. Wash the ligand-coupled medium thoroughly with two alternative pH buffers per cycle. Apply at least three cycles of washing.
8. Store the washed medium in 20% ethanol, and wash the medium with reaction buffer before use.

Residual reactive groups on the bead surface must be blocked with 1 M ethanolamine solution in the case of epoxy-activated groups to prevent nonspecific interaction with RNAs. Moreover, counter selection using the same agarose bead with no target molecule attached on its surface is highly recommended to eliminate unwanted RNAs with affinity to the bead surface rather than the target molecule.

3.3 RNA Aptamer SELEX

The initial diversity generation is followed by SELEX, which typically includes consecutive rounds of selection and amplification of RNAs that bind to the target molecule (*see* Fig. 2). The RNA pool is mixed with the ligand-coupled bead in a coupling solution containing sufficient monovalent and divalent cations (*see* **Note 9**).

Wash the beads with a ligand-free coupling buffer to remove weakly binding RNAs. The volume of the washing buffer or the number of washing steps should be carefully designed to ensure that target-specific aptamers with low equilibrium dissociation constant (K_d) are retrieved and enriched (*see* **Note 10**).

The initial selection round should be devised to maximize the probability of capturing the true aptamer candidates in the library pool (*see* **Note 11**).

1. Prepare the RNA library in 1 mL of 1× selection buffer, denature at 95 °C for 10 min, and cool to 25 °C at a ramp rate of 1 °C/s (*see* **Note 12**).
2. Supplement the RNA library with 10 mM $MgCl_2$ and incubate for 15 min at 25 °C.
3. Transfer the target-coupled beads in the storage solution to an empty column, and drain the buffer. The volume of the target-coupled bead solution must be chosen such that the number of moles of suspended ligands exceeds that of total RNAs.
4. Equilibrate the beads with 1 mL of 1× selection buffer, invert the capped column several times to mix, and drain the buffer. Repeat this step at least three times.
5. Add the solution in **step 2** to the bead-packed column, and the column was inverted for 30 min on a rotator mixer.
6. Drain the solution, and rinse the beads with washing buffer at a volume ten times that of drained beads (*see* **Note 13**).
7. Next, prepare 3 mL of preheated elution buffer (95 °C), and vigorously pipette the beads with preheated elution buffer three times (1 mL/repeat).
8. Collect the elution buffer from **step 7** in a 15 mL conical tube.

After each round, the enrichment index should be determined to adjust the selection condition for the next round. The parallel elution of RNA in ligand-coupled beads and ligand-free beads enables the relative quantification of the enrichment. Enrichment can be tracked based on the indirect index obtained from the following experiments:

9. Perform two separate SELEX applying identical methods and materials with ligand-coupled bead and intact bead.
10. Perform two parallel reverse transcription reactions (Table 5) (*see* **Note 14**).
11. Perform quantitative PCR (qPCR) for the eluted RNAs from the ligand-coupled bead and from intact bead.
12. Calculate the molar ratio of the eluted RNAs from the ligand-coupled bead vs. input RNA or the eluted RNAs from the ligand-coupled bead vs. the eluted RNAs from intact bead (*see* **Note 15**).

Table 5
Reverse transcription mix

Reagent	Final concentration
RNA sample	0.25 of total volume
Reverse primer (10 μM)	0.5 μM
dNTPs (10 mM)	2 mM
5× buffer	1×
DTT (0.1 M)	5 mM
RNase inhibitor (40 U/μL)	2 U/μL
Reverse transcriptase (200 U/μL)	10 U/μL

Subsequent amplification steps of eluted RNAs include reverse transcription (RNA to ssDNA), PCR (ssDNA to dsDNA with T7 promoter), and in vitro transcription (dsDNA to RNA). The same recipe and reaction conditions were applied as those used in the RNA library preparation process, except for the reverse transcription reaction (Table 5). After the amplification step, the RNA library with the target length was purified from a denaturing polyacrylamide gel and concentrated to prepare the RNA aptamer library for the next round.

Consistent chemical composition and reaction conditions were employed during the preparation of the reaction mixture in every round of amplification to avoid amplification bias. Moreover, the volume of amplification reaction was sufficiently prepared to cover the diversity of the estimated RNAs recovered from the selection.

3.4 Riboswitch Library Preparation

A ligand-activated translation regulating riboswitch, which regulates gene expression at the translation level, promotes translation initiation in response to ligand concentration by favoring ligand binding conformation, resulting in the exposure of RBS (ON state). In the absence of sufficient ligands, the aptamer adopts alternative intra-strand hybridization, which blocks the access of the ribosome to RBS (OFF state). The stable secondary structure of the 5′-UTR interferes with the access of the ribosome to RBS, which results in translation repression (*see* Fig. 1).

Therefore, the structure of the "OFF" state riboswitch should not only be stable to prevent any leaky expression of the downstream gene in the absence of the target molecule but also be flexible to allow switching to the ligand binding conformation when sufficient concentrations of ligands are introduced. The net free energy change of the 5′-UTR of an mRNA thermodynamically determines the rate of transition between the OFF and ON states [22]. However, the complex correlation between the sequence and the secondary and/or tertiary structure of mRNA cannot be predicted, which restricts the rational design of translational riboswitches. Moreover, the free energy of mRNA varies with the progression of transcription and the RNA elongates.

Currently, obtaining key information for riboswitch engineering depends on experiments, such as structure determination and footprinting of RNA interacting with ligands. In contrast, the evolutionary approach is relatively independent of background information. Huge mutational space is screened at high efficiency, which reduces the importance of the rational design of candidate RNA sequences. In the evolutionary approach, a plasmid library containing candidate riboswitches was transformed into bacterial cells (*see* Fig. 3a). Next, dual selection guided the direction of the

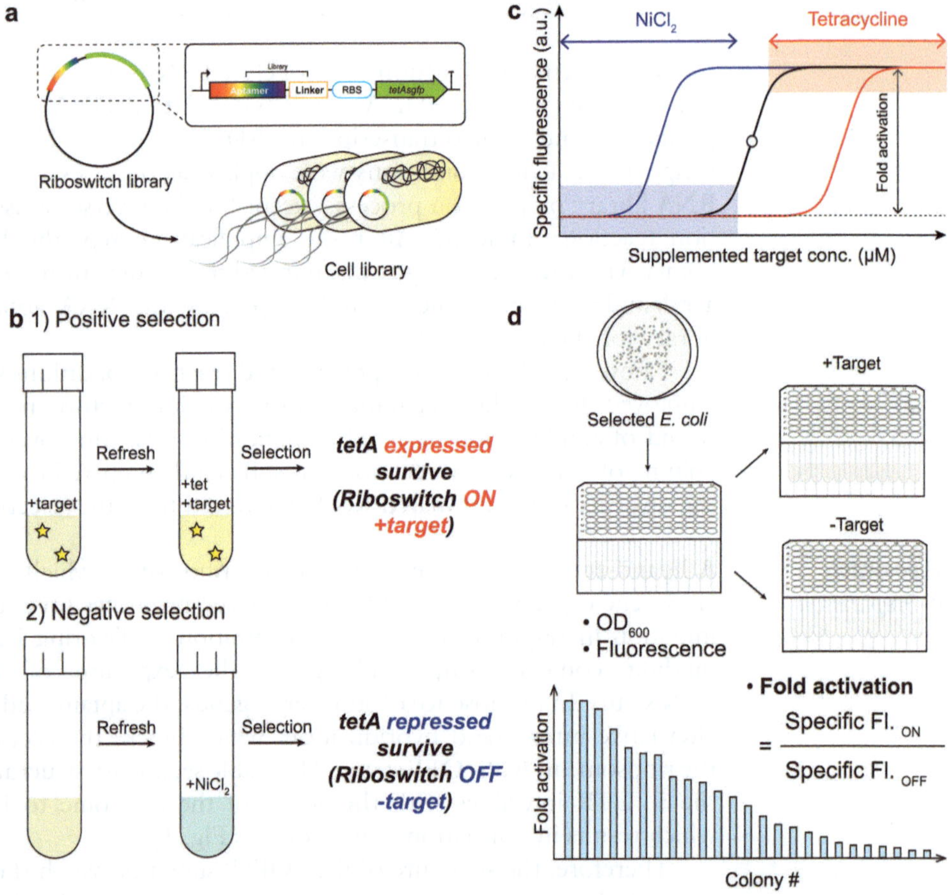

Fig. 3 Scheme of in vivo selection. (**a**) Library cloning and transformation steps for in vivo selection. The design of the synthetic riboswitch library plasmid was from Jang et al. (2017). (**b**) Positive and negative selections are illustrated in separate rows. (**c**) Dose-response curve of the riboswitch. Blue and red lines indicate the response curves of riboswitches with low and high half maximal effective concentration (EC_{50}) values, respectively. Blue and red shades indicate specific fluorescence or expression levels of *tetAsgfp* in the surviving population at each selection step (negative and positive selections). (**d**) Validation of selected riboswitch pools. An example of fold activation ranking of screened riboswitch variants is illustrated

enrichment toward target-sensing toggled switch by killing false positives that only retain a single status (ON or OFF). The *tetAsgfp*, which has been applied in previous studies [18, 25, 26], encodes a fusion protein of an antibiotic selection marker (*tetA*) and a green fluorescent protein (superfolder *gfp*). Dual selection can be performed with *tetAsgfp* using tetracycline and nickel (*see* Fig. 3b). The cells expressing *tetA* will survive in the presence of tetracycline and a target molecule, whereas those containing riboswitches that do not express the gene in the presence of the target molecule will not survive (positive selection). In the presence of nickel and the absence of ligand, the cells containing malfunctioning riboswitches, which express the gene even in the absence of the ligand, do not survive owing to their TetA-mediated sensitivity to nickel ions (negative selection). Serial rounds of positive and negative selections can enrich functional riboswitches for the target ligand from the riboswitch library.

Insert a random decamer called a linker between the aptamer and the RBS sequence to construct a riboswitch library. This is because the translation initiation rate is determined by the local sequence and structure around the 5′-UTR and start codon (*see* Fig. 1). A weak synthetic promoter and moderate strength of the 5′-UTR were used in the backbone plasmid. The promoter was chosen from the experiments examining the degree of growth rate retardation at various promoter strengths predicted from the Anderson promoter set and antibiotic concentrations [18].

Subclone the enriched aptamer pool from SELEX into the backbone plasmid (pACYC_J23113_tetAsgfp) using scarless golden gate cloning [27] to exclude any additional nucleotide sequence that may affect ligand-dependent conformation switching ability (*see* Fig. 3a). The reaction volume must be adjusted to include the maximum diversity of the combinations of aptamers and linker sequences in the riboswitch library.

The competent cells must be handled at ice-cold temperature and electroporation be performed immediately, as in the following:

1. Incubate the purified plasmids with electrocompetent *E. coli* DH10B T1^R in ice for 5 min (*see* **Note 16**).
2. Transfer the plasmid and cell mixture to pre-chilled 0.1 cm gapped e-cuvettes. The cavity contained less than 100 μL of the sample.
3. Remove the moisture on the surface of e-cuvettes with Kimwipes, and apply a single pulse at 18 kV/cm, 200 Ω, and 25 μF (*see* **Note 17**).
4. Collect transformants in 1 mL of pre-warmed S.O.C. medium for every e-cuvette, and incubate for 1 h in a shaking incubator at 37 °C and 220 rpm.

5. Spread the recovered transformants on at least five LB-agar plates (150 mm × 90 mm) supplemented with appropriate antibiotics. Also, spread serially diluted transformants on LB-agar plates to evaluate the transformation efficiency of the recovered *E. coli* library.
6. Incubate the LB-agar plates at 37 °C in an oven, and monitor the size of the colony to avoid overgrowth.
7. Harvest the colonies on the plates with sterilized 1× PBS using a flame-sterilized spreader.
8. Centrifuge the recovered colonies, resuspend in 10% glycerol solution, and store at −80 °C.

3.5 In Vivo Selection of Riboswitch Library

The concentrations of tetracycline hydrochloride and nickel chloride used in in vivo selection, which determine the overall stringency of selection, should be validated. Adopt the concentration at which the specific growth rate is neither severely affected nor unaffected by comparing the specific growth rates under various concentrations. As stated in 3.3, extreme conditions could eliminate positives in the selection pool, while mild conditions may delay the enrichment of positives. Adjust the supplemented antibiotic and ligand concentrations during the selection rounds (*see* Fig. 3c). To avoid enrichment of nonfunctional riboswitches, apply alternative positive and negative selections (*see* Fig. 3b). M9 minimal medium is recommended for in vivo selection to eliminate any possible interference or interactions in ligand-dependent sensor operation.

1. Inoculate the library cell into 3 mL of liquid medium, and incubate the culture at 37 °C with vigorous shaking.
2. Dilute the library cell in fresh medium to a final optical density at 600 nm (OD_{600}) of 0.05, and culture at 37 °C with vigorous shaking until the OD_{600} value reached 1.0. Supplement the target molecule in this step for positive selection (*see* **Note 18**).
3. Dilute the library cell in a fresh medium supplemented with appropriate antibiotics in the presence or absence of the target molecule depending on the selection. Adjust the final OD_{600} to 0.05, and culture the cells at 37 °C with vigorous shaking. Tetracycline or nickel chloride should be added in this step depending on the selection (*see* **Notes 18** and **19**).
4. Store the selected library after the OD_{600} value reached 1.0.
5. To continue the selection, harvest the cells by centrifugation, wash in fresh medium to remove residual ligands and antibiotics, and return to **step 2**. Store the selected cells in 10% glycerol stock at −80 °C.

3.6 *In Vivo* Riboswitch Library Validation and Characterization

Spread the selected pool on an LB-agar plate to isolate a single colony. To compare the response of the riboswitch to the target, perform a test along with the negative control, which shares the common structure with the riboswitch plasmid except the aptamer, under the same culture conditions (*see* Fig. 3d).

1. Inoculate 95 colonies from the selection and control colonies to the test medium (500 μL) in a deep well plate. Incubate the culture with vigorous shaking at 800 rpm and 37 °C overnight.
2. Re-culture the cells in 500 μL of fresh medium in a deep well plate with vigorous shaking at 800 rpm and 37 °C for 3 h.
3. Inoculate the culture in 500 μL of fresh medium in the presence or absence of ligands in deep well plates with vigorous shaking at 800 rpm and 37 °C for 6 h.
4. Sample each well with a multichannel pipette in a multiwell plate for fluorescence measurement.
5. Measure the OD_{600} value and the fluorescence intensity (excitation, 488 nm; emission, 513 nm).
6. Calculate the specific fluorescence and fold activation of each variant (*see* **Note 20**).

4 Notes

1. The optimal concentration of template single-stranded DNA should be pretested. An excessive amount of input DNA results in the early enrichment of unwanted heterodimers derived from nonspecific hybridization between random sequences. The initial generation of heterodimers decreases the quality of the library because unamplified template DNA compromises the initial diversity. In contrast, limited input DNA may delay the by-product enrichment and impair the amplification efficiency in each reaction volume. An appropriate concentration of initial library DNA must be selected in the reaction mixture by performing PCR amplification and polyacrylamide gel electrophoresis with serially diluted input DNA to ensure successful diversity generation.
2. The purification kit should be able to recover appropriate sizes of DNA to exclude primers while including the double-stranded target DNA. A kit with an excellent recovery yield must be selected to conserve diversity.
3. In cases where organic solvents are used for improving solubility during the preparation of the stock solution, the working concentration of target molecules and the inhibitory concentration of organic solvents must be considered before application in cell culture.

4. To ensure a consistent composition of the reaction mixture, which may affect the quality of library amplification, the bulk volume of the PCR mix (over 2 mL) must be mixed first. A pilot PCR amplification must be performed with a 50 μL aliquot of the PCR mix. Pilot PCR enables the identification of the optimal amplification cycle, which avoids amplification of byproducts and results in a high amplification yield. The same reaction conditions must be maintained for both the pilot and final amplifications. A 6 μL aliquot of the reaction mix must be sampled every two cycles during the last 10 s of elongation step (72 °C). To compare each PCR sample according to the rounds, the sample must be retrieved at the termination of the elongation stage at 72 °C. The sampling time should not be missed because the rising temperature significantly affects the band pattern in the polyacrylamide gel.
5. The amplification cycle must be selected such that the target band is fully amplified and the band of byproducts is barely identifiable.
6. The purity of DNA or RNA should always be validated before subsequent procedures. Residual organic solvents or target molecules have the potential to inhibit the reaction. The ratio of absorbance at 260 nm to that at 230 nm must be used to assess contamination of extraction reagents.
7. Reference RNA with the same size or single-stranded template DNA must be analyzed with in vitro transcription mixtures. The cavity of the wells must be pipetted with loading buffer before loading the sample, and the sample must be mixed thoroughly with 8 M urea.
8. To maximize the coupling efficiency, excess ligands, about 5–10 times concentration of active groups, should be applied in the coupling buffer. The composition and final pH of the buffer depend on the coupling chemistry to be applied. Organic solvent is allowed to use up to 50% of coupling solution to dissolve the ligand. The ratio of medium to coupling buffer can be ranged from 2:1 to 1:1. The coupling condition (e.g., temperature, incubation time, agitation speed) also depends on the coupling chemistry to be applied. The coupling efficiency of the molecule and activated reactive groups on the bead can be validated by performing a colorimetric reaction of unreacted beads and ligand-coupled beads or by comparing the concentration of the ligand in the coupling buffer before and after the reaction.
9. Monovalent and divalent cations assist secondary structure formation and stabilization of aptamer-ligand interactions. Gradually reduce the concentration of cations as SELEX progresses to provide increased selection pressure to remove

nonspecifically binding RNA aptamers from the pool. Therefore, the aim of selection is to decrease the false positives that coincidently pass through the designed selection filter by increasing the selection threshold while maintaining the true positives.

10. Stringent washing at early rounds may wash away the candidate aptamer pool, whereas mild washing may delay the enrichment. Therefore, the washing stringency must be gradually increased, or the cation concentration must be decreased.
11. Excessive input RNA and target-coupled beads along with a mild washing process are preferred in the first round. After the first two rounds of SELEX, the input RNA and bead concentration were maintained at a fixed concentration (e.g., a tenfold dilution of the first round).
12. Gradual cooling should be performed in the thermocycler to regenerate the intrinsic secondary structure of RNA aptamers.
13. Washing stringency, the number of washing repeats, and salt concentration should be adjusted at every round of selection.
14. The first three components (RNA, reverse primer, and dNTP) were preincubated at 65 °C for 5 min, followed by incubation on ice for 1 min. The other reagents were mixed thoroughly and incubated at 55 °C for 1 h. Heat inactivation was performed at 75 °C for 15 min.
15. Enriched rounds of selection exhibited a significant difference between the amounts of RNA released from eluents with and without target molecules. For example, the amount of recovered RNA in the eluent from ligand-coupled bead was at least eightfold higher than that from the ligand-free bead in our experiments.
16. During the preparation of competent cells, the volume of the cell culture should be estimated by considering the target library size and transformation efficiency. The quality of the plasmids should be validated before electroporation. The salts remaining in the DNA pellet must be completely removed during the purification and precipitation steps. The ratio of absorbance at 230 nm to that at 260 nm must be examined to detect contamination with residual organic chemicals.
17. The mixture must be divided into multiple e-cuvettes to prevent unexpected arcs and improve the electro-transformation efficiency. Fresh e-cuvettes should be used for library cell preparation.
18. The supplemented concentrations of antibiotics and target molecules should be completely soluble in the culture medium. Pretests must be performed to identify the threshold concentration at which sufficient intracellular levels are achieved to

promote active interactions between the riboswitches without affecting cell viability. The concentrations of antibiotics and target molecules should be adjusted to control the selection pressure (*see* Fig. 3c).

19. Freshly prepared and filtered tetracycline and nickel chloride stock solutions must be used as they are prone to denaturation. The same standard ligand stock solution must be used throughout the process of in vivo selection.

20. Specific fluorescence was the detected fluorescence subtracted by basal fluorescence divided by the OD_{600} of the culture subtracted by basal OD_{600}. This index reflects the fluorescence of the riboswitch independent of cell growth. Fold activation was defined as the ratio of specific fluorescence obtained under ligand-rich conditions to that obtained under ligand-free conditions (*see* Fig. 3d).

Acknowledgments

This work was supported by the National Research Foundation of Korea (NRF) grant funded by the Korean government (MSIT) (No. 2021R1F1A1062778) and Global Research Laboratory Program (NRF-2016K1A1A2912829).

References

1. Mandal M, Lee M, Barrick JE et al (2004) A glycine-dependent riboswitch that uses cooperative binding to control gene expression. Science 306:275–279. https://doi.org/10.1126/science.1100829
2. Garst AD, Héroux A, Rambo RP, Batey RT (2008) Crystal structure of the lysine riboswitch regulatory mRNA element. J Biol Chem 283:22347–22351. https://doi.org/10.1074/jbc.C800120200
3. Batey RT, Gilbert SD, Montange RK (2004) Structure of a natural guanine-responsive riboswitch complexed with the metabolite hypoxanthine. Nature 432:411–415. https://doi.org/10.1038/nature03037
4. Sudarsan N, Lee ER, Weinberg Z et al (2008) Riboswitches in eubacteria sense the second messenger cyclic di-GMP. Science 321: 411–413. https://doi.org/10.1126/science.1159519
5. Winkler W, Nahvi A, Breaker RR (2002) Thiamine derivatives bind messenger RNAs directly to regulate bacterial gene expression. Nature 419:952–956. https://doi.org/10.1038/nature01145
6. Nahvi A, Barrick JE, Breaker RR (2004) Coenzyme B12 riboswitches are widespread genetic control elements in prokaryotes. Nucleic Acids Res 32:143–150. https://doi.org/10.1093/nar/gkh167
7. Ames TD, Rodionov DA, Weinberg Z, Breaker RR (2010) A eubacterial riboswitch class that senses the coenzyme tetrahydrofolate. Chem Biol 17:681–685. https://doi.org/10.1016/j.chembiol.2010.05.020
8. Mironov AS, Gusarov I, Rafikov R et al (2002) Sensing small molecules by nascent RNA: a mechanism to control transcription in bacteria. Cell 111:747–756. https://doi.org/10.1016/s0092-8674(02)01134-0
9. Garst AD, Edwards AL, Batey RT (2011) Riboswitches: structures and mechanisms. Cold Spring Harb Perspect Biol 3. https://doi.org/10.1101/cshperspect.a003533
10. Kushwaha M, Rostain W, Prakash S et al (2016) Using RNA as molecular code for programming cellular function. ACS Synth Biol 5:

795–809. https://doi.org/10.1021/acssynbio.5b00297

11. Rogers JK, Taylor ND, Church GM (2016) Biosensor-based engineering of biosynthetic pathways. Curr Opin Biotechnol 42:84–91. https://doi.org/10.1016/j.copbio.2016.03.005
12. Rogers JK, Church GM (2016) Genetically encoded sensors enable real-time observation of metabolite production. Proc Natl Acad Sci U S A 113:2388–2393. https://doi.org/10.1073/pnas.1600375113
13. Varma A, Palsson BO (1994) Metabolic flux balancing: basic concepts, scientific and practical use. Nat Biotechnol 12:994–998. https://doi.org/10.1038/nbt1094-994
14. Stephanopoulos G, Sinskey AJ (1993) Metabolic engineering – methodologies and future prospects. Trends Biotechnol 11:392–396. https://doi.org/10.1016/0167-7799(93)90099-U
15. Leonard E, Ajikumar PK, Thayer K et al (2010) Combining metabolic and protein engineering of a terpenoid biosynthetic pathway for overproduction and selectivity control. Proc Natl Acad Sci U S A 107:13654–13659. https://doi.org/10.1073/pnas.1006138107
16. Lim HG, Jang S, Jang S et al (2018) Design and optimization of genetically encoded biosensors for high-throughput screening of chemicals. Curr Opin Biotechnol 54:18–25. https://doi.org/10.1016/j.copbio.2018.01.011
17. Dietrich JA, McKee AE, Keasling JD (2010) High-throughput metabolic engineering: advances in small-molecule screening and selection. Annu Rev Biochem 79:563–590. https://doi.org/10.1146/annurev-biochem-062608-095938
18. Jang S, Jang S, Xiu Y et al (2017) Development of artificial riboswitches for monitoring of naringenin in vivo. ACS Synth Biol 6:2077–2085. https://doi.org/10.1021/acssynbio.7b00128
19. Jang S, Jang S, Im DK et al (2019) Artificial caprolactam-specific riboswitch as an intracellular metabolite sensor. ACS Synth Biol 8:1276–1283. https://doi.org/10.1021/acssynbio.8b00452
20. Dwidar M, Seike Y, Kobori S et al (2019) Programmable artificial cells using histamine-responsive synthetic riboswitch. J Am Chem Soc 141:11103–11114. https://doi.org/10.1021/jacs.9b03300
21. Boussebayle A, Torka D, Ollivaud S et al (2019) Next-level riboswitch development-implementation of capture-SELEX facilitates identification of a new synthetic riboswitch. Nucleic Acids Res 47:4883–4895. https://doi.org/10.1093/nar/gkz216
22. Espah Borujeni A, Mishler DM, Wang J et al (2016) Automated physics-based design of synthetic riboswitches from diverse RNA aptamers. Nucleic Acids Res 44:1–13. https://doi.org/10.1093/nar/gkv1289
23. Serganov A, Nudler E (2013) A decade of riboswitches. Cell 152:17–24. https://doi.org/10.1016/j.cell.2012.12.024
24. Stoltenburg R, Reinemann C, Strehlitz B (2007) SELEX--a (r)evolutionary method to generate high-affinity nucleic acid ligands. Biomol Eng 24:381–403. https://doi.org/10.1016/j.bioeng.2007.06.001
25. Jang S, Yang J, Seo SW, Jung GY (2015) Riboselector: riboswitch-based synthetic selection device to expedite evolution of metabolite-producing microorganisms. Methods Enzymol 550:341–362. https://doi.org/10.1016/bs.mie.2014.10.039
26. Muranaka N, Sharma V, Nomura Y, Yokobayashi Y (2009) An efficient platform for genetic selection and screening of gene switches in Escherichia coli. Nucleic Acids Res 37:e39. https://doi.org/10.1093/nar/gkp039
27. Engler C, Kandzia R, Marillonnet S (2008) A one pot, one step, precision cloning method with high throughput capability. PLoS One 3:e3647. https://doi.org/10.1371/journal.pone.0003647

Chapter 10

Efficient Method to Identify Synthetic Riboswitches Using RNA-Based Capture-SELEX Combined with In Vivo Screening

Janice Kramat and Beatrix Suess

Abstract

Synthetic riboswitches are a promising tool for conditional gene expression. In vitro selected aptamers used as binding domains for the design of RNA-based switches have to exhibit excellent binding affinity as well as ligand binding-induced structural changes. Selection via Capture-SELEX favors the enrichment of aptamers which exhibit both characteristics. For the Capture-SELEX, an RNA pool is used that gets immobilized onto a capture oligonucleotide by hybridization. Addition of the ligand frees the aptamers by their binding to the ligand, resulting in the release from the capture oligonucleotide through structural changes. These sequences get reverse transcribed, PCR amplified, and used for the following selection rounds. In this publication, we present a detailed protocol for Capture-SELEX, followed by screening in yeast to identify aptamers suitable for the design of synthetic riboswitches.

Key words Capture-SELEX, RNA, Synthetic riboswitches, Aptamer, In vivo screening

1 Introduction

Systematic evolution of ligands by exponential enrichment (SELEX) has been used for the in vitro selection of aptamers for three decades. Aptamers are single-stranded nucleic acids capable of selectively binding a wide variety of ligands, including ions [1], small compounds [2, 3], proteins [4–6], and even whole cells [7–10]. Thus, potential applications for aptamers are diverse, ranging from sensing and diagnostics to therapeutic applications to the use of aptamers as regulatory devices to control gene expression. The original SELEX developed in 1990 [11, 12] has since been adapted to fit a myriad of such purposes [13]. One of these adaptations was described as Capture-SELEX. It is based on a capture oligonucleotide, which binds to a complementary docking sequence within a semi-randomized DNA or RNA pool. Consequently, the entire nucleic acid pool is immobilized to a solid support via docking to

James Chappell and Melissa K. Takahashi (eds.), *Riboregulator Design and Analysis*,
Methods in Molecular Biology, vol. 2518, https://doi.org/10.1007/978-1-0716-2421-0_10,

the capture oligonucleotide (e.g., via biotin to streptavidin-coated beads). After incubation with a target compound, some of the immobilized pool sequences are able to recognize the target. If this binding event leads to a structural rearrangement of the RNA, the sequence is released from the capture oligonucleotide, which allows their elution and further processing [14–16]. Due to the refolding upon ligand binding, Capture-SELEX-derived aptamers are ideally suited for the development of synthetic riboswitches.

Natural riboswitches are highly structured RNA elements that are mostly located in the untranslated region of various bacterial mRNAs, thereby controlling a plethora of metabolic genes [17]. They are composed of a binding domain (also called aptamer domain) and an expression platform. Ligand recognition by the binding domain is directly interpreted by the expression platform, mostly by inherent conformational changes within the RNA structure, resulting in changes in gene expression [18]. The advantage of riboswitch regulation is the combination of a sensor and a regulator in one physical element. Thus, the ligand is recognized directly by the RNA, and no further cofactors such as regulatory proteins are required. The regulatory unit is therefore considered as a simple and defined element, which makes it an interesting blueprint for the design of synthetic riboswitches.

Even though a lot of aptamers were selected via a classical SELEX, having excellent binding properties in vitro, only few of them were able to regulate gene expression as synthetic riboswitches in vivo. One feature all of these riboswitches have in common is that they undergo structural changes triggered by the binding event [19–21]. So, one hypothesis is that a structural rearrangement is needed to turn an aptamer into a working riboswitch. As Capture-SELEX does not only select for binding but also for structural switching, it is an ideal method for the selection of aptamers intended to be used as synthetic riboswitches. Note that it is necessary to follow up the Capture-SELEX with in vivo screening to isolate those aptamers that can be engineered into riboswitches.

We have established a protocol that combines RNA Capture-SELEX with an in vivo screening approach in yeast that allows the identification of synthetic riboswitches responding, for example, to paromomycin [22, 23]. A selection cycle of the RNA Capture-SELEX is illustrated in Fig. 1. It starts with the in vitro transcription of a randomized DNA pool. The resulting RNA pool is hybridized to a biotinylated DNA capture oligonucleotide that is subsequently bound to streptavidin-coated magnetic beads. After washing, the beads are incubated with the target molecule. The eluted sequences are reversely transcribed and amplified by PCR. This pool is then in vitro transcribed again, starting with the next selection cycle.

To find active aptamers, the resulting pool must be screened after successful enrichment. The used screening method relies on the most basic concept of a synthetic riboswitch. In this system, an

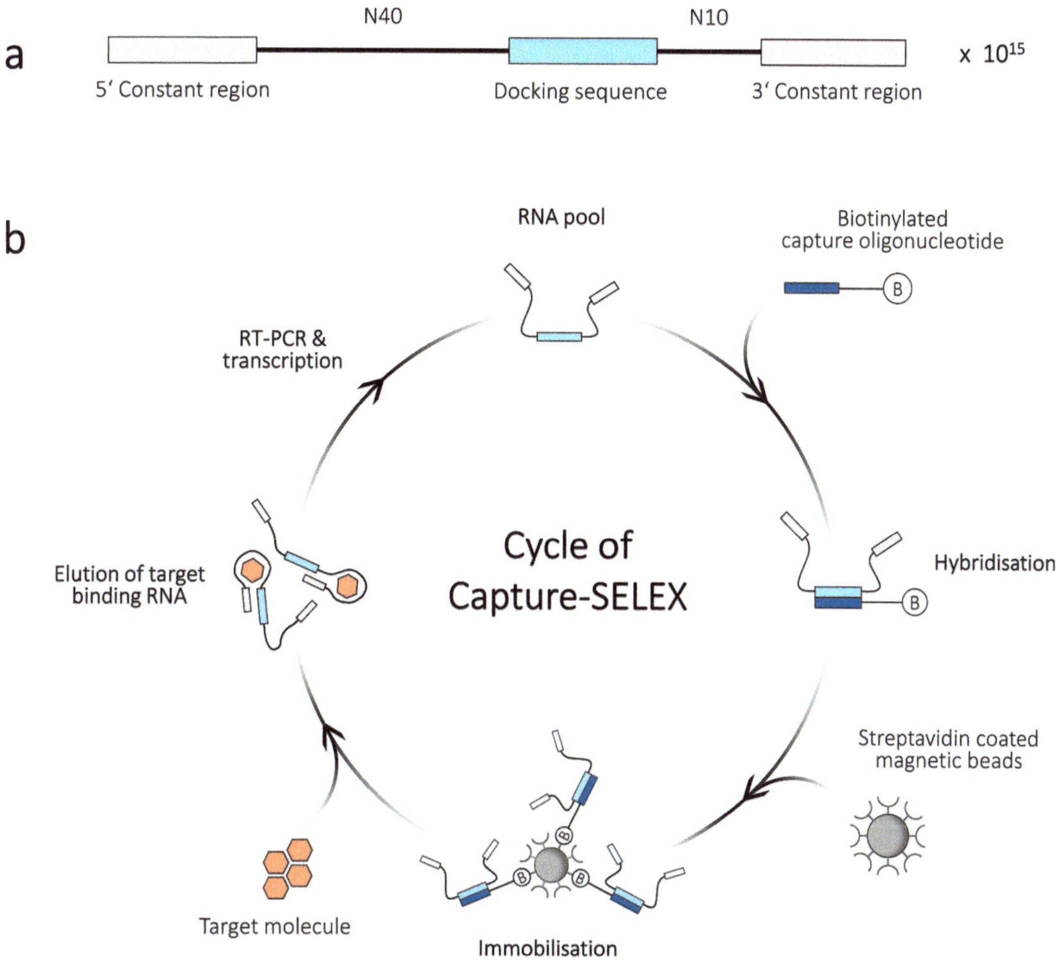

Fig. 1 The RNA Capture-SELEX cycle: (**a**) The RNA pool is transcribed in vitro from a purchased DNA pool. It consists of constant regions at the 5′ and 3′ ends and a docking sequence located between the randomized regions, which comprise 40 (N40) and 10 (N10) randomized bases, respectively. For the first round of selection, approximately 10^{15} RNA molecules are used to cover a high number of potential three-dimensional structures. (**b**) The RNA pool is hybridized to a biotinylated DNA capture oligonucleotide. The pool/capture oligonucleotide hybrid is immobilized on streptavidin magnetic beads via the biotin group. A target molecule solution (hexagons) is added. Aptamers which bind a target molecule can be eluted when they detach from the capture oligonucleotide. These sequences are reversely transcribed, amplified by PCR, and in vitro transcribed. The next cycle continues

aptamer is inserted into the 5′ untranslated region (UTR) of a *GFP* reporter gene in yeast. In the ligand bound state, the aptamer serves as a roadblock interfering with scanning of the small ribosomal subunit (*see* Fig. 2). This is done by cloning the entire enriched aptamer pool into the 5′ UTR of a *GFP* reporter gene and testing the resulting individual clones to determine whether the addition of the ligand decreases GFP expression. The *GFP* gene is located on a plasmid that constitutively co-expresses mCherry for

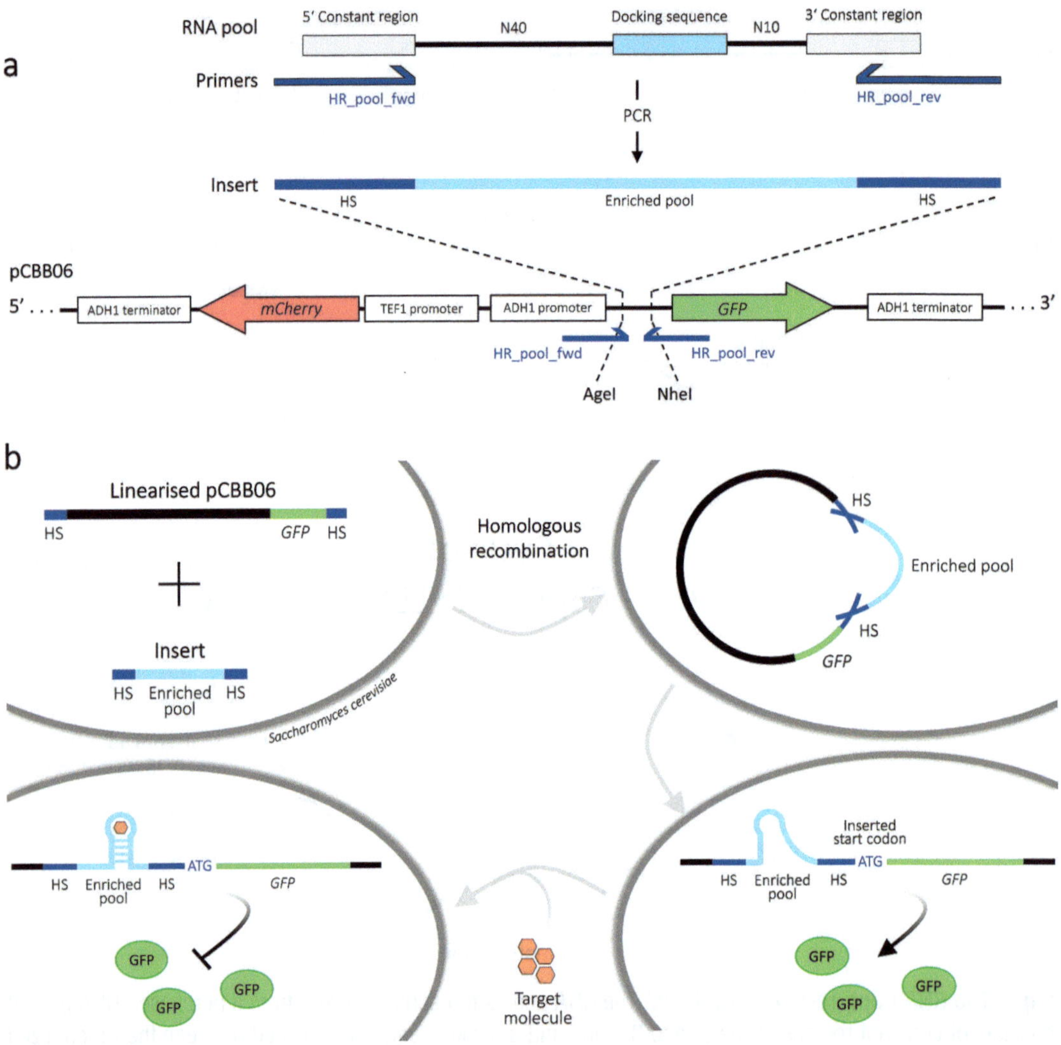

Fig. 2 Homologous recombination and riboswitch-dependent GFP expression. (**a**) Displayed is the plasmid pCBB06 (ColE1/pMB1/pBR322/pUC ori, an ampicillin resistance gene, a CEN6/ARS4 ori, an URA3 auxotrophic selection marker, an ADH1-driven *GFP*, and a TEF1-driven *mCherry*). The two reporter genes *GFP* and *mCherry* are displayed as arrows, and the corresponding promoter and terminator sequences are shown as boxes. The primers HR_pool_fwd and HR_pool_rev (dark blue half arrows) are complementary to specific regions in the ADH1 promoter and *GFP* gene of pCBB06 as well as to the end regions of the aptamer sequences from the enriched pool of the Capture-SELEX (RNA pool). They are used to extend the 5′ and 3′ ends of the enriched pool via PCR (insert). (**b**) The resulting homologous sequence regions (HS) of insert and plasmid enable homologous recombination in yeast cells. The successful insertion of an aptamer sequence from the enriched pool creates a circular plasmid. In addition, the insert located upstream of the *GFP* gene contains the missing start codon (ATG) for GFP expression. Thus, the expression of GFP can only take place after successful and correct insertion of the aptamer sequence. Ideally, the secondary structure of the aptamer causes a weakening of the expression. However, strong inhibition of GFP expression by a too rigid structure (no fluorescence) or unaffected gene expression due to a too flexible secondary structure (very strong fluorescence) is undesirable. An inhibitory structure of the aptamer should only be formed through the binding of the target molecule (hexagon in stem loop). Consequently, GFP expression is dependent on the presence of the target molecule

normalization. The protocol for the introduction of the enriched pool into yeast cells is based on the repair of a linearized plasmid by homologous recombination [24–27] (*see* Fig. 2). The linearized plasmid lacks the 5′ UTR and the start codon of *GFP* which is introduced together with the aptamer sequence. Consequently, GFP is expressed only after successful recombination. The individual yeast colonies then need to be screened for ligand-dependent changes in GFP fluorescence. Figure 2 illustrates the described concept. The described protocol within this section provides a detailed description of the selection of small molecule-binding RNA aptamers using Capture-SELEX together with the subsequent in vivo screening for riboswitches. A series of notes highlights critical issues and provides practical advice.

2 Materials

Prepare all buffers and solutions with ultrapure water, make sure to prevent RNase contamination by using RNase-free materials, and autoclave.

3 Capture-SELEX

3.1 Buffers and Solutions

1. 5× Capture-SELEX buffer (5× CSB): 200 mM HEPES, 1.25 M KCl, 100 mM NaCl, and 25 mM $MgCl_2$. Adjust pH to 7.4 with HCl.
2. 1× Capture-SELEX buffer (1× CSB): Dilute 5× CSB and add 0.01% (v/v) Tween-20.
3. Binding and washing buffer (B&W): 5 mM Tris-HCl, 0.5 mM EDTA, 1 M NaCl, and 0.01% (v/v) Tween-20. Adjust pH to 7.5 with HCl.
4. Target solution: 1 mM in 1× CSB.
5. Optional counter target solution (for the elution of nonspecific binders): 1 mM solution of a compound similar to your target molecule in 1× CSB.
6. Scintillation counter.
7. 3 M sodium acetate (NaAc) with pH 6.5 (adjust with HCl).
8. 7.5 M ammonium acetate (NH_4Ac).
9. Absolute ethanol (absolute EtOH).
10. 70% (v/v) ethanol (70% (v/v) EtOH).
11. GlycoBlue.
12. 50× TAE buffer: 2 M Tris, 1 M acetic acid, 50 mM Na-EDTA, and pH 8.3.

13. 2× RNA loading buffer: deionized formamide, 25 mM EDTA, and bromophenol blue.
14. 6× DNA loading dye: 125 mL glycerine, 5 mL 50× TAE buffer, 250 mg bromophenol blue, and 250 mg xylene cyanol.

3.2 Initial RNA Pool Preparation

The following sequences are based on our established protocol. For pool design considerations (*see* **Note 1**):

1. Oligonucleotide pool_fwd (only desalted): 5′-CCAAGTAATA CGACTCACTATAGGGCAACTCCAAGCTAGATC TACCGGT-3′.
2. Oligonucleotide pool_rev (only desalted): 5′-AGTGAAAAGT TCTTCTCCTTTGCTAGCCATTTT-3′.
3. Randomized pool_oligo_rev (PAGE purified): 5′-AGTGAAAA GTTCTTCTCCTTTGCTAGCCATTTT -N10-TAGAAGCC AGTAG-N40-ACCGGTAGATCTAGCTTGGAGTTGCCC-3′.
4. Syringe-filter (preferably 0.4 μm pore size).

3.3 Hybridization of RNA Pool and Capture Oligonucleotide

1. Biotinylated capture oligonucleotide (100 μM solution in water): 5′-TAGAAGCCAGTAG-[biotin tag].
2. RNA pool: Initial RNA pool for first round, enriched RNA pool for all following rounds.

3.4 Magnetic Bead Preparation

1. Streptavidin-coated magnetic beads: 4 mg/mL stored in buffer provided by manufacturer.
2. Magnetic rack for bead separation.

3.5 Reverse Transcription (RT) PCR

1. Taq DNA polymerase.
2. Superscript II reverse transcriptase.
3. RT PCR mix: 40 mM Tris-HCl pH 9.0, 130 mM KCl, 4.2 mM $MgCl_2$, 4 mM DTT, 0.2% (v/v) Triton X-100, 2 μM oligonucleotide pool_rev, and 0.6 mM each dNTP (store as 47 μL aliquots).

3.6 In Vitro Transcription and Radioactive Labelling

1. T7 RNA polymerase.
2. RNase inhibitor.
3. α-^{32}P-ATP solution (3000 Ci/mmol, 10 mCi/mL)
4. Transcription mix: 40 mM Tris-HCl pH 8.0, 5 mM DTT, 2.5 mM each NTP, and 15 mM $MgCl_2$ (store as 87 μL aliquots).

4 In Vivo Screening in *Saccharomyces cerevisiae*

4.1 Buffers, Solutions, and Media

All buffers, solutions, or media should be used in a sterile condition.

1. 1 M lithium acetate (LiAc) in ultrapure water
2. 1 M dithiothreitol (DTT) in ultrapure water
3. 1 M sorbitol in ultrapure water
4. 100 mM $CaCl_2$ in ultrapure water
5. Electroporation buffer (prepare on the same day as use, cool on ice): 1 M sorbitol and 0.1 mM $CaCl_2$.
6. LiAc/DTT solution (prepare on the same day as use): 100 mM LiAc and 10 mM DTT.
7. LB medium/plates: 1% (w/v) tryptone, 1% (w/v) NaCl, 0.5% (w/v) yeast extract, and 1.5% (w/v) agar (only for plates).
8. 50× amino acid mix: 1.05 g/L arginine, 0.6 g/L cysteine, 0.4 g/L histidine, 1.3 g/L isoleucine, 1.3 g/L leucine, 1.85 g/L lysine, 0.38 g/L methionine, 0.83 g/L phenylalanine, 1.2 g/L threonine, 0.2 g/L tryptophan, 0.9 g/L tyrosine, and 1.18 g/L valine
9. SCD-Ura medium/plates (synthetic complete dropout mix without uracil): 2% (w/v) glucose, 0.55% (w/v) $(NH_4)_2SO_4$, 0.2% (w/v) yeast nitrogen base without ammonium sulfate and amino acids, 4% (v/v) 300 μg/mL adenine solution, 2% (v/v) 50× amino acid mix, and 1.5% (w/v) agar (only for plates).
10. YEPD medium/plates (yeast extract peptone dextrose): 1% (w/v) yeast extract, 2% (w/v) peptone, 2% (w/v) glucose, and 1.5% (w/v) agar (only for plates).
11. YEPD sorbitol mix (prepare on the same day as use): 50% (v/v) sorbitol solution and 50% (v/v) YEPD medium.
12. 50 mM ligand in ultrapure water
13. SOC medium/plates (super optimal culture mix): 2% (w/v) tryptone, 0.5% (w/v) yeast extract, 10 mM NaCl, 2.5 mM KCl, 10 mM $MgCl_2$, 10 mM $MgSO_4$, 20 mM glucose, and pH 7.0, 2% (w/v) agar (only for plates).

4.2 Consumables

1. 24 well plate (u-bottom, sterile)
2. 96 deep well plate (u-bottom, sterile)
3. 96 well plate (flat bottom, sterile, clear)
4. Electroporation cuvette (0.2 cm electrode gap).

4.3 Plasmids and Primers

1. pCBB05: ColE1/pMB1/pBR322/pUC ori, AmpR, CEN6/ARS4 ori, URA3 auxotrophic selection marker, ADH1-driven

GFP, TEF1-driven *mCherry*. (Can be provided by our laboratory upon request) [23].

2. pCBB06: same as pCBB05, but it does not contain start codon in front of the *GFP* reporter gene. (Can be provided by our laboratory upon request) [23].
3. HR_pool_fwd (only desalted): 5′-CTCGTCATTGTTCTCGT TCCCTTTCTTCCTTGTTTCTTTTTCTGCACAATATTT-CAAGCTATACCAAGCATACAATGGGCAACTCCAAGC-TAGATCTACCGGT-3′.
4. HR_pool_rev (only desalted): 5′-CACCCTCTCCACTGACA GAAAATTTGTGCCCATTAACATCACCATCTAATTCAA-CAAGAATTGGGACAACTCCAGTGAAAAGTTCTTCTCC TTTGCTAGC-3′.
5. GFP_rev (only desalted): 5′-CCACTGACAGAAAATTTGT GC-3′.
6. Purification kit for plasmid isolation.

4.4 Enzymes

1. AgeI.
2. NheI.
3. High-fidelity DNA polymerase (e.g., Q5-Polymerase NEB).

4.5 Strains

1. *Escherichia coli* TOP 10: *F- mcrA*, *Δ(mrr-hsdRMS-mcrBC)*, *φ80lacZΔM15*, *ΔlacX74*, *nupG*, *recA1*, *araD139*, *Δ(ara-leu) 7697*, *galE15*, *galK16*, *rpsL(StrR)*, and *endA1 λ-*.
2. *S. cerevisiae* RS453α (yeast): *MATα*, *ade2-1*, *trp1-1*, *can1-100*, *leu2-3*, *his3-1*, and *ura3-52* [28].

5 Methods

5.1 Capture-SELEX

The first SELEX cycle is carried out with volumes that differ from the following rounds. This is done in order to make use of the greatest possible diversity of the randomized starting pool. Where appropriate, volumes for the first cycle are provided in brackets [].

5.2 Initial RNA Pool Preparation

Decide on a pool design beforehand. The described protocol is optimized for a design successfully applied in our lab: N_{40}-docking sequence-N_{10}. For design considerations (*see* **Note 1**).

1. Large-scale PCR: 50 mM KCl, 10 mM Tris-HCl pH 9.0, 0.1% (v/v) Triton X-100, 0.2 mM dNTPs (each), 1.5 mM $MgCl_2$, 24 nM pool_oligo_rev, 2.4 μM pool_fwd oligo, 2.4 μM randomized pool_oligo_rev, and 50 U/mL Taq DNA polymerase; fill to 50 mL with ultrapure water.
2. PCR program: 3 min 95 °C, 5× (1 min 95 °C, 1 min 54 °C, 2 min 72 °C), and 3 min 72 °C (*see* **Note 2**).

3. Precipitate DNA: add 5 mL 3 M NaAc pH 6.5, mix and add 125 mL absolute EtOH, mix well, incubate for 30 min at −20 °C, and centrifuge at 17,000 × *g* for 30 min.
4. Wash once with 20 mL fresh 70% (v/v) EtOH, centrifuge at 17,000 × *g* for 10 min, remove supernatant, let dry at room temperature, and dissolve pellet in 2 mL of ultrapure water (*see* **Note 3**).
5. Verify the size (145 base pairs if randomized pool_oligo_rev is used) and quality of the PCR product using a 3% (w/v) agarose gel.
6. Large-scale transcription: 200 mM Tris-HCl pH 8.0, 20 mM $MgAc_2$, 20 mM DTT, 10 mM spermidine, 4 mM NTPs (each), 1/10 of DNA pool from previous step, and 250 U/mL T7 RNA polymerase; fill to 10 mL with ultrapure water.
7. 16 h at 37 °C. If solution appears opaque, pyrophosphate has precipitated. Centrifuge at 8400 × *g* for 1 min and discard pyrophosphate pellet (*see* **Note 4**).
8. Precipitate RNA: Add EDTA to 80 mM and mix, add NaAc to 400 μM and 2.5 volumes of −20 °C absolute ethanol, mix thoroughly, incubate for 1 h at −20 °C, and centrifuge at 8400 × *g* for 1 h at 4 °C. Discard the supernatant (*see* **Note 5**).
9. Wash the RNA pellet twice with 10 mL 70% (v/v) EtOH, centrifuge at 8400 × *g* for 15 min at 4 °C, remove supernatant, and let dry at room temperature for maximum 5 min (*see* **Note 5**).
10. Dissolve RNA in 8 mL RNA loading dye; distribute onto four 20 × 20 cm gels for a denaturing PAGE with 8% PAA, 8 M urea, and 1X TBE buffer. Run the gels at 80 watts (watch for heat generation); let bromophenol blue almost run out.
11. Visualize RNA by UV shadowing and cut bands out of gels. The cutouts are sliced into 0.5 cm pieces. For each individual RNA band, the gel pieces are placed in a 50 ml falcon. Cover gel slices with 300 mM NaAc pH 6.0; place in slow mixer or rotor for 16 h at 4 °C.
12. Push supernatants through syringe filter to remove remaining gel debris.
13. Precipitate RNA of each filtrate: Add 1/10 volume of NaAc solution (3 M, pH 6.5) and 2.5 volumes of absolute ethanol, mix well, incubate for 1 h at −20 °C, centrifuge at 8400 × *g* for 15 min at 4 °C, and discard supernatant (*see* **Note 6**).
14. Wash pellets once with 10 mL fresh 70% (v/v) EtOH, centrifuge at 8400 × *g* for 15 min at 4 °C, remove supernatants, and let dry at room temperature for maximum 5 min.

15. Dissolve each RNA pellet in 500 μL ultrapure water, and pool them all in one reaction tube.
16. Determine OD_{260} to calculate RNA concentration, and run a denaturing PAGE with 8% PAA, 8 M urea, and 1× TBE buffer as control.

5.3 Hybridization of RNA Pool and Capture Oligonucleotide

1. Mix 10 μL 5× CSB [40 μL], 1.5 μL biotinylated-capture oligonucleotide [15 μL], and 5 μL RNA pool [volume equivalent to approximately 10^{15} RNA sequences ≈1 nmol RNA], and add 33.5 μL of ultrapure water [add until total volume of 200 μL].
2. Heat to 65 °C for 5 min. Incubate at 21 °C for 60 min (*see* **Note 7**).

5.4 Preparation of Magnetic Beads and Pool Immobilization

1. Take 150 μL of magnetic bead suspension [1000 μL].
2. Wash three times: place reaction tube in magnetic rack, discard supernatant, and mix thoroughly with 500 μL B&W [1000 μL].
3. Discard supernatant with magnetic rack and resuspend beads in 150 μL 1× CSB [200 μL].
4. Add the 50 μL [200 μL] from the previous hybridization reaction (*see* Subheading 5.3).
5. Leave in slow mixer or rotor for 30 min at room temperature.
6. Separate beads with magnetic rack and keep supernatant for later measurement. This is the "flow-through" fraction (FT).

5.5 Selection Process: Removal of Weak Binders and Elution of Ligand Binding Aptamers

1. Wash 1 (W1): suspend beads in 200 μL 1× CSB [400 μL], incubate on slow mixer or rotor for 5 min at room temperature, separate beads, and collect supernatant for later measurement (*see* **Note 8**).
2. Wash 2 (W2) with heat separation: suspend beads in 200 μL 1× CSB [400 μL], 2.5 min at 30 °C, vortex, 2.5 min at 30 °C, separate beads, and collect supernatant for later measurement (*see* **Note 8**).
3. Wash 3 (W3): suspend beads in 200 μL 1X CSB [400 μL], incubate on slow mixer or rotor for 5 min at room temperature, separate beads, and collect supernatant for later measurement.
4. Optional counter elution (C): suspend your beads in 200 μL counter target solution [400 μL], incubate on slow mixer or rotor for 5 min at room temperature, separate beads, and collect supernatant for later measurement.
5. Elution (E): suspend beads in 200 μL 1 mM target solution [400 μL], incubate on slow mixer or rotor for 5 min at room

temperature, separate beads, and collect supernatant for later measurement.

6. Bead fraction (B): Suspend the beads in 200 μL of 1× CSB [not required] and use for later measurement.

5.6 Quantification (Radioactive Measurement Using a Scintillation Counter)

Not necessary for first round as it is not performed with radiolabelled RNA.

1. Use 4 μL of each fraction you collected in Subheadings 5.4 and 5.5: FT, W1, W2, W3, C (optional), E, and B.
2. Prepare and measure your samples as specified by the manufacturer of your scintillation counter.
3. Calculate the percentages of your samples: the sum of fractions W1, W2, W3, C (if present), E, and B is the total amount of RNA pool bound in **step** 5.3. Calculate the percentages for each fraction (*see* **Note 9**).

5.7 Amplification of the Enriched Pool

1. Precipitate the RNA from the whole elution fraction: Add 20 μL NaAc solution (3 M, pH 6.5) [40 μL]; add 500 μL absolute EtOH [1000 μL]; add 1 μL GlycoBlue; mix well, 30 min at −20 °C; centrifuge at maximum speed for 30 min at 4 °C; and discard supernatant (*see* **Note 6**).
2. Wash pellet once with 800 μL fresh 70% (v/v) EtOH, centrifuge at 17,000 × *g* for 10 min at 4 °C, remove supernatant, and let dry at room temperature for maximum 5 min.
3. Dissolve RNA pellet in 50 μL ultrapure water.
4. Add 47 μL RT PCR mix and incubate at 65 °C for 5 min.
5. Add 1 μL of superscript II reverse transcriptase and 1 μL Taq DNA polymerase.
6. Reverse transcription: 10 min at 54 °C.
7. Reverse transcriptase inactivation: 1 min at 95 °C.
8. Add 1 μL of 100 μM oligonucleotide pool_fwd.
9. PCR: 10 s [1 min] 95 °C, 10 s [1 min] 58 °C, and 20 s [1 min] 72 °C. Number of PCR cycles varies (*see* **Note 10**).
10. Mix 10 μL of PCR amplicon with 2 μL 6× DNA loading dye and check on a 3% (w/v) agarose gel.

5.8 In Vitro Transcription and Radioactive Body Labelling

1. Add 10 μL of the PCR amplicon to 87 μL transcription mix.
2. Add 1 μL of α-^{32}P-ATP solution (3000 Ci/mmol, 10 mCi/mL), add 100 U T7 RNA polymerase, add 40 U RNase inhibitor (or amount specified by manufacturer), and fill to 100 μL with ultrapure water.

3. 37 °C for 6 h. If solution appears opaque, pyrophosphate has precipitated. Centrifuge at 17,000 × *g* for 1 min and discard pyrophosphate pellet.
4. Precipitate RNA: add 50 μL NH_4Ac solution (7.5 M); add 250 μL absolute EtOH, 10 min at −20 °C; centrifuge at 17,000 × *g* for 15 min at 4 °C; and discard supernatant (*see* **Note 6**).
11. Wash pellet once with 200 μL fresh 70% (v/v) EtOH, centrifuge at maximum speed for 10 min at 4 °C, remove supernatant, and let dry at room temperature for maximum 5 min.
12. Dissolve RNA pellet in 50 μL of ultrapure water.
13. Start next selection cycle from Subheading 5.3.

6 In Vivo Screening in *S. cerevisiae*

6.1 Preparation of the Pool for Use as Insert

1. Test PCR: Mix 10 μL of RT PCR product from enriched pool with reaction buffer and enhancer of used DNA polymerase according to manufacturer's instructions, 200 μM dNTPs, 0.3 μM HR_pool_fwd, 0.3 μM HR_pool_rev, high-fidelity DNA polymerase, and ultrapure water (fill up to 100 μL).
2. PCR program: 1 min 98 °C, 15× (30 s 98 °C, 30 s 58 °C, 30 s 72 °C), and 5 min 72 °C. Number of PCR cycles could vary (*see* **Note 11**). Check amplification on 3% (w/v) agarose gel (expected size of the PCR product: 272 bp).
3. Large-scale PCR: Prepare at least five tubes of PCR with optimized PCR conditions (each 100 μL reaction volume). Check amplification on 3% (w/v) agarose gel (expected size of the PCR product: 272 bp).
4. Precipitate products of big PCR with 2.5 volume of absolute ethanol and 1/10 volume of NaAc (3 M, pH 6.5). Incubate for 30 min at −20 °C, centrifuge at maximum speed for 30 min at 4 °C, and discard supernatant.
5. Wash with 200 μL fresh 70% (v/v) EtOH, centrifuge at maximum speed for 10 min at 4 °C, discard supernatant, and let pellet dry for maximum 5 min at 37 °C.
6. Resuspend all pellets in a total volume of 50 μL ultrapure water.
7. Determine concentration (ng/μL) (e.g., by UV/Vis spectrometry using a NanoDrop spectrophotometer).

6.2 Preparation of Plasmid

1. Perform a large-scale digestion (3 μg) of the plasmid pCBB06 (*see* **Note 12**).
2. Digest the plasmid with AgeI and NheI according to manufacturer's specifications.

3. Mix digested plasmid with equal amount of 6× DNA loading dye, run 1% (w/v) agarose gel, and cut DNA out of gel.
4. Use favored purification kit to isolate plasmid (*see* **Note 13**).
5. Determine concentration (ng/µL) (*see* **Note 13**).

6.3 Preparation of Electrocompetent Yeast Cells and Transformation

To be prepared in advance are the following:

1. Spread *S. cerevisiae* RS453α on YEPD plate and incubate for 2 days at 30 °C.
2. Pre-chill four electroporation cuvettes at −20 °C.
3. Several SCD-Ura plates (*see* **Note 14**).

Day 1

4. Inoculate 4 mL YEPD preculture with one colony of yeast.
5. Incubate overnight at 30 °C (shaking) (*see* **Note 15**).

Day 2

6. Inoculate 100 mL YEPD (prepared 1 l flask) with the preculture. Chose a start $OD_{600\ nm}$, which leads to an $OD_{600\ nm} = 1.6$ on the next day (*see* **Note 16**).
7. Incubate overnight at 30 °C (shaking) (*see* **Note 15**).

Day 3

8. Chill electroporation buffer on ice.
9. Collect cells in 2× 50 mL falcons by centrifugation at 1000 × *g* for 3 min at RT; discard supernatant.

   Do the next three steps with each falcon:

10. Resuspend cells in 25 mL ultrapure water, centrifuge at 1000 × *g* for 3 min at RT, and discard supernatant.
11. Repeat wash step from 10.
12. Resuspend cells in 10 mL LiAc/DTT.
13. Mix both resuspensions into sterile 250 mL culture flask; incubate for 30 min at 30 °C (shaking) (*see* **Note 15**).

    Keep cells on ice from now on.

14. Centrifuge cells at 1000 × *g* for 3 min at 4 °C; discard supernatant.
15. Resuspend cells in 50 mL ice-cold electroporation buffer, centrifuge at 1000 × *g* for 3 min at 4 °C, and discard supernatant.
16. Resuspend cells in electroporation buffer to reach a final volume of 1 mL (*see* **Note 17**).
17. Keep cells on ice until electroporation.

18. Gently mix 100 μL (transformation controls) or 400 μL (samples) electrocompetent cells with DNA mixtures, and transfer to pre-chilled electroporation cuvettes. Keep on ice until electroporation (at least 5 min) (*see* **Note 18**).
19. Electroporate the cells: 2.5 kV (and 25 μF). Typically, time constants range from 3.0 to 4.5 ms.
20. Add 1 mL YEPD sorbitol and mix immediately after electroporation.
21. Transfer electroporated cells from each reaction into separate 100 mL culture flask containing 10 mL YEPD sorbitol mix. Use for transformation controls only 2 mL reaction tubes with 2 mL YEPD sorbitol mix.
22. Incubate cells for 1 h at 30 °C (shaking) (*see* **Note 15**).
23. Plate 100 μL per transformation control directly onto SCD-Ura plate.
24. Collect cells of samples (electroporation reactions can now be pooled) by centrifugation at 1000 × *g* for 3 min at RT, and discard supernatant.
25. Resuspend cells in 3 mL SCD-Ura.
26. Determine CFU with dilution series on SCD-Ura plates, as in the following:

    10^1 dilution: plate 100 μL cell resuspension onto SCD-Ura plate.

    10^2 dilution: mix 10 μL cell resuspension with 90 μL SCD-Ura, and plate onto SCD-Ura plate.

    10^3 dilution: mix 10 μL cell resuspension with 990 μL SCD-Ura, and plate 100 μL mixture onto SCD-Ura plate.

    10^4 dilution: mix 10 μL 10^3 dilution with 90 μL SCD-Ura, and plate onto SCD-Ura plate.

    Incubate plates for 72 h at 30 °C. Count colonies after 3 days (*see* **Notes 14** and **19**).
27. Plate remaining volume onto several SCD-Ura plates (100 μL per plate) (*see* **Note 19**).
28. Incubate for 72 h at 30 °C.

6.4 Screening Process

Check on binocular microscope for fluorescence distribution to discard non-fluorescent colonies. Such colonies indicate a failed homologous recombination. The start codon for *GFP* is only included after successful integration of the insert. It is also possible that an aptamer and thus the start codon are inserted correctly, but the aptamer is too stable and blocks the translation already in the absence of the ligand. Moreover, an AUG in the aptamer sequence that is not in frame with the start codon of the *GFP* open reading frame will lead to failure of the GFP fluorescence. Colonies with very strong GFP fluorescence should also not be considered as there is a risk that no aptamer structure is formed at all. Consequently, only colonies with a moderate fluorescence should be selected. Controls are also required: positive control with pCBB05 (colonies with very strong fluorescence) and negative control with pCBB06 (colonies with no fluorescence, *GFP* lacks start codon). In every 96 well plate, both controls should be carried as triplicates.

Preculture (Day 1)

1. Put 200 μL of fresh SCD-Ura medium into each well of 96 deep well plates (u-bottom).
2. Use only one single colony of transformed cells (pool/controls) to inoculate each well, let one well empty (only medium) for background subtraction (*see* **Note 20**).
3. Incubate plates at 1200 rpm for 24 h at 30 °C.

   Splitting culture (Day 2)

   Split one plate into two new 96 deep well plates (u-bottom).
4. Mix 180 μL fresh medium with 20 μL preculture, respectively, 180 μL fresh medium with 250 μM ligand and 20 μL preculture.
5. Incubate plates at 1200 rpm for 24 h at 30 °C.

   Measuring by cytometer and selection of possible candidates (Day 3)

   Use 96 well plates for measurement (flat bottom).
6. Mix 180 μL 1× PBS with 20 μL splitting culture.
7. Read the plate in a cytometer (GFP/mCherry fluorescent, GFP: 488 nm/fluorescent channel 510/20; mCherry: 561 nm/fluorescent channel 610/20) or a comparable device.
8. Calculate switch factor (SW): $SW = \frac{\frac{\text{median GFP (without ligand)}}{\text{median mCherry (without ligand)}}}{\frac{\text{median GFP (with ligand)}}{\text{median mCherry (with ligand)}}}$ (*see* **Note 21**).
9. Strike out potential candidates (SW $\geq$ 1.4) on SCD-Ura plates (*see* **Note 22**).
10. Incubate for 72 h at 30 °C.
11. Start screening process (*see* Subheading 6.4) again but use four clone colonies with medium GFP fluorescence per candidate; do not forget controls (*see* **Note 20**).
12. Use 30 μL from the positive candidates for the inoculation of 4 mL fresh SCD-Ura.

13. Incubate overnight at 30 °C (shaking) (*see* **Note 15**).
14. Use favored kit for plasmid isolation from appropriate candidates (e.g., PureYield™ Plasmid Miniprep System from Promega). Add glass beads to the step with lysis buffer and vortex for 5 min, allow to settle for 5 min, transfer supernatant into a new tube (without glass beads), then continue as instructed in the kit.

6.5 Passage Through *Escherichia coli* and Sequencing of Candidates

1. Transform $CaCl_2$ competent *E. coli* TOP 10 with obtained plasmids of potential candidates from in vivo screening.
2. Plate 100 μL per potential candidate onto LB plate.
3. Incubate overnight at 37 °C.
4. Inoculate 4 mL of LB medium with a single colony of each candidate.
5. Incubate overnight at 37 °C (shaking).
6. Use favored kit for plasmid isolation (e.g., PureYield™ Plasmid Miniprep System from Promega).
7. Reuse plasmids for new transformation of yeast; store them at −20 °C for subsequent sequencing.

6.6 Retransformation of *S. cerevisiae* with Pure Clones of Candidates and Remeasurement

1. Transform chemo-competent yeast with obtained plasmids from passage through *E. coli* (*see* **Note 23**).
2. Plate 100 μL per candidate onto SCD-Ura plate.
3. Incubate for 72 h at 30 °C.
4. Repeat the screening process from Subheading 6.4; stop at calculation of switch factor.
5. Sequence all candidates that still switch successfully (switching factor greater than or equal to 1.4, *see* **Note 22**) with the oligo GFP_rev; use corresponding plasmids obtained from *E. coli* in Subheading 6.5.

6.7 Further Steps

Candidates of interest now need to be characterized and optimized. For example, point mutations can be introduced to potentially increase the switching factor. At the same time, specific sequence regions involved in ligand binding or switching can be identified and non-required regions deleted. Read about these and other ways to study and optimize a riboswitch in the following publications [23, 27, 29].

7 Notes

1. The most important considerations for pool design are the docking sequence and the amount of randomization. When designing the docking sequence, it is key to strike a balance

that allows for sufficient hybridization with the capture oligonucleotide and the detachment upon ligand binding. We strongly recommend placing the docking sequence in between two randomized parts of the sequence and designing the flanking constant regions to be partially compatible with the docking sequence. Both will aid in the formation of secondary structures upon ligand binding. An important consideration for any SELEX experiment is also the length of the randomized region; shorter regions allow for better coverage, while longer ones allow for the formation of more complex RNA structures. For switching aptamers, we rather recommend a longer randomized region, for example, as described in the methods.

2. Large-scale PCR can be done by distributing your reaction into several reaction tubes that fit for your thermocycler. Afterward, the reaction can be pooled in one larger tube.
3. Overdrying an RNA/DNA pellet can make it hard to dissolve. Leaving the tube open to dry for 5 min is usually sufficient if supernatant was thoroughly removed.
4. Pyrophosphate does not always visibly precipitate during the transcription. Missing pyrophosphate is no indicator for a failed reaction.
5. After NaAc/EtOH precipitation, the pellet is often porous and fragile. Be careful when removing the supernatant. For a better separation of supernatant, centrifuge again briefly after the first discard, and remove the remaining supernatant with a pipette.
6. The RNA pellet will often be hardly visible, especially during precipitation after a SELEX elution step. A co-precipitant such as GlycoBlue™ can help with that.
7. In our experience, the hybridization time was critical for a successful Capture-SELEX. Increasingly it appeared to improve efficiency of the protocol. If you wish to reduce the time, we recommend you do so very carefully. We observed an improvement in hybridization when the time at 21 °C was increased from 30 to 60 min.
8. Washing for a longer time is more stringent than using a larger buffer volume and leads to harsher washing. The heat step is meant to remove weak capture oligonucleotide/RNA hybrids. If you find it difficult to enrich any sequences, you may replace this step with a simple wash step, until enrichment starts.
9. You are looking for an increase of the measured radioactively labelled RNA of the elution fraction with increasing numbers of SELEX cycles. If the elution fraction is consistently and at least twice as high as the last washing step (W3), you can consider your pool enriched. You may increase the stringency from now on. Measures to increase stringency include reducing the target concentration during elution, starting with counter elution, or increasing the wash time.

10. Unspecific amplifications occur easily, especially when primers are depleted. We suggest starting with four PCR cycles, check 10 μL on an agarose gel, and increase if necessary. As the pool gets more enriched and more RNA is eluted, less and less cycles will be required. When in doubt, stop the cycles while there are still some unused primers visible below your product.
11. If you get overamplifications or the resulting signal is too week, adjust the number of cycles of the PCR program (less or more cycles, respectively).
12. Large-scale preparation of plasmid should lead to a high concentration of approximately 1 μg/μL. Use favored plasmid kit (e.g., QIAGEN Plasmid Maxi Kit).
13. For optimal results, distribute excised DNA on several columns (if you use a column-based purification kit). If the concentration of salts is too high (detectable by inaccurate measuring in nanodrop with a low 260/230 ratio), perform an ethanol or butanol precipitation to clean the plasmid.
14. Sufficient SCD-Ura plates should be available, as they are required for the determination of the colony-forming units (CFU) and the plating of the transformed yeast cells. The determination of the CFU is necessary to check which dilution is useful if the plates are densely overgrown.
15. Shake with appropriate rpm, as the yeast cells settle quickly. Mild shaking is recommended, which prevents the cells from sedimentation, but also does not stress them.
16. Use the following formulas to calculate the appropriate starting $OD_{600\ nm}$:

$$N_0 = \frac{N}{2^n}$$

$$n = \frac{t}{1.5\ \mathrm{h}}$$

N_0: appropriate starting $OD_{600\ nm}$
N: desired OD600 nm after n cell divisions
n: number of cell divisions (1.5 h)
t: time between inoculation and reaching N in h

17. Approximately 600 μL are required. The 1 mL obtained is sufficient for 400 μL for electroporation reaction with your sample and 2× 100 μL transformation controls.
18. DNA mixture for samples should be in a plasmid/insert ratio between 1:2 and 1:3 with minimum 1 μg of plasmid. Rule of thumb: More is more. Use for negative transformation control 300 ng precut plasmid and for positive transformation control 300 ng intact plasmid. Every DNA mixture in water should be less than 50 μL.

19. The plating can be done with a Drigalski spatula as well as with glass beads. With a total number of approximately 30 plates, it is better to work with glass beads to save time.
20. If it is the first time that the enriched pool is screened, approximately 50 colonies as well as colonies of the controls could be tested in advance. This will give you an overview of the population.
21. Take care that mCherry expression is constant in all cultures (with and without ligand). If the cells are not shaken sufficiently (visible sedimentation), the mCherry values drop drastically.
22. Possible candidates should have a switching factor greater than or equal to 1.4. This seemingly low value is related to the applied cloning conditions. Since the entire pool sequence including the constant regions was cloned, the aptamer is in an unfavorable context. This can influence the regulation. Subsequent optimization usually results in a significant increase in regulation. However, it is important to include the constant regions for the initial screening as they are sometimes involved in aptamer structure formation.
23. The use of a suitable kit to produce chemo-competent yeast cells is possible at this point. The use of pure plasmids eliminates the need for electroporation.

Acknowledgments

The authors thank Dr. Marc Vogel for critical reading the manuscript. This work was supported by the DFG (SFB902 A2, SU402/9-1) and ONR (grant #12997118).

References

1. Ono A, Togashi H (2004) Highly selective oligonucleotide-based sensor for mercury (II) in aqueous solutions. Angewandte Chemie (International Ed. in English) 43:4300–4302. https://doi.org/10.1002/anie.200454172
2. Hanson S, Berthelot K, Fink B et al (2003) Tetracycline-aptamer-mediated translational regulation in yeast. Molecular Microbiology 49:1627–1637. https://doi.org/10.1046/j.1365-2958.2003.03656.x
3. Wrist A, Sun W, Summers RM (2020) The theophylline aptamer: 25 years as an important tool in cellular engineering research. ACS Synthetic Biology 9:682–697. https://doi.org/10.1021/acssynbio.9b00475
4. Min K, Cho M, Han S-Y et al (2008) A simple and direct electrochemical detection of interferon-gamma using its RNA and DNA aptamers. Biosensors & Bioelectronics 23:1819–1824. https://doi.org/10.1016/j.bios.2008.02.021

5. Potty ASR, Kourentzi K, Fang H et al (2009) Biophysical characterization of DNA aptamer interactions with vascular endothelial growth factor. Biopolymers 91:145–156. https://doi.org/10.1002/bip.21097
6. Long SB, Long MB, White RR et al (2008) Crystal structure of an RNA aptamer bound to thrombin. RNA 14:2504–2512. https://doi.org/10.1261/rna.1239308
7. Hamula CLA, Zhang H, Le Guan L et al (2008) Selection of aptamers against live bacterial cells. Analytical Chemistry 80: 7812–7819. https://doi.org/10.1021/ac801272s
8. Ye X, Shi H, He X et al (2015) Iodide-responsive Cu-Au nanoparticle-based colorimetric platform for ultrasensitive detection of target cancer cells. Analytical Chemistry 87: 7141–7147. https://doi.org/10.1021/acs.analchem.5b00943
9. Wang K, Fan D, Liu Y et al (2015) Highly sensitive and specific colorimetric detection of cancer cells via dual-aptamer target binding strategy. Biosensors & Bioelectronics 73:1–6. https://doi.org/10.1016/j.bios.2015.05.044
10. Cao X, Li S, Chen L et al (2009) Combining use of a panel of ssDNA aptamers in the detection of Staphylococcus aureus. Nucleic Acids Research 37:4621–4628. https://doi.org/10.1093/nar/gkp489
11. Tuerk C, Gold L (1990) Systematic evolution of ligands by exponential enrichment: RNA ligands to bacteriophage T4 DNA polymerase. Science:505–510
12. Ellington AD, Szostak JW (1990) In vitro selection of RNA molecules that bind specific ligands. Nature:818–822
13. Zhang Y, Lai BS, Juhas M (2019) Recent advances in aptamer discovery and applications. Molecules 24. https://doi.org/10.3390/molecules24050941
14. Nutiu R, Li Y (2005) In vitro selection of structure-switching signaling aptamers. Angewandte Chemie (International Ed. in English) 44:1061–1065. https://doi.org/10.1002/anie.200461848
15. Rajendran M, Ellington AD (2003) In vitro selection of molecular beacons. Nucleic Acids Research 31:5700–5713. https://doi.org/10.1093/nar/gkg764
16. Stoltenburg R, Nikolaus N, Strehlitz B (2012) Capture-SELEX: Selection of DNA aptamers for aminoglycoside antibiotics. Journal of Analytical Methods in Chemistry 2012:415697. https://doi.org/10.1155/2012/415697
17. Winkler WC, Nahvi A, Roth A et al (2004) Control of gene expression by a natural metabolite-responsive ribozyme. Nature:281–286
18. Sashital DG, Butcher SE (2006) Flipping off the riboswitch: RNA structures that control gene expression. ACS Chemical Biology:341–345
19. Duchardt-Ferner E, Gottstein-Schmidtke SR, Weigand JE et al (2016) What a difference an OH makes: Conformational dynamics as the basis for the ligand specificity of the neomycin-sensing riboswitch. Angewandte Chemie (International Ed. in English) 55: 1527–1530. https://doi.org/10.1002/anie.201507365
20. Weigand JE, Schmidtke SR, Will TJ et al (2011) Mechanistic insights into an engineered riboswitch: A switching element which confers riboswitch activity. Nucleic Acids Research 39: 3363–3372. https://doi.org/10.1093/nar/gkq946
21. Bauer G, Suess B (2006) Engineered riboswitches as novel tools in molecular biology. Journal of Biotechnology 124:4–11. https://doi.org/10.1016/j.jbiotec.2005.12.006
22. Boussebayle A, Groher F, Suess B (2019) RNA-based Capture-SELEX for the selection of small molecule-binding aptamers. Methods 161:10–15. https://doi.org/10.1016/j.ymeth.2019.04.004
23. Boussebayle A, Torka D, Ollivaud S et al (2019) Next-level riboswitch development-implementation of Capture-SELEX facilitates identification of a new synthetic riboswitch. Nucleic Acids Research 47:4883–4895. https://doi.org/10.1093/nar/gkz216
24. Weigand JE, Sanchez M, Gunnesch E-B et al (2008) Screening for engineered neomycin riboswitches that control translation initiation. RNA 14:89–97. https://doi.org/10.1261/rna.772408
25. Hamilton R, Watanabe CK, de Boer HA (1987) Compilation and comparison of the sequence context around the AUG startcodons in Saecharvmyces cerevisiae mRNAs. Nucleic Acids Research:3581–3593
26. Ma H, Kunes S, Schatz PJ et al (1987) Plasmid construction by homologous recombination in yeast. Gene:201–206
27. Schneider C, Suess B (2016) Identification of RNA aptamers with riboswitching properties. Methods 97:44–50. https://doi.org/10.1016/j.ymeth.2015.12.001
28. Sauer N, Stadler R (1993) A sink-specific H+/monosaccharide co-transporter from Nicotiana tabacum: Cloning and heterologous expression in baker's yeast. The Plant Journal:601–610

29. Groher F, Bofill-Bosch C, Schneider C et al (2018) Riboswitching with ciprofloxacin-development and characterization of a novel RNA regulator. Nucleic Acids Research 46: 2121–2132. https://doi.org/10.1093/nar/gkx1319

[illegible]

Chapter 11

RNA Design Principles for Riboswitches that Regulate RNase P-Mediated tRNA Processing

Anna Ender, Peter F. Stadler, Mario Mörl, and Sven Findeiß

Abstract

Riboswitches are an attractive target for the directed design of RNA-based regulators by in silico prediction. These noncoding RNA elements consist of an aptamer platform for the highly selective ligand recognition and an expression platform which controls gene activity typically at the level of transcription or translation. In previous work, we could successfully apply RNA folding prediction to implement a new riboswitch mechanism regulating processing of a tRNA by RNase P. In this contribution, we present detailed information about our pipeline consisting of in silico design combined with the biochemical analysis for the verification of the implemented mechanism. Furthermore, we discuss the applicability of the presented biochemical in vivo and in vitro methods for the characterization of other artificial riboswitches.

Key words RNase P riboswitch, RNA design, Ligand-dependent tRNA processing, Riboswitch characterization, Design score calculation

1 Introduction

In the field of synthetic biology, there is a strong demand for various types of orthogonal regulation systems. For this purpose, synthetic riboswitches came into focus during the last decades. Riboswitches are RNA elements which are mostly located in the 5′-untranslated region (5′-UTR) of bacterial messenger RNAs. Natural riboswitches regulate gene expression in a ligand-dependent manner at the level of transcription or translation [1–3] and are composed of two overlapping domains – an aptamer domain and an expression platform [4–6]. The highly sensitive interaction of the ligand with the aptamer domain triggers conformational refolding of the expression platform, leading to a change in gene activity. This modular composition of riboswitches makes them attractive molecular regulators which can be also applied to address different RNA-based cellular processes [7–11]. Previously, we successfully combined in silico design with biochemical

James Chappell and Melissa K. Takahashi (eds.), *Riboregulator Design and Analysis*,
Methods in Molecular Biology, vol. 2518, https://doi.org/10.1007/978-1-0716-2421-0_11,

characterization to investigate a novel, presumably completely artificial riboswitch-based mechanism – the regulation of the processing of a tRNA by Ribonuclease P (RNase P) [12]. Before a tRNA fulfills its function as an adapter in the translation machinery, it undergoes a multistep maturation process [13–15], which involves cleavage of a 5′-overhang (5′-leader) of the tRNA by RNase P in nearly all organisms [16, 17]. Due to its universal function in all kingdoms of life, RNase P represents an attractive target for a riboswitch-based regulation which can be initially designed and tested in more simple organisms and upon success might be transferred to more complex targets. The implemented mechanism of our RNase P riboswitches relies on the ligand-dependent sequestering of a tRNA 5′-leader [12]. We describe in this chapter the parameters and constraints that should be considered when addressing an enzyme reaction through folding of its RNA substrate. Further, we present and discuss our computational and biochemical methods, which were applied, on the one hand, to design riboswitch candidates de novo and, on the other hand, to monitor the switching activity and for the investigation of the postulated switching mechanism.

The implementation of such a regulatory system represents a new riboswitch-based tool to regulate tRNA maturation depending on the presence or absence of a specific ligand, in our case theophylline, a xanthine-related purine. Our riboswitches were applied to control two different suppressor tRNAs – tRNA supF from *Escherichia coli* [12] and $tRNA^{Pyl}$ from *Methanosarcina mazei* [16]. Using our pipeline, other orthogonal tRNAs could in principle be regulated in a ligand-dependent fashion and applied for genetic code expansion. We have shown that our RNase P riboswitches, initially designed to regulate *E. coli* RNase P, are also able to control tRNA maturation of other RNase P enzymes in a ligand-dependent way [18]. As it is quite difficult to convert eukaryotic RNA polymerase III promoters into regulable elements [19], a possible riboswitch application is to render tRNA maturation inducible in these species. Additionally, RNase P riboswitches have the potential to reveal new insights into the structural requirements of the 5′-leader and its efficient recognition by diverse RNase P enzymes. Utilizing our riboswitches, we were able to support the hypothesis of Lin et al. [20] that the structure of the tRNA's 5′-leader has a direct impact on RNase P processing. Whereas Lin and colleagues demonstrated this behavior in vitro, we confirmed it by ligand-dependent control of the 5′-leader accessibility in vivo.

We are confident that our design strategy is applicable to regulate maturation of different types of tRNA transcripts in general. However, more data would be essential to further improve the prediction accuracy of the in silico design approach. Furthermore, it might be possible to transfer our pipeline and the applied analysis to other enzyme-based reactions where an RNA molecule is the recognized substrate.

1.1 Design Strategy

Riboswitch design typically starts with the selection of an RNA aptamer which is utilized as sensory domain. Using SELEX (Systematic Evolution of Ligands by EXponential Enrichment) [21–23], an RNA aptamer could be selected for any ligand at least in principle. However, in the riboswitch design context, requirements such as a membrane-permeable and nontoxic ligand as well as a structured and highly specific aptamer are essential. This holds only for a couple of currently available ligand-aptamer pairs. A prominent example frequently used to design ligand-triggered switches is the well-established theophylline-dependent aptamer [10, 24–29]. Its binding mechanism has been studied in great detail and important structural features as well as the K_D of 320 nM are known for this aptamer [24, 30]. Whenever a novel aptamer is to be used for the design, such information need to be acquired before the implementation of a regulatory mechanism becomes possible. Specifically, the proposed design strategy requires a priori knowledge of the binding-competent secondary structure as well as the K_D of the aptamer-ligand complex.

1.1.1 Model Development

To regulate an enzyme-based reaction by RNA folding, substantial knowledge about the enzyme and especially the enzyme-substrate interaction is required. Bacterial RNase P is composed of two moieties – an RNA subunit and a protein subunit, which are named M1 RNA and C5 protein in *E. coli* [31, 32]. Whereas the RNA subunit is catalytically active in vitro, the protein subunit is essential for the in vivo activity of the enzyme [33]. Various structural and sequence-based features of the tRNA interact with RNase P and serve for the substrate recognition, for example, parts of the T-stem [34], the 3′-CCA end [35], the 5′-leader [36, 37], the identity of the N-1 position, and the acceptor-stem [38]. The minimal substrate for the bacterial enzyme is composed of the acceptor-stem in combination with the 3′-CCA end [39, 40].

The mechanism of our RNase P riboswitches relies on the hypothesis that RNase P prefers a single-stranded 5′-leader as substrate (*see* Fig. 1). This hypothesis was already postulated based on previous results [20], but there was no explicit evidence for this assumption. In our model, the 5′-leader should be sequestered in a double-stranded conformation by the formation of a hairpin structure together with the theophylline-dependent aptamer. Addition of theophylline should trigger an opening of this sequestor hairpin, releasing the interaction site for RNase P. For the development of such a mechanistic model, it is important to clearly define the region which should change its conformation. In our case, this is the 5′-leader that interacts with the enzyme. *E. coli* RNase P interacts with up to eight nucleotides upstream of the tRNA [41–44]. It was shown that shortening the 5′-leader to less than four nucleotides leads to a clearly decreased cleavage activity in the case of *Bacillus subtilis* RNase P [42]. Based on this information, the first

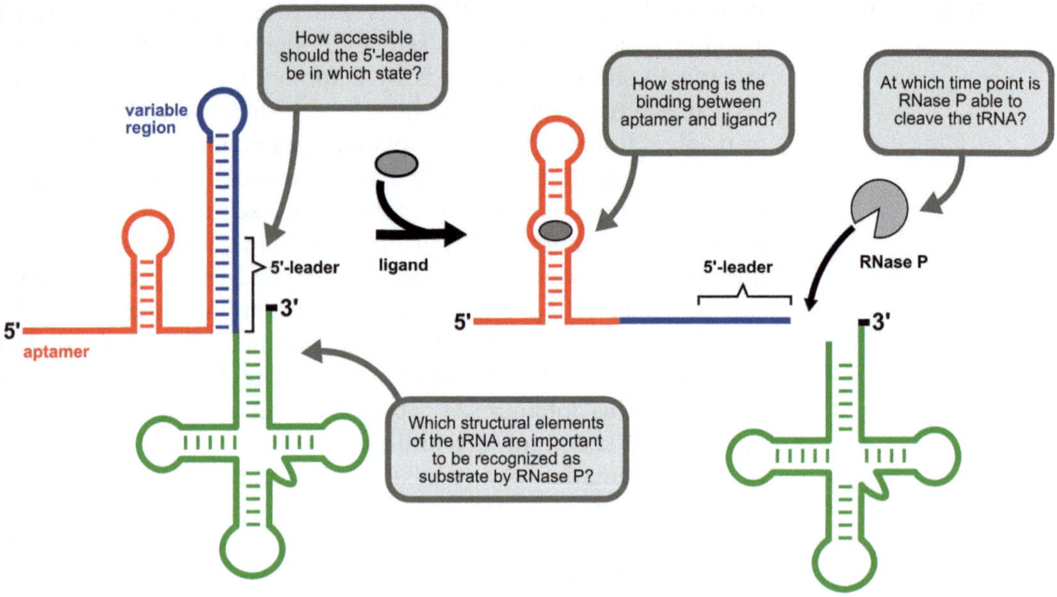

Fig. 1 Schematic representation of the regulatory principles of RNase P riboswitches and the most important parameters to be considered. The aptamer sequence (red) is combined with a designable variable region (blue) and a tRNA sequence (green). In the absence of the ligand, the 5′-leader should be sequestered within a hairpin structure formed between the aptamer's 3′-part and the variable region (left). In the presence of the ligand, refolding of the aptamer sequence is triggered and the ligand bound state (right) is populated. In this conformation, the 5′-leader becomes accessible for RNase P processing. Important aspects to be considered during the design process are indicated (gray boxed questions) and described in the main text

seven nucleotides upstream of the tRNA are defined as 5′-leader and enforced to be unpaired for the score calculation in our setting.

1.1.2 Model Implementation, Sequence Sampling, and Scoring

For the herein described RNase P riboswitches, we combine the sequence of the theophylline aptamer TCT8-4 [24] with a variable region and the sequence of *E. coli's* suppressor tRNA^{Tyr} supF [45] (*see* **Note 1**). In principle, any well-characterized ligand binding RNA-aptamer and any tRNA sequence could be used as long as a readout system exists and the ligand is able to enter the cell for in vivo application. To be recognized by RNase P, the full-length primary transcript including the tRNA at its 3′-end needs to be transcribed and folded correctly as the acceptor-stem, the CCA tail, and the 5′-leader sequence represent the minimal substrate [46]. Our in silico design is therefore based on the assumption that the RNA's structure ensemble is equilibrated and is sufficiently modeled by thermodynamic structure prediction methods such as the routines implemented in RNAfold [47]. The developed design model comprises two structural states. In absence of the ligand, the formation of the sequestor hairpin renders the leader inaccessible, whereas in the presence of the ligand, the bound aptamer structure gets stabilized upon interaction and thereby the sequestor hairpin is

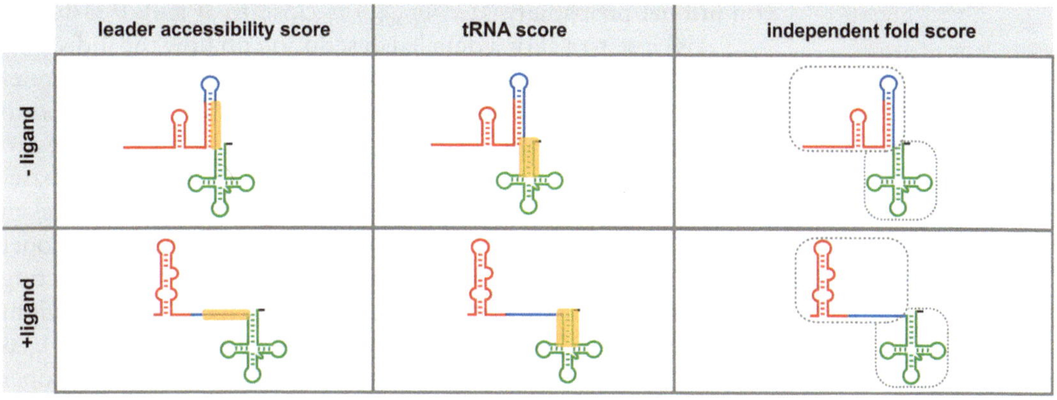

Fig. 2 Graphical representation of scores applied either to optimize or subsequently filter and select candidate sequences. The *leader accessibility score* evaluates whether the 5′-leader of the tRNA is in a double-stranded conformation in the absence and in a single-stranded conformation in the presence of the ligand. The *tRNA score* estimates whether the tRNA acceptor-stem is correctly formed in the absence as well as in the presence of the ligand. The *independent fold score* evaluates whether the tRNA and the riboswitch form independent substructures such that RNase P binding is presumably only regulated by the accessibility of the tRNA leader. Regions of interest are highlighted in yellow or gray for each score

opened, resulting in an accessible leader sequence (*see* Fig. 1). To be able to regulate (at least in principle) different tRNA sequences by the same riboswitch construct, the tRNA sequence should ideally fold into its native structure independent of the structural state of the upstream sequence. The hard and the soft constraint framework of RNAfold is well suited to model ligand binding and structure constraints at the level of secondary structure prediction [48]; for details, *see* Subheading 3.1. The predefined components, i.e., the aptamer and the tRNA sequence, are linked by the variable region which should be able to form a sequestor hairpin with the 3′-part of the aptamer sequence. In principle, any sequence could work as variable region resulting in 4^n possible sequences, with n being the length of the variable region. However, the intended partial complementarity to a subsequence of the aptamer limits the number of candidates to those that are able to form the sequestor hairpin. RNAblueprint [49] can be applied to fairly sample all sequences that are compatible to a predefined set of sequence and structure constraints; *see* Subheading 3.1. In order to quantify how well a sampled candidate sequence meets the design criteria described above, we developed three scores which are graphically summarized in Fig. 2. The *leader accessibility score*

$$s_{\text{accessibility}}(x) = (1 - P(x|\Phi_{\text{leader}}))P(x|\Phi_{\text{leader}}, c_{\text{ligand}}) \qquad (1)$$

combines the probability of an accessible leader region in the absence ($P(x|\Phi_{\text{leader}})$) and presence ($P(x|\Phi_{\text{leader}}, c_{\text{ligand}})$) of the ligand. Sequence x implements the intended design model if the conditional probability $P(x|\Phi_{leader})$ is close to 0 and $P(x|\Phi_{leader}, c_{ligand})$ is close to 1. For a detailed description on how the individual probabilities are estimated based on RNAfold predictions, we refer to Subheading 3.1. All sequences with an accessibility score close to 1 are likely to sequester and present the leader region according to the intended design model (*see* Fig. 1). However, unintended base-pairing between the riboswitch and the tRNA sequence is not taken into account so far. Such interactions might interfere with proper RNase P recognition and subsequent processing. When predicting the secondary structure of a single tRNA sequence, the result might not be similar to the expected clover-leaf structure (*see* **Note 2**). As the complete tRNA sequence is predefined, we assumed proper tRNA folding if the acceptor-stem is formed (*see* Fig. 2). The resulting *tRNA score*

$$s_{tRNA}(x) = \exp\left(-1\left|P(x|\Phi_{stem}, c_{ligand}) - P(x|\Phi_{stem})\right|\right) \quad (2)$$

becomes 1 when the acceptor-stem is present independent of ligand availability. Finally, our *tRNA independent fold score*

$$s_{indepfold}(x) = \frac{G(x_{switch}) + G(x_{tRNA})}{G(x)} \quad (3)$$

estimates the impact of structures that form unintended base pairs between the riboswitch and tRNA subsequences. Ideally, the sum of the ensemble free energies of both subsequences, i.e., $G(x_{switch})$ for the riboswitch and $G(x_{tRNA})$ for the tRNA sequence, is equal to the ensemble free energy $G(x)$ of the concatenated sequences, and the $s_{indepfold}(x)$ becomes 1. In [12], the leader accessibility score was applied to select optimal sequences that implement the desired design model. However, further in silico analysis indicated the problem of base pairs between riboswitch and tRNA, and the two tRNA scores have therefore been added as additional filter criteria.

1.2 Biochemical Evaluation of Riboswitch Functionality and the Switching Mechanism

1.2.1 In Vivo Analysis

To allow fast and convenient detection of tRNA maturation in vivo, a fluorescence-based reporter system was established. For this purpose, the open reading frame for *c3gfp* (cycle 3 green fluorescent protein) carrying an amber stop codon at position Y39 was used. Transcription of the reporter was regulated by an arabinose promoter on plasmid pACYC. The riboswitch tRNA fusion was expressed from a second plasmid (pBAD). To monitor the release of the mature tRNA, *E. coli* suppressor $tRNA^{Tyr}$ supF was chosen, which is able to translate the amber stop codon in the *c3gfp* reporter mRNA. Hence, the c3GFP-induced fluorescence of the cells indicates the maturation of the riboswitch-controlled suppressor tRNA. When using such a stop-codon suppression, it should be considered that protein expression can be only monitored if an appropriate aminoacyl-tRNA synthetase is present in the cell. As tRNA supF differs only in one nucleotide from the native *E. coli* $tRNA^{Tyr}$, no additional aminoacyl-tRNA synthetase is needed, since the tyrosyl-tRNA synthetase recognizes tRNA supF as substrate. A positive control (identical tRNA with a single-stranded 5′-leader) should be analyzed in parallel, as fluorescence measurements are difficult to quantify in absolute values. The fluorescence intensity can then be quantified relative to the positive control. Additionally, the positive control allows to examine the maximal possible fluorescence signal for the tested riboswitch candidates.

If acceptable switching activities are observed with the chosen candidates, tRNA processing can be further characterized through Northern blot analysis. To avoid unspecific background hybridization signals, it is important to use an orthogonal tRNA with a sequence sufficiently different from endogenous tRNAs in the analyzed organism. As tRNA supF differs only in one nucleotide from native *E. coli* $tRNA^{Tyr}$, a tRNA hybrid was constructed by replacing the anticodon arm and the variable loop by the corresponding region of $tRNA^{Tyr}$ from *Thermus thermophilus.* The introduced sequence was used for the probe design, as it is unique in *E. coli* so that no background hybridization signals were obtained. Furthermore, monitoring tRNA maturation by Northern blot analysis does not require the presence of a corresponding aminoacyl-tRNA synthetase and represents a more direct way to investigate riboswitch functionality, as no additional layer of gene expression (translation) is needed. Hence, in subsequent studies, the functionality of our riboswitches was analyzed only by Northern blot analysis.

When riboswitches of interest show promising switching activities at the level of tRNA maturation, the next step is to verify that the constructs indeed regulate 5′-leader processing by RNase P.

1.2.2 Kinetic Analysis According to Michaelis-Menten

To check whether theophylline has an impact on the riboswitch constructs concerning RNase P-dependent cleavage, kinetic analysis was performed according to Michaelis-Menten. In order to set the correct reaction parameters for the switching analysis, it is important to consider the expected regulation mechanism. In case of tRNA processing by RNase P, the 5′-leader is mainly involved in enzyme-substrate binding and has less effects on the cleavage catalysis [20]. Thus, studies under single turnover conditions are not suitable, as the accessibility of the 5′-leader will probably address substrate binding, which has no impact under those reaction conditions. The experiment was performed under steady-state conditions to include substrate binding. Additionally, a positive control with a single-stranded leader was used to exclude artificial effects of theophylline addition to the enzyme reaction. The concentration of the ligand should be adjusted to the known binding efficiencies to avoid artifacts by oversaturating ligand concentrations. In our analysis, the observed activation rate in vitro showed a higher background (OFF-state) compared to the in vivo investigation, leading to a lower activation rate. Such differences can occur, as different parameters can influence the activation rates in different experimental settings (*see* Subheading 1.2.4).

1.2.3 Structure Analysis by In-Line Probing

Finally, the postulated ligand-dependent refolding of the riboswitch should be analyzed to verify the connection between RNA folding and the observed switching activities of the enzyme reaction. For this purpose, folding dynamics should be considered. In our case, a thermodynamically stable refolding of the riboswitch was expected upon theophylline addition. Hence, in-line probing can be performed to analyze the secondary structure in the presence and the absence of theophylline. However, this procedure can only be applied if the process is thermodynamically stable, as the RNA of interest is incubated for up to 40 h at room temperature. Hence, differences in the folding kinetics cannot be revealed by this method. If the mechanism depends on the folding time (e.g., in the case of transcription-regulating riboswitches), other methods must be applied. In some cases, lead probing could improve the structural analysis in a shorter time frame [50]. Taken together, the method for structure analysis must be chosen according to the underlying riboswitch mechanism.

1.2.4 Interpretation of the Biochemical Data

Taken together, riboswitch functionality and the switching mechanism should be investigated by a combination of in vivo and in vitro experiments. Each experiment offers new insights and shows qualitative connections for each riboswitch construct, but the quantitative activation rates will differ depending on the experiment, as varying parameters may influence the switching activity in different settings [12]. Figure 3 sums up different parameters which can impact the output of the chosen experiment. For correct data interpretation, it is important to consider the parameters in the context of the respective experiment.

2 Materials

2.1 Applied Software

1. RNAblueprint version 1.2.2.
2. RNAfold routines of the ViennaRNA package version 2.4.11.

2.2 Fluorescence Measurements

1. *E. coli* TOP10 cells carrying riboswitch plasmid (pBAD-ampR-riboswitch-tRNA) and reporter plasmid (pACYC/tetM/c3GFPY39TAG); see Fig. 4.
2. 20% (w/v) (+) arabinose.
3. Cultivation medium: LB medium with 100 μg/ml ampicillin and 10 μg/ml tetracycline.

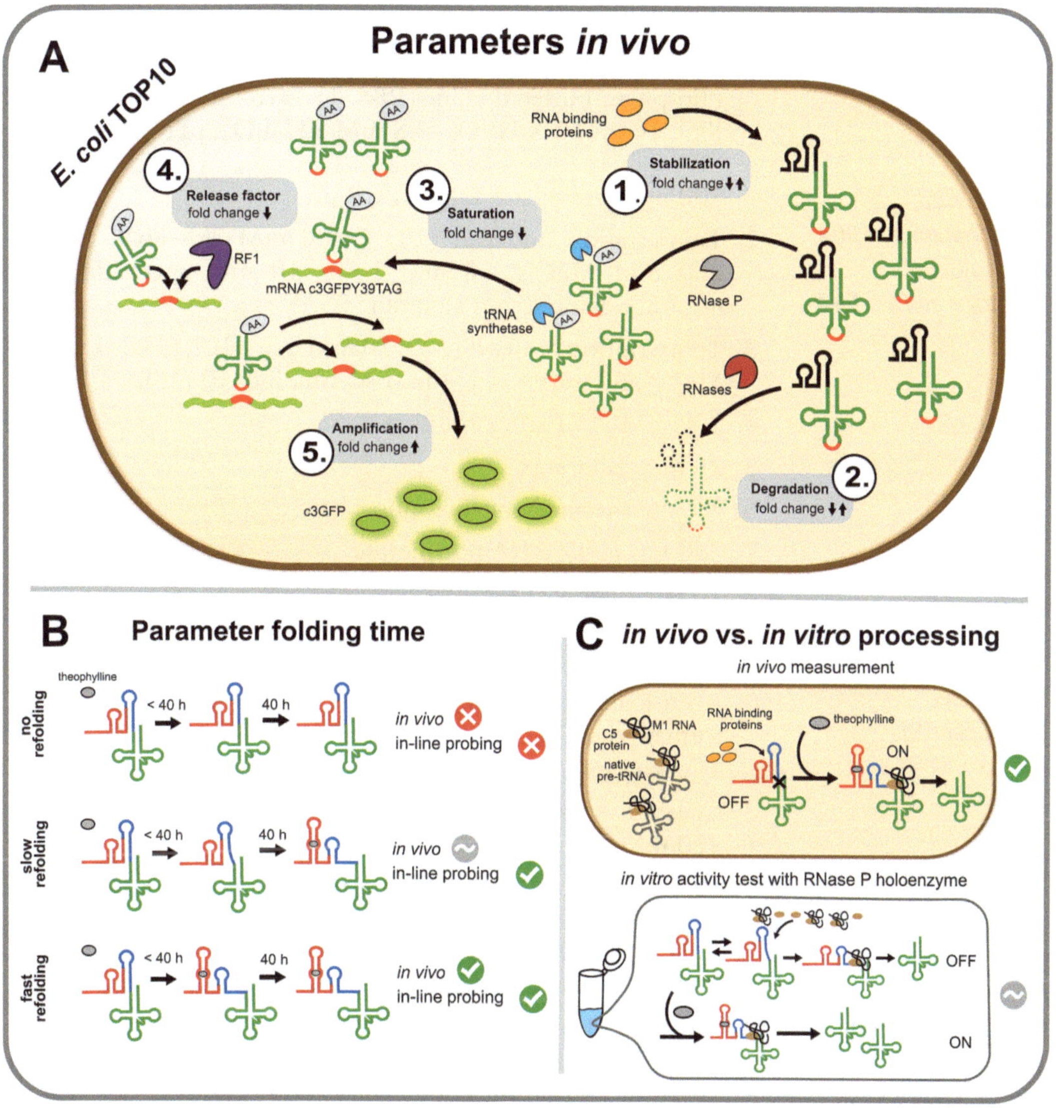

Fig. 3 Possible parameters influencing the final determined fold changes depending on the chosen experiment. (**a**) Parameters in vivo. 1. RNA-binding proteins may stabilize the sequestor hairpin structure (lower fold change ↓) or enhance the processing by RNase P (higher fold change ↑). 2. Slowly processed riboswitch tRNA precursors can be rapidly degraded by other RNases. This could reduce the background (higher fold change ↑), but in parallel, it may reduce the ON-state if RNase P processing needs too much time (lower fold change ↓).3. Oversaturation of the translation system due to a high amount of processed tRNA leads to a constant ON-state, while background fluorescence still can increase (lower fold change ↓). 4. Competition of the release factor 1 (RF1) and the mature suppressor tRNA lowers the translation efficiency (lower fold change ↓). 5. On protein level, the signal may be amplified, as the concentration of tRNA has a positive impact on the rate of protein synthesis (higher fold change ↑). (**b**) Folding time has a strong impact on the fold change. No refolding upon theophylline addition: No activation can be detected in vivo or through in-line probing in vitro. Slow refolding: only a low activation can be detected in vivo (due to rapid degradation of unprocessed tRNA precursors), but a clear opening is visible by in-line probing analysis. Fast refolding: a clear activation is visible by in vivo analysis and by in-line probing experiments. (**c**) Comparison of RNase P-dependent processing in vivo and in vitro. In vivo: in the OFF-state, the riboswitch construct will be rarely cleaved by RNase P, as it competes with native and more accessible tRNA precursors for processing. Additionally, the sequestor hairpin may be

4. 40 mM theophylline stock solution.
5. Spectrofluorometer.
6. Phosphate-buffered saline (PBS, 1×): 137 mM NaCl, 2.7 mM KCl, 10 mM Na_2HPO_4, 1.8 mM KH_2PO_4, pH 7.4.

2.3 In Vitro Transcription and RNA Purification of Substrate and M1 RNA

1. 500 ng/μl template-DNA (PCR product).
2. Transcription buffer A (10×): 800 mM Tris-HCl, pH 7.5, 200 mM $MgCl_2$, 50 mM DTT, 10 mM spermidine, 1.2 mg/ml bovine serum albumin (BSA).
3. Transcription buffer B (2×): 80 mM Tris-HCl, pH 8.0, 16 mM $MgCl_2$, 50 mM NaCl, 4 mM spermidine, 10 mM DTT.
4. 200 mM DTT stock solution.
5. 25 mM NTP mix.
6. 2 U/μl thermostable pyrophosphatase (TIPP).
7. 250 ng/μl T7 RNA polymerase.
8. 10 μCi/μl α-[^{32}P]-ATP.

2.4 Purification of RNA by Gel Electrophoresis

1. Denaturing polyacrylamide (PAA) gel (10%).
2. RNA-loading dye (3×): 10 mM Tris-HCl, pH 7.6, 80% (v/v) formamide, 0.25% (w/v) bromophenol blue, 0.25% (w/v) xylene cyanol.
3. Crushed soak solution (1×): 10 mM Tris-HCl, pH 7.5, 200 mM NaCl, 1 mM EDTA.
4. 10× Tris/borate/EDTA (TBE) buffer: 890 mM Tris-HCl, pH 8.0, 890 mM borate, 20 mM EDTA.

2.5 5′-Dephosphorylation of RNA

1. 5 U/μl antarctic phosphatase.
2. Antarctic phosphatase buffer (10×).

2.6 5′-Labeling of DNA Probes for Northern Blot or RNA for In-Line Probing

1. 10 U/μl T4 polynucleotide kinase (T4 PNK).
2. T4 PNK buffer (10×).
3. 100 pmol/μl DNA oligonucleotide or 100 pmol/μl RNA in vitro transcript.
4. 10 μCi/l [γ-^{32}P]-ATP.
5. NucAway™ Spin-Column.

Fig. 3 (continued) stabilized by RNA-binding proteins. In vitro: even in the OFF-state, a part of the riboswitch tRNA precursor is processed, since no other substrate is present (higher background, lower fold change) Red cross: no activation detected; gray wave: low activation detected; green checkmark: strong activation

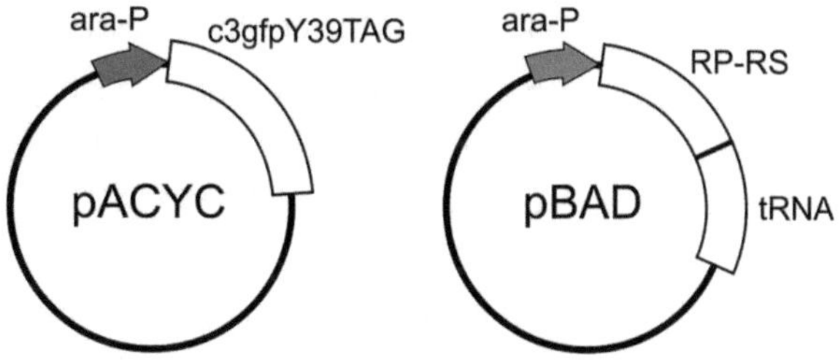

Fig. 4 Plasmids for the in vivo analysis of RNase P riboswitches (RP-RS). The reporter plasmid pACYC/tetM/c3GFPY39TAG was used to monitor maturation of the suppressor tRNA by fluorescence measurement. The fusion of riboswitch and suppressor tRNA was encoded on a second plasmid (pBAD). Both the reporter gene and the riboswitches were regulated by arabinose induction

2.7 RNA Extraction and Northern Blot

1. *E. coli* TOP10 cells carrying riboswitch plasmid (pBAD-ampR-riboswitch-tRNA-supF) and reporter plasmid (pACYC/tetM/c3GFPY39TAG).
2. LB medium with 100 μg/ml ampicillin and 10 μg/ml tetracycline.
3. 40 mM theophylline stock solution.
4. Zirconia glass beads 0.1 mm.
5. FastPrep®-24 Classic bead beating grinder lysis system.
6. TRIzol™.
7. 10× TBE buffer: 890 mM Tris-HCl, pH 8.0, 890 mM borate, 20 mM EDTA.
8. Filter paper.
9. BrightStar®-Plus Nylon Membrane.
10. Trans-Blot Turbo Transfer System.
11. Pre-hybridization buffer (1×): 6× saline sodium citrate (SSC), 10× Denhardt's solution, 0.5% sodium dodecyl sulfate (SDS).
12. Hybridization buffer (1×): 6× SSC, 0.1% SDS
13. 10 pmol/μl 5′-labeled DNA probe.
14. Stripping solution (1×): 0.5% SDS.

2.8 Kinetic Analysis

1. RNase P reaction buffer (2×): 100 mM 4-(2-hydroxyethyl)-1-piperazineethanesulfonic (HEPES, pH 7.4), 200 mM NH_4Cl, 20 mM $MgCl_2$ [51].
2. 200 nM freshly diluted *E. coli* RNase P C5 protein in 1× concentrated RNase P reaction buffer (store concentrated protein in 50 mM Tris-HCl (pH 8,0) buffer with 50 mM NaCl) (*see* **Note 3**).
3. 100 nM M1 RNA stock solution prepared by in vitro transcription and RNA purification (*see* Subheadings 3.3 and 3.4).

4. 300 μM theophylline stock solution.
5. 2 μM stock solution of riboswitch in vitro transcript (substrate) internally radioactively labeled with α-[^{32}P]-ATP.
6. Denaturing PAA gel (10%) – ROTIPHORESE®DNA sequencing system kit.
7. Loading dye (3×): 10 mM Tris-HCl, pH 7.6, 80% (v/v) formamide, 0.25% (w/v) bromophenol blue, 0.25% (w/v) xylene cyanol.

2.9 Structural Analysis by In-Line Probing

1. 300 μM theophylline.
2. In-line probing buffer (10×): 500 mM Tris-HCl, pH 8.5, 200 mM $MgCl_2$, 1 M KCl.
3. RNase T1 buffer (10×): 100 mM Tris-HCl, pH 7.5, 300 mM NaCl, 5 mM EDTA.
4. Alkaline hydrolysis buffer (10×): 500 mM $NaCO_3$, pH 9.7.
5. Denaturing polyacrylamide gel (10%).
6. RNA-loading dye (3×): 10 mM Tris-HCl, pH 7.6, 80% (v/v) formamide, 0.25% (w/v) bromophenol blue, 0.25% xylene cyanol.
7. Colorless loading dye (2×): 10 M urea, 1.5 mM EDTA, pH 8.0.

3 Methods

3.1 In Silico Design of a New Riboswitch Mechanism

The following step-by-step instructions exemplify sequence sampling with RNAblueprint and score calculation utilizing structure prediction routines of RNAfold. For an efficient screening of thousands of sampled sequences, we recommend to use the Python bindings available for both tools. Both software packages are available on GitHub and equipped with detailed install instructions. Here we assume a successful setup of both tools.

Sequence Sampling

1. Create input file, `RNaseP_RS_design.in`, containing the following sequence and structure constraints:

```
AAGUGAUACCAGCAUCGUCUUGAUGCCCUUGGCAGCACUUCANNNNNNNNNNNNNNNNNN
NNGGUGGGGUUCCCGAGCGGCCAAAGGGAGCAGACUCUAAAUCUGCCGUCAUCGACUUCG
AAGGUUCGAAUCCUUCCCCCACCACCA
. . . . . . . . . . . . . . . . . . . . . . . . . . . . . .
(((((((((((((((....))))))))))))))).........................-
.........................................................
```

which correspond to the intended design model (*see* Fig. 1). Please note, for the given design, that the structural state containing the sequestor hairpin is sufficient, as the aptamer and the tRNA sequences are fixed.

2. Sample compatible sequences:

   Apply RNAblueprint to fairly sample ten compatible sequences from the solution space of all compatible sequences:

```
# RNAblueprint -i RNaseP_RS_design.in
```

Please note that the resulting sequences change every time the command is executed. All returned sequences are compatible to the design model, but they meet the design criteria not equally well. For the subsequent score calculation, the following sequence has been selected:

```
AAGUGAUACCAGCAUCGUCUUGAUGCCCUUGGCAGCACUUCAACCCCUGUUGGGGUGCU-
GUUAGGUGGGGUUCCCGAGCGGCCAAAGGGAGCAGACUCUAAAUCUGCCGUCAUCGA-
CUUCGAAGGUUCGAAUCCUUCCCCCACCACCA
```

This sequence is high scoring with respect to the leader accessibility score.

Secondary Structure Prediction and Energy Evaluation

3. Partition function calculation:

   By default, RNAfold predicts for a given RNA sequence the most stable structure and returns a text-based representation of it as well as the corresponding minimum free energy (MFE). Utilizing the partition function calculations implemented in RNAfold, the energy of the thermodynamic ensemble can be computed.

   Execute the command

```
# RNAfold -i RNaseP_RS_example.in --noPS -p0
```

to predict the MFE and the corresponding secondary structure in dot-bracket notation for each sequence in the input file (specified by `-i`). Additionally, the ensemble free energy $G(x)$ and the frequency of the MFE structure in the ensemble are calculated (`-p0`). To save disc space, no postscript drawings are generated (`--noPS`). For the subsequent score calculations, we focus on the ensemble free energy.

4. Hard constraint folding:

   To calculate ensemble free energies of a specified subset of the structure ensemble, the hard constraint framework of RNAfold is applied.

   First, add the hard structure constraint

```
...................................................................
x x x x x x -
x.................................................................-
............................
```

to the input file, `RNaseP_RS_example.in`, on a separate line. In this example, the structure constraint that forces a seven nucleotide long leader sequence, marked with x symbols, to be completely unpaired is used. Apply the program call

```
# RNAfold -i RNaseP_RS_example.in --noPS -p0 -C
```

containing the `-C` option to predict the constraint ensemble free energy $G(x|\, \Phi_{\text{leader}}) = -54.81$ kcal mol^{-1}.

5. Soft constraint folding:

   To simulate ligand binding on the secondary structure level, an energy bonus of -9.22 kcal mol^{-1} is added to every structure in the ensemble that contains the correctly folded theophylline binding pocket. Specify the respective interior and bulge loop substructure and the energy bonus, by the `--motif` option. The program call

```
# RNAfold -i RNaseP_RS_example.in --noPS -p0 --motif "GAUAC-
CAG&CCCUUGGCAGC,(...((((&)...)))...),-9.22"
```

now returns the ligand binding corrected ensemble free energy $G(x|\, c_{\text{ligand}}) = -64.09$ kcal mol^{-1}.

Add the `-C` option to the above program call to predict the leader constraint and ligand binding corrected ensemble free energy $G(x|\, \Phi_{\text{leader}}, c_{\text{ligand}}) = -63.91$ kcal mol^{-1}.

Score Calculation

6. *Leader accessibility score:*

   Use the partition function calculation described above to estimate the conditional probability

$$
\begin{aligned}
P(x|\Phi_{\text{leader}}) &= \exp\left(-\frac{G(x|\Phi_{\text{leader}}) - G(x)}{RT}\right) \\
&= \exp\left(-\frac{-54.81 - -59.96}{RT}\right) \\
&= 2.3 \times 10^{-4}
\end{aligned}
\tag{4}
$$

of observing sequence x, here the selected example sequence, in a conformation where all seven tRNA leader nucleotides are unpaired. R is the universal gas constant of 1.98717×10^{-3} kcal K^{-1}mol^{-1} and T the temperature with a default value of 310.15 K. All energy values are in kcal mol^{-1}. For simplicity we omit the units within the formulas. Estimate the conditional probability $P(x|\,\Phi_{\text{leader}}, c_{\text{ligand}})$ of an unpaired leader sequence in the presence of the ligand similarly by using $G(x|\,\Phi_{\text{leader}}, c_{\text{ligand}})$ and $G(x|\,c_{\text{ligand}})$.

For the example sequence, the resulting leader accessibility score is

$$\begin{aligned} s_{\text{accessiblity}}(x) &= (1 - P(x|\Phi_{\text{leader}}))P(x|\Phi_{\text{leader}}, c_{\text{ligand}}) \\ &= (1 - 2.3 \times 10^{-4})(0.75) \\ &= 0.75 \end{aligned} \tag{5}$$

7. *tRNA score:*

The fraction of structures compatible with the tRNA acceptor-stem can be estimated by utilizing the corresponding structure constraint. Replace the structure constraint applied to calculate the leader accessibility score by

```
...........................................................-
....
((((((.....................................................-
.............)))))).....
```

Estimate the conditional probabilities $P(x|\,\Phi_{\text{stem}})$ and P-$(x\,|\Phi_{\text{stem}}, c_{\text{ligand}})$ of observing sequence x in a structure compatible to the tRNA acceptor-stem in the presence and absence of theophylline as described above.

For the example sequence, the resulting tRNA score is

$$\begin{aligned} s_{\text{tRNA}}(x) &= \exp\left(-\left|P(x|\Phi_{\text{stem}}, c_{\text{ligand}}) - P(x|\Phi_{\text{stem}})\right|\right) \\ &= \exp(-|1.00 - 0.98|) \\ &= 0.98 \end{aligned} \tag{6}$$

8. *Independent fold score:*

To estimate the number of structures that form unintended base pairs between the riboswitch (AAGUGAUACCAGCAUCGU-CUUGAUGCCCUUGGCAGCACUUCAACCCCUGUUGGGGUGCUGUUA) and tRNA (GGUGGGGUUCCCGAGCGGCCAAAGGGAGCAGACU-CUAAAUCUGCCGUCAUCGACUUCGAAGGUUCGAAUCCUUCCCC-CACCACCA) subsequences, the respective ensemble free energies can be used. If the riboswitch and the tRNA sequence fold independently, the sum of the individual ensemble free energies, $G(x_{\text{switch}})$ and $G(x_{\text{tRNA}})$, is equal to the ensemble free energy of the full-length construct, $G(x)$.

Apply RNAfold to predict the $G(x_{\text{switch}})$ and $G(x_{\text{tRNA}})$ in order to calculate the tRNA independent fold score for the example sequence

$$\begin{aligned} s_{\text{indepfold}(x)} &= \frac{G(x_{\text{switch}}) + G(x_{\text{tRNA}})}{G(x)} \\ &= \frac{-26.73 + -32.83}{-59.96} \\ &= 0.99 \end{aligned} \tag{7}$$

3.2 Fluorescence Measurements

1. Use one colony per sample (*E. coli* TOP10 cells carrying riboswitch plasmid and reporter plasmid) to generate overnight cultures in 15 ml cultivation tubes with 5 ml cultivation medium and the appropriate antibiotics.
2. For the preparation of the main culture, dilute a pre-culture to an OD_{600} of 0.06 in 5 ml cultivation medium supplemented with 0.01% (w/v) arabinose and optional 2 mM theophylline (final concentration).
3. Cultivate cells at 37 °C and 200 rpm for 4.5 h (*see* **Note 4**).
4. Adjust OD_{600} in LB medium to 0.8 and wash the cell pellet (spin 3 min once with 1 ml of 1× PBS buffer).
5. Resuspend cell pellet in 500 μl 1× PBS buffer.
6. Determine fluorescence intensity using fluorescence spectrometer (mode: emission; extinction bandwidth: 2.5 nm; emission bandwidth: 5 nM; response: 1 s; sensitivity: medium; extinction wavelength: 497; measurements: 503–600 nm). For relative quantification, a positive control should be included for each experiment.
7. The quantification of the fluorescence intensity (signal intensity at 509 nm emission) is done relative (in percent) to the positive control (tRNA with a single-stranded 5′-leader).

3.3 In Vitro Transcription

1. Set up the transcription reaction in 30 μl reaction volume (in a PCR tube) in 30 μl reaction volume. Standard conditions: 3 μl of 10× transcription buffer A, 2 μl of 500 ng/μl PCR product, 5 μl of 25 mM NTP, 0.5 μl of 200 mM DTT, 0.2 μl of 2 U/μl TIPP, 5 μl of 250 ng/ml T7 polymerase, 14.3 μl dH_2O, 1 μl of 10 μCi/μl α-[^{32}P]-ATP (if radioactively labeled RNA is needed).

   For the transcription of M1 RNA, use the following modification of the protocol: use 15 μl of 2 x transcription buffer B instead of transcription buffer A, and use 2.3 μl dH_2O instead of 14.3 μl dH_2O.
2. Incubate the reaction for 5–7 h at 37 °C.
3. Stop reaction by the addition of 15 μl 3 x RNA-loading dye.

3.4 Purification of RNA by Gel Electrophoresis

1. Load in vitro transcripts on a 10% denaturing PAA gel and run the gel in 1× TBE.
2. Pre-run: 40 W for 1 h (without sample loaded).
3. Run: 30 W for 3–4 h.
4. Detection of the RNA transcript. In case of radioactively labeled RNA, detect RNA of interest by autoradiography. In case of unlabeled RNA, detect RNA of interest by UV shadowing.
5. Cut out gel slices containing RNA bands. Perform gel extraction in 400 μl crushed soak solution for 1 h at 65 °C and 700 rpm in a thermomixer.
6. Transfer the solution into a new tube. Avoid the transfer of gel slices. Add 800 μl of ice-cold 100% ethanol and perform an ethanol precipitation of the RNA.

3.5 5′-Dephosphorylation and 5′-Labeling of RNA

Dephosphorylation

1. Set up dephosphorylation reaction: 1 μl of 10× antarctic phosphatase buffer, 5 μl of 100 pmol/μl RNA in vitro transcript, 1 μl of 5 U/μl antarctic phosphatase, and 3 μl dH_2O.
2. Incubate the reaction for 30 min at 37 °C.
3. Perform heat inactivation of the phosphatase by incubation at 80 °C for 20 min.

5′-Labeling

4. Set up 5′-labeling reaction: 10 μl of dephosphorylated RNA (50 pmol/μl) (reaction mixture from above), 1 μl of 10 U/μl T4 PNK, 5 μl of 10× T4 PNK buffer, 1 μl of 10 μCi/μl γ-[^{32}P]-ATP, and 33 μl dH_2O.
5. Incubate reaction for 1 h at 37 °C.
6. Purify labeled RNA by gel electrophoresis (*see* Subheading 1.2 **step 3**).

3.6 5′-Labeling of DNA Probes

1. Set up 5′-labeling reaction: 5 μl of 100 pmol/μl DNA, 1 μl of 10 U/μl T4 PNK, 5 μl of 10× T4 PNK buffer, 1 μl of 10μCi/μl γ-[P32]-ATP, and 38 μl dH_2O.
2. Incubate the reaction solution for 1 h at 37 °C.
3. Remove unincorporated γ-[^{32}P]-ATP by NucAway Spin Columns according to the manufacturer.

3.7 RNA Extraction for Northern Blot Analysis

Preparation of the Cells

1. Use one colony per sample (*E. coli* TOP10 cells carrying riboswitch plasmid and reporter plasmid) to generate overnight cultures in 15 ml cultivation tubes with 5 ml cultivation medium and the appropriate antibiotics.

2. For the preparation of the main culture, dilute a pre-culture to an OD_{600} of 0.06 in 2× 5 ml cultivation medium supplemented with 0.01% (w/v) arabinose and optional 2 mM theophylline (final concentration).
3. Cultivate cells at 37 °C and 200 rpm for 3.5 h.

RNA Extraction

4. Combine 2× 5 ml cultures of the same clone type and harvest cell pellet.
5. Resuspend pellet in 500 μl TRIzol™, and transfer the suspension into a FastPrep® tube (2 ml) with 300 mg zirconia beads.
6. Homogenize the suspension for 40 s with 6 m/s using FastPrep® system.
7. Transfer the supernatant without zirconia beads into a new tube.
8. Follow the protocol for RNA extraction using TRIzol™ according to the manufacturer.

3.8 Northern Blot

Gel Run

1. Prepare a 10% PAA gel.
2. Load 10 μg of the extracted total RNA per sample and 20 ng of in vitro transcripts (size marker) with 1× RNA-loading dye, and run the gel in 1× TBE (total volume 9 μl) (*see* **Note 5**).
3. Pre-run: 5 W for 1 h (without sample loaded).
4. Run: 150 V or 3–4 h.

Transfer to the Nylon Membrane

5. Equilibrate two pieces of filter paper and one piece of the nylon membrane (8× 8 cm) for 10 min in 1× TBE buffer.
6. When the gel run is finished, place a wet filter paper into the cassette of the Trans-Blot Turbo Transfer System. Place the membrane on the filter. Place the PAA gel on the membrane. Place the second filter paper on the gel. Remove air bubbles.
7. Perform the transfer for 35 min at 0.1 A.

Crosslinking

8. Perform crosslinking of the membrane at 254 nm UV light (120 mJ/cm^2).

Pre-hybridization

9. Incubate the nylon membrane in 5 ml of 1× pre-hybridization buffer for 14–20 h at 37 °C in a rotating 50 ml Falcon tube.

Hybridization

10. Discard the pre-hybridization buffer.
11. Add 6 ml of 1× hybridization buffer into the Falcon tube.
12. Add 10 μl of 10 pmol/μl of the 5′-labeled DNA probe into the Falcon tube.
13. Incubate for 14–20 h at 42 °C while rotating.

Wash and Detection

14. Discard the hybridization buffer.
15. Wash the membrane for 10 min with 6× SSC and repeat, and then wash with 4× SSC and finally with 2× SSC.
16. Detection of the samples is accomplished by autoradiography.

Quantification

17. Signal intensities of the processed tRNA are quantified through ImageQuant^{TL}. The band intensity was normalized to the appropriate 5S rRNA signal, and final results were calculated relatively to processed tRNA of the positive control (tRNA with single-stranded 5′-leader).

3.9 Kinetic Analysis

Preparation of Substrate/Ligand/Buffer Mixture

1. Prepare substrate in 10 μl of a 2× concentrated reaction mixture (different substrate concentrations from 20 to 300 nM): 3–0.4μl of 2000 nM substrate, 2 μl of 300 μM theophylline stock solution (60 μM), 5μl of 2× RNase P reaction buffer, dH_2O to 10 μl.
2. Incubate the solution for 1 min at 90 °C.
3. Refold the substrate for at least 5 min at room temperature (25 °C).

Preparation of RNase P Holoenzyme Solution

4. Prepare 2× concentrated mixture per reaction (*see* **Note 6**): 0.4 μl of 200 nM C5 protein solution (8 nM), 0.4 μl of 100 nM M1 RNA solution (4 nM), 5 μl 2× RNase P reaction buffer, 4.2 μl dH_2O.
5. Assemble C5 protein and M1 RNA by incubation for 5 min at 37 °C. Start the cleavage reaction immediately after the assembly of both enzyme components to avoid loss of activity.

RNase P Reaction

6. Start reaction by combining 10 μl of the substrate solution with 10 μl of the enzyme solution (final concentrations: 2 nM M1 RNA, 4 nM C5 protein, 10–150 nM substrate).
7. Incubate reaction for 8 min at 37 °C.

8. Stop reaction by the addition of 20 µl of ice-cold RNA-loading dye and store samples on ice or at −20 °C.
9. Analyze cleavage products by gel electrophoresis using a 10% PAA gel.

Quantification of the Cleavage Products and Linear Regression

10. Quantification of the percentual product formation using ImageQuant.
11. Calculation of the kinetic parameters according to Michaelis-Menten using nonlinear regression (e.g., using GraphPad Prism7).

3.10 Structural Analysis by In-Line Probing

1. Perform in-line probing according to Regulski and Breaker [52] with minor deviations below.
2. 5′-labeled RNA constructs were denatured at 90 °C for 1 min followed by 5 min incubation at room temperature. Different samples were prepared afterwards.

Negative Control

3. Set up reaction: 1 µl of 5′-labeled RNA (8000–15,000 counts per minute (cpm)), 19 µl dH_2O.
4. Add directly 20 µl of colorless loading dye and store sample at −20 °C.

In-line Probing Reaction

5. Set up reaction: 1 µl of 5′-labeled RNA (8000–15,000 cpm), 2 µl of 10× in-line probing buffer, 2 µl of 300 µM theophylline (optional). Add dH_2O to 20 µl reaction volume
6. Incubate reaction for 40 h at 25 °C.
7. Add 20 µl of colorless loading dye and store sample at −20 °C.

Partial Hydrolysis Ladder

8. Set up reaction: 1 µl of 5′-labeled RNA (8000–15,000 cpm), 10 µl of $NaCO_3$ buffer (2×), 9 µl dH_2O.
9. Incubate for 2 min at 90 °C.
10. Add 20 µl of colorless loading dye and store sample at −20 °C.

RNase T1 Digestion

11. Set up reaction: 1 µl of 5′-labeled RNA (8,000–15,000 cpm), 2 µl RNase T1 buffer (10×), 1 µl RNase T1 (0.5 U/µl), 16 µl dH_2O.
12. Add 20 µl of colorless loading dye and store sample at −20 °C

Analysis of the Samples by Gel Electrophoresis (see **Note** *7)*

13. Load all samples on a 10% PAA sequencing gel.
14. Pre-run: 2 h at 30 W.
15. Run: 6–8 h at 30 W.
16. Detection of the cleavage pattern by autoradiography.

4 Notes

1. For the published RNase P riboswitch designs [12], the variable sequence was composed of a given spacer sequence – resulting from a previous transcription regulating riboswitch design – and an 8 to 16 nucleotide designable long random sequence. In principle, the spacer sequence could also be designed. Hence, we extended the variable region in this contribution by the length of the spacer. This increases the solution space from 4^8–4^{16} in our previous study to 4^{13}–4^{25} possible sequences depending on the actual length of the variable region.
2. When analyzing the secondary structure of the $tRNA^{Tyr}$ supF sequence with RNAfold, the acceptor-stem and the T arm are correctly predicted. However, the remaining enclosed base pairs do not fit the classical clover-leaf conformation as depicted in Fig. 2.
3. For the expression and purification of the RNase P protein subunit, we recommend to use methods described in [12]
4. Depending on the activity of the tested riboswitch, the fluorescence signal may reach 100% of the single-stranded positive control. We figured out that at this point an appropriate quantification of the tRNA maturation is not possible, as the translation machinery may be oversaturated and thus the GFP signal does not represent the amount of processed suppressor tRNA in a linear way. If the riboswitch is working at capacity, reducing the cultivation time may help to evaluate the riboswitch activity (e.g., to 3.5 h).
5. If needed, adjust the amount of the in vitro transcript depending on the amount of processed tRNA of the in vivo sample.
6. It is important to prepare a fresh solution of RNase P holoenzyme for each experiment and to use the solution immediately after assembly of the two subunits. For one experiment, an RNase P holoenzyme Master Mix can be prepared and used for up to ten reactions. This should be done rapidly to avoid loss of enzyme activity.
7. To estimate the sample position in the gel during electrophoresis, RNA-loading dye (with bromophenol blue and xylene cyanol) without any RNA sample is recommended to be used in an additional lane.

Acknowledgments

We thank Dr. Stefan Hammer for his ongoing support although he left academia. Furthermore, we thank Dr. Manuela Geiß for fruitful discussion on all mathematical aspects especially of the applied scoring. Special thanks to Dr. Christina Weinberg for valuable discussions about the experimental setup and Felix Kühnl for fruitful discussions about the project in general.
This work was supported by the Deutsche Forschungsgemeinschaft (grant numbers MO 634/9-2 and STA 850/15-2).

References

1. Mironov AS, Gusarov I, Rafikov R et al (2002) Sensing small molecules by nascent RNA: a mechanism to control transcription in bacteria. Cell 111(5):747–756
2. Nahvi A, Sudarsan N, Ebert MS et al (2002) Genetic control by a metabolite binding mRNA. Chem Biol 9(9):1043–1049. https://doi.org/10.1016/S1074-5521(02)00224-7
3. Winkler WC, Breaker RR (2003) Genetic control by metabolite-binding riboswitches. Chembiochem 4(10):1024–1032. https://doi.org/10.1002/cbic.200300685
4. Mandal M, Breaker RR (2004) Gene regulation by riboswitches. Nat Rev Mol Cell Biol 5(6):451–463. https://doi.org/10.1038/nrm1403
5. Nudler E, Mironov AS (2004) The riboswitch control of bacterial metabolism. Trends Biochem Sci 29(1):11–17. https://doi.org/10.1016/j.tibs.2003.11.004
6. Winkler WC, Breaker RR (2005) Regulation of bacterial gene expression by riboswitches. Annu Rev Microbiol 59:487–517. https://doi.org/10.1146/annurev.micro.59.030804.121336
7. Berschneider B, Wieland M, Rubini M et al (2009) Small-molecule-dependent regulation of transfer RNA in bacteria. Angew Chem Int Ed Engl 48(41):7564–7567. https://doi.org/10.1002/anie.200900851
8. Groher F, Suess B (2014) Synthetic riboswitches – a tool comes of age. Biochim Biophys Acta 1839 10:964–973. https://doi.org/10.1016/j.bbagrm.2014.05.005
9. Strobel B, Düchs MJ, Blazevic D et al (2020) A small-molecule-responsive riboswitch enables conditional induction of viral vector-mediated gene expression in mice. ACS Synth Biol 9(6):1292–1305. https://doi.org/10.1021/acssynbio.9b00410
10. Suess B, Fink B, Berens C et al (2004) A theophylline responsive riboswitch based on helix slipping controls gene expression in vivo. Nucleic Acids Res 32(4):1610–1614. https://doi.org/10.1093/nar/gkh321
11. Wachsmuth M, Findeiss S, Weissheimer N et al (2013) De novo design of a synthetic riboswitch that regulates transcription termination. Nucleic Acids Res 41(4):2541–2551. https://doi.org/10.1093/nar/gks1330
12. Ender A, Etzel M, Hammer S et al (2021) Ligand-dependent tRNA processing by a rationally designed RNase P riboswitch. Nucleic Acids Res 49(3):1784–1800. https://doi.org/10.1093/nar/gkaa1282
13. Betat H, Long Y, Jackman JE et al (2014) From end to end: tRNA editing at 5′- and 3′-terminal positions. Int J Mol Sci 15(12):23975–23998. https://doi.org/10.3390/ijms151223975
14. Deutscher MP (1984) Processing of tRNA in prokaryotes and eukaryotes. CRC Crit Rev Biochem 17(1):45–71
15. Schedl P, Roberts J, Primakoff P (1976) In vitro processing of E. coli tRNA precursors. Cell 8(4):581–594. https://doi.org/10.1016/0092-8674(76)90226-9
16. Klemm BP, Wu N, Chen Y et al (2016) The diversity of ribonuclease P: protein and RNA catalysts with analogous biological functions. Biomolecules 6(2). https://doi.org/10.3390/biom6020027
17. Robertson HD, Altman S, Smith JD (1972) Purification and properties of a specific Escherichia coli ribonuclease which cleaves a tyrosine transfer ribonucleic acid presursor. J Biol Chem 247(16):5243–5251 https://doi.org/10.1016/S0021-9258(19)44963-6
18. Ender A, Grafl N, Kolberg T et al (2022) Synthetic riboswitches for the analysis of tRNA processing by eukaryotic RNase P enzymes.

RNA 28(4): 551–567. https://doi.org/10.1261/rna.078814.121
19. Henriksen JR, Løkke C, Hammerø M et al (2007) Comparison of RNAi efficiency mediated by tetracycline-responsive H1 and U6 promoter variants in mammalian cell lines. Nucleic Acids Res 35(9):e67. https://doi.org/10.1093/nar/gkm193
20. Lin H-C, Zhao J, Niland CN et al (2016) Analysis of the RNA binding specificity landscape of C5 protein reveals structure and sequence preferences that direct RNase P specificity. Cell Chem Biol 23(10):1271–1281. https://doi.org/10.1016/j.chembiol.2016.09.002
21. Aquino-Jarquin G, Toscano-Garibay JD (2011) RNA aptamer evolution: two decades of SELEction. Int J Mol Sci 12(12):9155–9171. https://doi.org/10.3390/ijms12129155
22. Gold L (2015) SELEX: how It happened and where it will go. J Mol Evol 81(5-6):140–143. https://doi.org/10.1007/s00239-015-9705-9
23. Gold L, Janjic N, Jarvis T et al (2012) Aptamers and the RNA world, past and present. Cold Spring Harb Perspect Biol 4(3): a003582–a003582. https://doi.org/10.1101/cshperspect.a003582
24. Jenison RD, Gill SC, Pardi A et al (1994) High-resolution molecular discrimination by RNA. Science 263(5152):1425–1429
25. Ma AT, Schmidt CM, Golden JW (2014) Regulation of gene expression in diverse cyanobacterial species by using theophylline-responsive riboswitches. Appl Environ Microbiol 80(21):6704–6713. https://doi.org/10.1128/AEM.01697-14
26. Nakahira Y, Ogawa A, Asano H et al (2013) Theophylline-dependent riboswitch as a novel genetic tool for strict regulation of protein expression in Cyanobacterium Synechococcus elongatus PCC 7942. Plant Cell Physiol 54(10):1724–1735. https://doi.org/10.1093/pcp/pct115
27. Pu Q, Zhou S, Huang X et al (2020) Intracellular selection of theophylline-sensitive hammerhead aptazyme. Mol Ther Nucleic Acids 20:400–408. https://doi.org/10.1016/j.omtn.2020.03.001
28. Shanidze N, Lenkeit F, Hartig JS et al (2020) A theophylline-responsive riboswitch regulates expression of nuclear-encoded genes. Plant Physiol 182(1):123–135. https://doi.org/10.1104/pp.19.00625
29. Wrist A, Sun W, Summers RM (2020) The theophylline aptamer: 25 years as an important tool in cellular engineering research. ACS Synth Biol 9(4):682–697. https://doi.org/10.1021/acssynbio.9b00475
30. Zimmermann GR, Jenison RD, Wick CL et al (1997) Interlocking structural motifs mediate molecular discrimination by a theophylline-binding RNA. Nat Struct Biol 4(8):644–649. https://doi.org/10.1038/nsb0897-644
31. Altman S, Baer M, Guerrier-Takada C et al (1982) Transfer RNA processing and gene regulation. In: Gene regulation. Elsevier, pp 207–217
32. Kole R, Baer MF, Stark BC et al (1980) E. coli RNAase P has a required RNA component in vivo. Cell 19(4):881–887. https://doi.org/10.1016/0092-8674(80)90079-3
33. Guerrier-Takada C, Gardiner K, Marsh T et al (1983) The RNA moiety of ribonuclease P is the catalytic subunit of the enzyme. Cell 35(3 Pt 2):849–857
34. Loria A, Pan T (1997) Recognition of the T stem-loop of a pre-tRNA substrate by the ribozyme from Bacillus subtilis ribonuclease P. Biochemistry 36(21):6317–6325. https://doi.org/10.1021/bi970115o
35. Oh BK, Pace NR (1994) Interaction of the 3′-end of tRNA with ribonuclease P RNA. Nucleic Acids Res 22(20):4087–4094
36. Loria A, Pan T (1998) Recognition of the 5′ leader and the acceptor stem of a pre-tRNA substrate by the ribozyme from Bacillus subtilis RNase P. Biochemistry 37(28):10126–10133. https://doi.org/10.1021/bi980220d
37. Niland CN, Anderson DR, Jankowsky E et al (2017) The contribution of the C5 protein subunit of Escherichia coli ribonuclease P to specificity for precursor tRNA is modulated by proximal 5′ leader sequences. RNA 23(10):1502–1511. https://doi.org/10.1261/rna.056408.116
38. Svard SG, Kirsebom LA (1993) Determinants of Escherichia coli RNase P cleavage site selection: a detailed in vitro and in vivo analysis. Nucleic Acids Res 21(3):427–434
39. Davies-Sala C, Soler-Bistué A, Bonomo RA et al (2015) External guide sequence technology: a path to development of novel antimicrobial therapeutics. Ann N Y Acad Sci 1354: 98–110. https://doi.org/10.1111/nyas.12755
40. McClain WH, Guerrier-Takada C, Altman S (1987) Model substrates for an RNA enzyme. Science 238(4826):527–530. https://doi.org/10.1126/science.2443980
41. Koutmou KS, Zahler NH, Kurz JC et al (2010) Protein–precursor tRNA contact leads to sequence-specific recognition of 5′ leaders by

bacterial ribonuclease P. J Mol Biol 396(1):195–208. https://doi.org/10.1016/j.jmb.2009.11.039

42. Niranjanakumari S, Stams T, Crary SM et al (1998) Protein component of the ribozyme ribonuclease P alters substrate recognition by directly contacting precursor tRNA. Proc Natl Acad Sci U S A 95(26):15212–15217

43. Reiter NJ, Osterman AK, Mondragón A (2012) The bacterial ribonuclease P holoenzyme requires specific, conserved residues for efficient catalysis and substrate positioning. Nucleic Acids Res 40(20):10384–10393. https://doi.org/10.1093/nar/gks744

44. Sun L, Campbell FE, Zahler NH et al (2006) Evidence that substrate-specific effects of C5 protein lead to uniformity in binding and catalysis by RNase P. EMBO J 25(17):3998–4007. https://doi.org/10.1038/sj.emboj.7601290

45. Kraemer KH, Seidman MM (1989) Use of supF, an Escherichia coli tyrosine suppressor tRNA gene, as a mutagenic target in shuttle-vector plasmids. Mutat Res/Rev Genet Toxicol 220(2–3):61–72. https://doi.org/10.1016/0165-1110(89)90011-0

46. Hansen A, Pfeiffer T, Zuleeg T et al (2001) Exploring the minimal substrate requirements for trans-cleavage by RNase P holoenzymes from Escherichia coli and Bacillus subtilis. Mol Microbiol 41(1):131–143

47. Lorenz R, Bernhart SH, Zu H, Siederdissen C et al (2011) ViennaRNA package 2.0. Algorithms Mol Biol 6:26. https://doi.org/10.1186/1748-7188-6-26

48. Lorenz R, Hofacker IL, Stadler PF (2016) RNA folding with hard and soft constraints. Algorithms Mol Biol 11:8. https://doi.org/10.1186/s13015-016-0070-z

49. Hammer S, Tschiatschek B, Flamm C et al (2017) RNAblueprint: flexible multiple target nucleic acid sequence design. Bioinformatics 33(18):2850–2858. https://doi.org/10.1093/bioinformatics/btx263

50. Lindell M, Romby P, Wagner EGH (2002) Lead (II) as a probe for investigating RNA structure in vivo. RNA 8(4):534–541. https://doi.org/10.1017/s1355838201020416

51. Kazakov S, Altman S (1991) Site-specific cleavage by metal ion cofactors and inhibitors of M1 RNA, the catalytic subunit of RNase P from Escherichia coli. Proc Natl Acad Sci 88(20):9193–9197. https://doi.org/10.1073/pnas.88.20.9193

52. Regulski EE, Breaker RR (2008) In-line probing analysis of riboswitches. Methods Mol Biol 419:53–67. https://doi.org/10.1007/978-1-59745-033-1_4

Chapter 12

Design, Characterization, and Application of Targeted Gene Activation in Bacteria Using a Modular CRISPRa System

Maria Claudia Villegas Kcam and James Chappell

Abstract

CRISPR-Cas regulators have provided an excellent toolbox to control gene expression due to the versatility of its components and the easy programming of the single guide RNA (sgRNA) to target DNA sequences. Included in this are CRISPR activation (CRISPRa) systems. These systems allow users to activate transcription of a target gene through the localization of transcription activation domains (ADs) near promoter elements, which in turn recruit RNA polymerase (RNAP) to turn on transcription. A variety of different CRISPRa systems have been described that vary in AD type, recruitment strategies, and CRISPR-Cas systems. Recently, a highly modular CRISPRa system was described that allows for facile exchange of ADs and CRISPR-Cas components. This allows for the creation of CRISPRa systems with unique properties, for example, ability to activate from specific positions upstream of a gene of interest. Here, we describe a protocol for designing, characterizing, and applying the modular CRISPRa system for gene activation in *E. coli*. We first focus on how to identify activating sites upstream of promoters and the cloning of the targeting sgRNA. We then describe how to perform a fluorescence experiment to evaluate activation of a single target site. Finally, we explain how to adapt the system to expand the target range and how to characterize the activation pattern obtained from different CRISPRa designs.

Key words CRISPR activation, CRISPR-Cas regulators, Transcription activation

1 Introduction

The discovery of CRISPR-Cas systems has transformed many aspects of biological sciences, biotechnology, and medicine [1–3]. Engineering CRISPR-Cas systems for synthetic control of gene expression has allowed the development of a powerful suite of *trans*-acting gene regulatory tools able to precisely program gene expression in a diversity of cellular contexts. Applications of these tools are diverse, including creating synthetic gene circuits [4–6], uncovering gene function, and inducing valuable phenotypes and cell fates [7–10].

The first CRISPR-Cas regulator developed was the CRISPR interference (CRISPRi) system that targets either the promoter or

James Chappell and Melissa K. Takahashi (eds.), *Riboregulator Design and Analysis*,
Methods in Molecular Biology, vol. 2518, https://doi.org/10.1007/978-1-0716-2421-0_12,

coding region of a gene to block transcription initiation or elongation [11, 12]. The two main elements of this system are a catalytically dead Cas9 (dCas9) protein that can bind the DNA and a single guide RNA (sgRNA) that is easily programmed to determine target specificity. The sgRNA can be codified to target the system to any sequence with the NGG protospacer adjacent motif (PAM) at the 3′ of the target sequence, which is recognized by the dCas9 [13]. The ability to engineer both elements of the system led to further innovations where the CRISPR-dCas9 complex was utilized as a localization platform for protein effector domains, such as transcription activator domains (ADs), to create CRISPR activator (CRISPRa) systems.

Bacterial CRISPRa systems work by localizing an AD at a precise upstream position relative to the promoter of the gene of interest. The first bacterial CRISPRa system was obtained by the direct fusion of the ω subunit of the RNAP to dCas9 via a two alanine linker (*see* Fig. 1a) [11]. An interesting potential advantage of the ω subunit is its ability to work across hosts [14, 15]; however, this system required the deletion of the endogenous omega-encoding gene (*rpoZ*) from the host, requiring additional strain engineering. Following this first-generation system, alternative bacterial CRISPRa designs have been explored harnessing different ADs and recruitment strategies. For example, the transcription factor SoxS has been recruited through a modified sgRNA strategy called scaffold RNAs (scRNAs) (*see* Fig. 1b) [16]; the AsiA_m2.1, derived from the AsiA anti-sigma factor from T4 bacteriophage, has been covalently fused to dCas9 through a flexible linker (*see* Fig. 1c) [17], and the α N-terminal domain (αNTD), derived from the α subunit of the RNAP, has been recruited by dCas9 through non-covalent interactions in the modular CRISPRa system (*see* Fig. 1d) [18].

In this Methods chapter, we focus on the use of the modular CRISPRa system (see Fig. 1d), in which the dCas9 recruits the AD αNTD, through non-covalent SYNZIP interactions [19]. The AD and dCas9 are both fused to SYNZIP domains that form heterodimers when expressed together. To activate a gene (*see* Fig. 1e), the sgRNA is programmed to target a sequence upstream of the promoter where a PAM is present. When the CRISPRa system is bound to its target DNA sequence, the dCas9-SYNZIP forms a heterodimer with the AD-SYNZIP, and the localization of the AD results in the subsequent recruitment of the RNAP near the promoter element to stimulate transcription (*see* Fig. 2).

A major limitation of CRISPRa systems for bacteria has been the low availability and strict distance-dependent requirements for functional target sites. Specifically, it has been reported that independent of the recruitment mechanism and AD type, bacterial CRISPRa systems typically show activation patterns that are periodical, with activation only seen when the system is targeted within

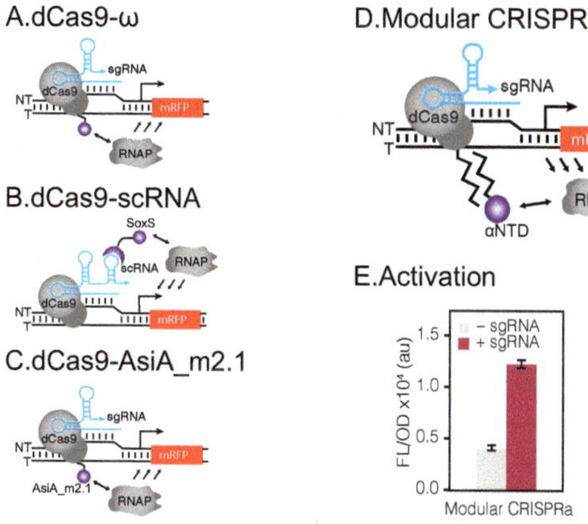

Fig. 1 CRISPRa variants. (**a**) Original CRISPRa design that uses an omega subunit as activation domain (AD). (**b**) CRISPRa design with a scaffold RNA (scRNA) that recruits SoxS. (**c**) CRISPRa system with direct fusion to the Asia_m2.1 AD. (**d**) Modular CRISPRa system that recruits the αNTD AD through non-covalent SYNZIP interactions. (**e**) Fluorescence activation targeting position 80bp upstream of the Transcription start site (TSS) on the template strand with the modular CRISPRa system

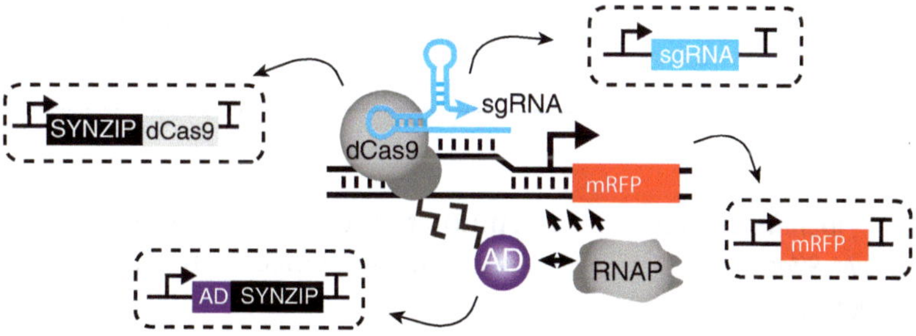

Fig. 2 Plasmid-encoded elements of the modular CRISPRa system. The modular CRISPRa system contains a sgRNA, a dCas9 fused to SYNZIP, and an AD fused to SYNZIP. Each element is expressed on a separate plasmid. A reporter plasmid containing mRFP is used to evaluate activation

a 2–4 bp window that repeats every 10–11 bp upstream of the promoter transcription start site (TSS) [18, 20]. This significantly reduces the versatility of CRISPRa systems, as only genes with correctly positioned PAM sites can be activated. It had been proposed that this periodicity is explained by the position of the AD relative to the promoter in the turn of the DNA double helix, which results in the co-localization of the AD and the promoter in the same helical plane only when the CRISPRa system is targeted to a specific upstream distance. The modular CRISPRa system addresses this limitation by allowing for facile exchange of different engineered dCas9 variants that can activate from distinct positions upstream of the TSS. Specifically, circular permutations of the dCas9 protein (cpdCas9) are used, where the original N and C termini are fused and new termini are created [21]. This allows fusion of the SYNZIP domain to different areas of the dCas9, resulting in the recruitment of the AD to distinct locations, which can uniquely position ADs to create shifted activation patterns.

CRISPR-Cas regulators have the main advantage of not requiring modifications in the target sequence, and its programming is relatively simple [12, 13]. The optimization of CRISPRa systems opens the door for future applications such as metabolic engineering and functional genomics. This includes modifying the expression of endogenous genes to uncover gene function and connectivity, gain-of-function screening, and optimization of metabolic flux.

In this chapter, we will provide a walkthrough of the design, characterization, and application of the modular CRISPRa system, starting from the identification of available PAM sites and measuring their distance to the transcription start site (TSS). We then describe how to obtain and clone the selected sgRNA sequence and the steps to evaluate the activation of an RFP reporter with the modular CRISPRa system. We also describe how to modify the dCas9 with engineered variants to expand the target range. Finally, we show how to characterize the activation pattern of different CRISPRa systems at a single nucleotide level resolution.

2 Materials

2.1 Identification of Target Sites and sgRNA Design

1. Annotated sequence of target gene including the promoter and upstream sequence up to 150 bp.

2.2 sgRNA Cloning

1. Spacer sgRNA sequence to be used.
2. Two oligonucleotides for spacer cloning.
3. NEB buffer 4 (*see* **Note 1**).
4. Golden Gate plasmid for sgRNA cloning (e.g., pJEC625 [Addgene plasmid #174368]).
5. 10x T4 ligase buffer.
6. 2,000,000 units/mL T4 ligase enzyme.
7. BbsI enzyme.
8. Thermocycler.
9. Competent *E. coli* cells.
10. LB media with 100 μg/mL carbenicillin (carb).
11. LB agar plates with 100 μg/mL carb.
12. DNA miniprep kit.
13. Sequencing primer (e.g., 5′ CACGAGGCAGAATTTCAG 3′).

2.3 Gene Activation Experiments

1. LB media with 100 μg/mL carb, 100 μg/mL chloramphenicol (cam), 100 μg/mL spectinomycin (spec), and 25 μg/mL kanamycin (kan).
2. LB agar plates with 100 μg/mL carb, 100 μg/mL cam, 100 μg/mL spec, and 25 μg/mL kan.
3. Deep 96-well plates and breathable sealing membrane.
4. Transparent 96-well plates.
5. 1X PBS solution.
6. Competent MG1655 *E. coli* cells.
7. Shaking well plate incubator.
8. Fluorescence plate reader.
9. Purified DNA plasmids: dCas9-encoding plasmids (e.g., pJEC578 [Addgene plasmid #174366]), sgRNA-encoding plasmids (e.g., pJEC625 [Addgene plasmid #174368]), AD-encoding plasmids (e.g., pJEC626 [Addgene plasmid #174367]), reporter-encoding plasmids (e.g., pJEC581 [Addgene plasmid #174369]), and blank plasmids containing only antibiotic resistance gene (e.g., pJBL001 [Addgene plasmid #71211], an empty backbone containing chloramphenicol resistance cassette).

2.4 Expanding Gene Activation Range

1. cpdCas9-encoding plasmids.

2.5 Evaluating Activation Patterns for CRISPRa Systems

1. Materials for gene activation experiment.
2. Reporter-encoding plasmid library with shifted sgRNA target sites.

3 Methods

In this Methods section, we will describe how to create and characterize CRISPRa systems. As an example, an RFP gene will be used as our target gene. The same design procedure can be applied to other genes.

3.1 Identification of Target Sites and sgRNA Design

To activate a gene in bacteria using a CRISPRa system, it is necessary to target precise distances upstream of the promoter transcription start site (TSS) (*see* **Note 2**). This should be considered additionally to the binding of the CRISPR-Cas system which requires a protospacer adjacent motif (PAM).

1. Identify PAM sites for the CRISPR-Cas system, which are 5′ NGG 3′ for *Streptococcus pyogenes* Cas9, where N represents any nucleotide (*see* Fig. 3a).

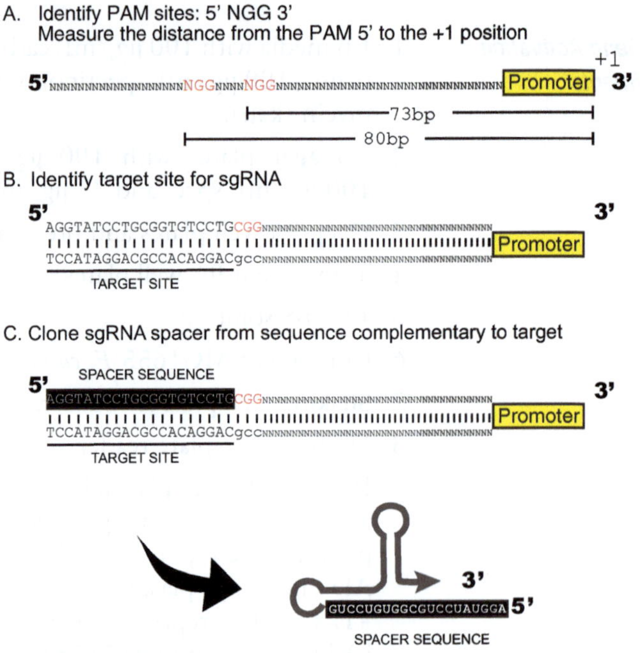

Fig. 3 Design of sgRNAs. (**a**) Identify PAM sequences in the target DNA, and measure the distance to the promoter transcription start site (TSS). (**b**) Identify the best available target site for CRISPRa. (**c**) Identify spacer sequence required for the sgRNA

2. Measure the distance between the PAM site and the promoter, and specifically measure from the 5′ of the NGG to the 3′ end of the promoter (*see* Fig. 3a) (*see* **Note 3**).
3. Choose the site whose distance allows for the highest activation (e.g., 80 base pairs upstream in the template strand [80T] or 81 base pairs upstream in the non-template strand [81NT]) (*see* **Note 2**) (*see* Fig. 3b).
4. Identify the DNA target site that corresponds to the 20 nt sequence upstream of the PAM site on the opposite strand (*see* Fig. 3b).
5. Identify the spacer sequence for the sgRNA that is the 20 nt sequence upstream of the PAM site on the same strand (*see* Fig. 3c).

3.2 sgRNA Cloning

Cloning of the sgRNA spacer into a suitable sgRNA expression plasmid can be achieved through a diversity of cloning methods. We describe cloning into a sgRNA Golden Gate vector pJEC625 (Addgene plasmid #174368) which has been designed to contain cutting sites for the restriction enzyme BbsI (*see* Fig. 4a) where the spacer sequence is inserted through ligation of annealed oligonucleotides (*see* Fig. 4b).

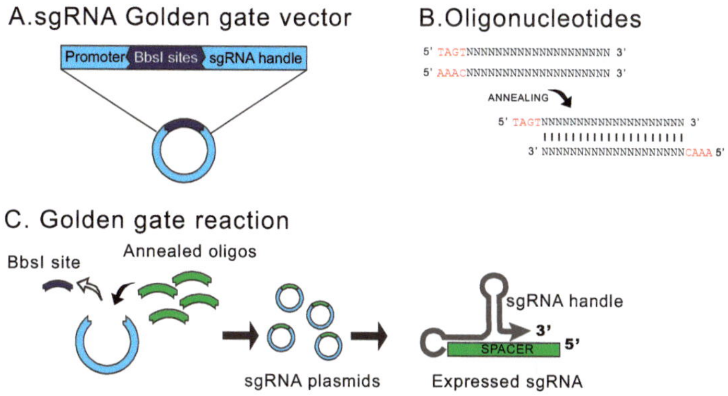

Fig. 4 Cloning of the sgRNA spacer. (**a**) sgRNA Golden Gate vector contains the BbsI sites between the promoter and sgRNA handle to allow for Golden Gate cloning of the spacer site. (**b**) Annealing of oligonucleotides is used to create the spacer region for the sgRNA. (**c**) Golden Gate reaction to insert the annealed oligonucleotides in the Golden Gate vector, obtaining the desired sgRNA plasmid

1. Identify spacer sequence as described in Subheading 1.2.
2. Obtain two oligonucleotides with the following design (*see* Fig. 4b): Oligo1: TAGT followed by the spacer sequence at the 3′. Oligo2: AAAC followed by sequence reverse complementary to the spacer at the 3′.
3. Resuspend the oligonucleotides to 100 μM.
4. To anneal the oligonucleotides, assemble the annealing reaction mix: 1 μL 100 μM Oligo1, 1 μL 100 μM Oligo2, 1 μL 10X NEB buffer 4, and 7 μL of H_2O.
5. Set the reaction mix into a thermocycler protocol: 95–12 °C with a slow ramp of 0.1 °C/s (*see* **Note 4**).
6. Set up the BbsI Golden Gate reaction (Table 1) (*see* **Notes 5 and 6**).
7. To assemble the sgRNA-encoding plasmid (*see* Fig. 4c), run the Golden Gate reaction in a thermocycler following protocol described in Table 2.
8. Transform 5 μL of the Golden Gate reaction into *E. coli* competent cells (*see* **Note 7**) and plate in LB agar with 100 μg/mL carb.
9. Incubate overnight at 37 °C.
10. Pick a single colony and inoculate in culture tubes containing LB media with 100 μg/mL carb. Grow cells at 37 °C with constant shaking at 250 rpm.
11. Perform DNA plasmid isolation using a miniprep kit according to the manufacturer's protocol.

Table 1
BbsI Golden Gate reaction

Reagent	Volume
Golden gate vector (40ng/μL)	1 μL
Annealed oligos (40fm)	0.5 μL
10X T4 ligase buffer	1μL
BbsI enzyme	0.75 μL
T4 ligase enzyme (2000000 u)	0.25 μL
Water	6.5 μL

Table 2
Golden Gate thermocycler protocol

Step	
1	37 °C, 2 min
2	16 °C, 5 min
3	GOTO 1, 24X
4	37 °C, 10 min
5	80 °C, 20 min
6	12 °C, forever

12. Perform Sanger sequencing on isolated DNA plasmid to confirm correct DNA assembly using the sequencing primer that binds upstream of the sgRNA promoter to sequence all the genes.

3.3 Gene Activation Experiments

Measuring gene activation by RFP fluorescence measurements requires all the elements of the CRISPRa system: dCas9-SYNZIP, sgRNA, AD-SYNZIP, and the RFP reporter containing the upstream target sequence. Include a control with all blank plasmids (plasmid that only contains antibiotic resistance genes) (*see* **Note 8**), and include a no sgRNA control for each reporter.

1. Co-transform into competent MG1655 *E. coli* cells the plasmids encoding the RFP reporter, the dCas9-SYNZIP, the AD-SYNZIP, and the sgRNA (*see* **Note 9**).
2. Plate transformed cells onto LB agar plates with 100 μg/mL carb, 100 μg/mL cam, 100 μg/mL spec, and 25 μg/mL kan, and grow overnight at 37 °C.

3. Pick four colonies from each transformation into a deep 96-well plate containing 300 μL of LB media with appropriate antibiotics. Include four wells with only media (i.e., no cells). Seal with a breathable sealing membrane.
4. Grow in a shaking well plate incubator at 1000 rpm at 37 °C.
5. After overnight growth, remove 4 μL of each well and inoculate a deep 96-well plate containing 196 μL of fresh LB media with the appropriate antibiotics. Include four wells with only media (i.e., no cells). Seal with a breathable sealing membrane.
6. Grow in a shaking well plate incubator at 1000 rpm at 37 °C.
7. After 8 h, remove 50 μL of culture from each well of the deep 96-well plate and add to a transparent 96-well plate. Then, add to each well 50 μL of PBS.
8. Measure the fluorescence (FL) (540 nm excitation and 600 nm emission) and optical density at 600 nm (OD) in the plate reader.
9. Calculate normalized FL/OD values. First, subtract the average media FL and OD values from each culture's FL and OD values. Second, divide FL values by corresponding OD for each culture. Normalize the FL/OD for each culture by subtracting FL/OD from cells transformed with blank plasmids.

3.4 Expanding Gene Activation Range

The modular CRISPRa system allows for easy exchange of each element to optimize the overall characteristics for the desired application (*see* **Note 10**). Particularly, the target range can be expanded using circularly permuted dCas9 (cpdCas9) with the SYNZIP domain fused to a different area in the dCas9 structure. This yields CRISPRa systems with distinct and shifted activation patterns compared to wild-type dCas9. Depending on the positions of the available PAM sites in the reporter plasmid, different cpdCas9 variants are optimal. The following steps can be used to determine which cpdCas9 variant is optimal.

1. Identify potential target sites.
2. Measure distance between PAM and promoter as described in Subheading 1.1.
3. Select dCas9-SYNZIP variants capable of activating transcription with CRISPRa from the available target sites using a previously described library [1].

3.5 Evaluating Activation Patterns for CRISPRa Systems

The characterization of activation patterns for a CRISPRa system provides information about periodicity and optimal target sites within a given reporter. For this, we suggest using a strategy described by Fontana et al. 2020 [2] where the initial reporter plasmid contains sgRNA target sites every 10 bp. Then, another nine reporters are cloned where each one has an additional

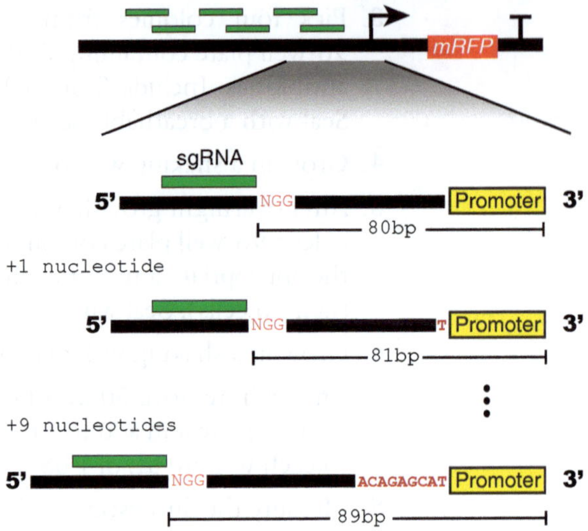

Fig. 5 Reporter strategy to characterize activation pattern at 1 nt resolution. Initial RFP reporter contains multiple sgRNA binding sites (green) spaced every 10 nt. To shift the sgRNA binding positions, additional nucleotides (+1 to +9 nucleotides) are inserted between the promoter and upstream sequence, creating nine additional reporters

nucleotide between the promoter and its upstream region, shifting the distance of the sgRNA target sites relative to the promoter TSS (*see* Fig. 5). This allows the same sgRNA to target up to ten different distances relative to the promoter and cover the target range at 1 nucleotide resolution. The following steps overview how to perform these experiments:

1. Identify the desired target range to obtain the activation pattern.
2. Clone the sgRNAs needed to target the specified area.
3. Perform the fluorescence experiment as indicated in Subheading 1.4. The FL/OD calculated for each sgRNA targeting the corresponding reporter (*see* Fig. 5) is shown next to the corresponding no sgRNA control (*see* Fig. 6a).
4. Calculate fold activation (*see* Fig. 6b). Each FL value (activating or non-activating) is divided by the no sgRNA control to obtain the fold activation. To calculate the propagated error, we use the following formula:

$$\text{Propagated error} = \text{fold activation} * \sqrt{\left(\frac{\text{Stdv}_{\text{no sgRNA}}}{\text{FL}_{\text{no sgRNA}}}\right)^2 + \left(\frac{\text{Stdv}_{\text{sgRNA}}}{\text{FL}_{\text{sgRNA}}}\right)^2} \quad (1)$$

where FL_x is the normalized FL/OD value and Stdv_x is the standard deviation calculated for each condition.

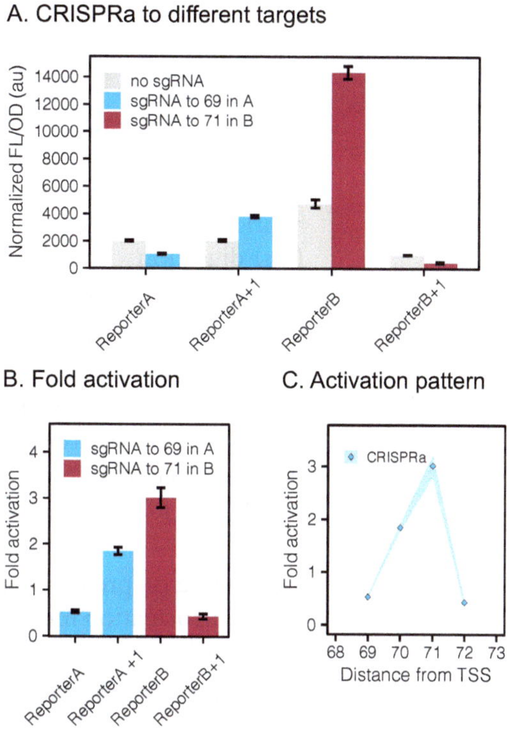

Fig. 6 Characterizing CRISPRa activation patterns. (**a**) Measured FL/OD obtained targeting four reporters. Each reporter allows a sgRNA to be targeted at specific distances to the TSS. (**b**) Calculation of the fold activation dividing the FL/OD from the sgRNA by the no sgRNA control for each condition. (**c**) Activation pattern with fold activation displayed as a function of the targeting distance to the TSS

5. Plot the activation pattern (*see* Fig. 6c). Fold activation values (y axis) are plotted against the distance to the promoter's TSS (y axis). This allows activation patterns to be displayed at 1 nt resolution.

4 Notes

1. NEB buffer 4 is used to perform the annealing reaction. This can be replaced by annealing buffer made in lab with its components. An example annealing buffer is 10 mM Tris-HCL (pH 7.5–8.0), 50 mM NaCl, and 1 mM EDTA.
2. Bacterial CRISPRa systems typically show activation patterns that are periodical, with activation only seen when the system is targeted within a 2–4 bp window that repeats every 10–11 bp upstream of the promoter transcription start site (TSS). Previous studies can be used to guide identification of appropriate target sites [18, 20].

3. To measure the distance of the target site to the TSS, we count from the 5′ of the PAM site to the last nucleotide at the 3′ of the promoter.
4. The annealing reaction results in dsDNA that have four nucleotides of overhang in the 5′ of each strand. These overhangs are used for the ligation in the Golden Gate reaction as they will hybridize with the complementary sequence in the vector after BbsI digestion.
5. The annealed oligonucleotides are diluted 1/20 before the Golden Gate reaction to have them at an approximate concentration of 40 fm. 0.5 μL of each dilution is used in the reaction.
6. The sgRNA Golden Gate vector is a 2.8 kb plasmid that contains BbsI sites between the promoter and the sgRNA handle. For the Golden Gate reaction, use 20 fm of each DNA part, which is 1 μL for our vector if the DNA stock is diluted to 40 ng/μL.
7. Depending on preference, different competent *E. coli* cell types and transformation methods can be used (e.g., electroporation, heat shock).
8. The purpose of the blank plasmids is to allow the autofluorescence of *E. coli* cells to be determined under the same growth conditions (i.e., media) as the experimental conditions.
9. Co-transforming four plasmids simultaneously can have reduced efficiencies and can be complex if multiple reporters and sgRNAs are being used to characterize the activation pattern. We suggest to co-transform the reporter, the dCas9-SYNZIP, and the AD-SYNZIP plasmids first and then transform the sgRNAs. To do this, perform the first transformation and plate the cells on LB agar with the three required antibiotics. The next day pick colonies, prepare competent cells, and transform the sgRNA.
10. Alternative dCas9 elements are not just limited to cpdCas9. For example, to work with an inducible system, the constitutive dCas9 can be replaced by an arabinose inducible dCas9, which has no CRISPRa activity in the absence of the arabinose inducer.

Acknowledgments

Thanks goes to Welch Foundation [C-1982–20190330 to J.C.] and Alfred P. Sloan Research Fellowship [FG-2018-10500 to J.C.]; J.C. is an Alfred P. Sloan Research Fellow.

References

1. Adli M (2018) The CRISPR tool kit for genome editing and beyond. Nat Commun 9: 1911
2. Ameruoso A, Gambill L, Liu B, Villegas Kcam MC, Chappell J (2019) Brave new 'RNA' world—advances in RNA tools and their application for understanding and engineering biological systems. Curr Opin Syst Biol 14: 32–40
3. Xu X, Qi LS (2019) A CRISPR–dCas toolbox for genetic engineering and synthetic biology. J Mol Biol 431:34–47
4. Kiani S, Beal J, Ebrahimkhani MR, Huh J, Hall RN, Xie Z, Li Y, Weiss R (2014) CRISPR transcriptional repression devices and layered circuits in mammalian cells. Nat Methods 11: 723–726
5. Nielsen AA, Voigt CA (2014) Multi-input CRISPR/Cas genetic circuits that interface host regulatory networks. Mol Syst Biol 10: 763
6. Kim H, Bojar D, Fussenegger M (2019) A CRISPR/Cas9-based central processing unit to program complex logic computation in human cells. Proc Natl Acad Sci USA 116: 7214–7219
7. Kampmann M (2018) CRISPRi and CRISPRa screens in mammalian cells for precision biology and medicine. ACS Chem Biol 13: 406–416
8. Boettcher M, Tian R, Blau JA, Markegard E, Wagner RT, Wu D, Mo X, Biton A, Zaitlen N, Fu H et al (2018) Dual gene activation and knockout screen reveals directional dependencies in genetic networks. Nat Biotechnol 36: 170–178
9. Peters JM, Colavin A, Shi H, Czarny TL, Larson MH, Wong S, Hawkins JS, Lu CHS, Koo B-M, Marta E et al (2016) A comprehensive, CRISPR-based functional analysis of essential genes in bacteria. Cell 165:1493–1506
10. Wang T, Guan C, Guo J, Liu B, Wu Y, Xie Z, Zhang C, Xing X-H (2018) Pooled CRISPR interference screening enables genome-scale functional genomics study in bacteria with superior performance. Nat Commun 9:2475
11. Bikard D, Jiang W, Samai P, Hochschild A, Zhang F, Marraffini LA (2013) Programmable repression and activation of bacterial gene expression using an engineered CRISPR-Cas system. Nucleic Acids Res 41:7429–7437
12. Qi LS, Larson MH, Gilbert LA, Doudna JA, Weissman JS, Arkin AP, Lim WA (2013) Repurposing CRISPR as an RNA-guided platform for sequence-specific control of gene expression. Cell 152:1173–1183
13. Jinek M, Chylinski K, Fonfara I, Hauer M, Doudna JA, Charpentier E (2012) A programmable dual-RNA–guided DNA endonuclease in adaptive bacterial immunity. Science 337: 816–821
14. Kurkela J, Fredman J, Salminen TA, Tyystjärvi T (2020) Revealing secrets of the enigmatic omega subunit of bacterial RNA polymerase. Mol Microbiol. https://doi.org/10.1111/mmi.14603
15. Lu Z, Yang S, Yuan X, Shi Y, Ouyang L, Jiang S, Yi L, Zhang G (2019) CRISPR-assisted multi-dimensional regulation for fine-tuning gene expression in Bacillus subtilis. Nucleic Acids Res 47:e40–e40
16. Dong C, Fontana J, Patel A, Carothers JM, Zalatan JG (2018) Synthetic CRISPR-Cas gene activators for transcriptional reprogramming in bacteria. Nat Commun 9:2489
17. Ho H, Fang JR, Cheung J, Wang HH (2020) Programmable CRISPR-Cas transcriptional activation in bacteria. Mol Syst Biol 16:e9427
18. Villegas Kcam MC, Tsong AJ, Chappell J (2021) Rational engineering of a modular bacterial CRISPR–Cas activation platform with expanded target range. Nucleic Acids Res 49: 4793–4802
19. Thompson KE, Bashor CJ, Lim WA, Keating AE (2012) SYNZIP protein interaction toolbox: *in Vitro* and *in Vivo* specifications of heterospecific coiled-coil interaction domains. ACS Synth Biol 1:118–129
20. Fontana J, Dong C, Kiattisewee C, Chavali VP, Tickman BI, Carothers JM, Zalatan JG (2020) Effective CRISPRa-mediated control of gene expression in bacteria must overcome strict target site requirements. Nat Commun 11:1618
21. Oakes BL, Fellmann C, Rishi H, Taylor KL, Ren SM, Nadler DC, Yokoo R, Arkin AP, Doudna JA, Savage DF (2019) CRISPR-Cas9 circular permutants as programmable scaffolds for genome modification. Cell 176: 254–267.e16

Chapter 13

Reprogramming TracrRNAs for Multiplexed RNA Detection

Chunlei Jiao and Chase L. Beisel

Abstract

CRISPR-based detection and recording technologies are gaining increasing attention in disease surveillance and prevention. In this chapter, we describe how our recent discovery of noncanonical crRNAs inspired the engineering of reprogrammed tracrRNAs and led to a powerful platform for multiplexed RNA detection. We provide detailed protocols regarding how to design reprogrammed tracrRNA and carry out assays in vitro and in vivo.

Key words CRISPR, Noncanonical crRNA, Reprogrammed tracrRNA, RNA detection

1 Introduction

The discovery of CRISPR-Cas immune systems not only revolutionized the genome editing field but also offered various tools to tackle challenges in disease control and prevention, especially amid the ongoing COVID-19 pandemic [1–3]. CRISPR-based diagnostics (CRISPR-Dx) are gaining more attention as faster, cheaper, and more portable technologies compared to the traditional gold standard quantitative PCR (qPCR). The highest profile CRISPR-Dx uses the collateral effect of Cas12 and Cas13 nucleases. As part of a diagnostic assay, Cas12 or Cas13 recognizes and cuts a double-stranded DNA substrate or single-stranded RNA, respectively, present in a sample. Each nuclease then proceeds to nonspecifically cleave single-stranded DNA or RNA reporter to release a detectable signal [4–6]. The nonspecific cleavage is catalytic, resulting in many more cleaved reporters than detected targets that improves sensitivity and the sampling-to-readout time. However, this nonspecific readout poses an obstacle for multiplexed detection [7, 8].

Beyond in vitro diagnostics, CRISPR holds potential to detect and record cellular behaviors, states, and responses to environmental stimuli. Although current omics-based technologies provide valuable insights about complex molecular events and profiles,

James Chappell and Melissa K. Takahashi (eds.), *Riboregulator Design and Analysis*,
Methods in Molecular Biology, vol. 2518, https://doi.org/10.1007/978-1-0716-2421-0_13,

they cannot provide dynamic cell profiles due to using methods that destroy the cell (e.g., disrupting cells to collect nucleic acids for sequencing), and they are not suitable for direct measurement in many environments (e.g., gut or brain) [9, 10]. CRISPR-based memory technologies in a cell-based way are providing promising solutions for the above shortcomings. Two main technical routes have been developed, with one depending on the DNA targeting activity of effector protein Cas9 [11–13] and the other depending on spacer integration of acquisition complex Cas1-Cas2 [14–16]. As an example of the first technical route, CAMERA (CRISPR-mediated analog multi-event recording apparatus) exploits base editors and Cas9 nucleases to record cellular events [11]. Environmental stimuli-responsive promoters were used to drive sgRNA expression, which guided the Cas9 base editor to the matching DNA target and then created nucleotide changes in the target. CAMERA could achieve analog recording of the stimulus amplitude and duration in an order-dependent manner. However, leaky expression of the stimuli-responsive promoters resulted in background recording in the absence of the stimulus [11]. As an example of the second technical route, SENECA (selective amplification of expanded CRISPR arrays) employs the CRISPR-acquisition machinery of reverse transcriptase (RT)-Cas1-Cas2 to randomly capture snippets of cellular transcripts, reverse-transcribe them into DNA fragments, and integrate them into the CRISPR array. Its nonspecific RNA capturing enables recording complex cellular behaviors on a transcriptome scale. However, the technique requires at least millions of cells to faithfully record an expression profile due to the low spacer integration efficiency, and the ends of AT-rich transcripts are preferentially recorded as RNA-derived spacers [14]. In addition, recording transcripts with a single-nucleotide polymorphism (SNP) is not achievable for current recording systems unless they capture transcripts directly.

Our recent discovery of noncanonical crRNAs inspired us to create unique paradigms for RNA detection in different contexts [17]. When interrogating Cas9-RNA interactions in *Campylobacter jejuni,* we found that the tracrRNA [18] can bind to some cellular RNAs and convert them into noncanonical crRNAs (ncrRNAs), which guide DNA targeting by Cas9. The discovery inspired us to employ a reprogrammed tracrRNA (Rptr) to sense the presence of any RNA-of-interest. Based on the sensing capability of Rptrs, we established a multiplexed RNA diagnostic platform called LEOPARD (Leveraging Engineered tracrRNAs and On-target DNAs for PArallel RNA Detection) [17]. With LEOPARD, we achieved multiplexed detection of RNAs with single-nucleotide specificity from different viruses including SARS-CoV-2 and other respiratory viruses in one reaction. We also ported LEOPARD into cells for the detection and recording of selected RNAs.

The aim of this chapter is to provide a detailed protocol for Rptr-based RNA detection. Using the *C. jejuni* gene CJ8421_04975 as a model, we describe how to design Rptrs and the associated target DNA, how to carry out in vitro RNA transcription and a DNA cleavage assay, and finally how to record transcription of this gene in vivo. These approaches should be broadly applicable to multiplexed in vitro diagnostics and the recording of cellular and environmental states in vivo.

2 Materials

2.1 Reagents and Consumables

1. DNA Clean and Concentration Kit
2. Plasmid Miniprep Kit
3. RNA Clean and Concentration Kit
4. 2% agarose gel
5. LB medium
6. SOC medium
7. Antibiotics: ampicillin (Amp), chloramphenicol (Cm), and kanamycin (Kan)
8. L-arabinose
9. IPTG
10. Q5® Site-Directed Mutagenesis Kit
11. Q5® High-Fidelity 2X Master Mix
12. Onetaq PCR 2X Master Mix
13. Electroporation cuvette
14. SpyCas9 nuclease
15. 1X diluent buffer: 300 mM NaCl, 10 mM Tris-HCl, pH 7.4, 1 mM DTT, 0.1 mM EDTA, 500 μg/ml BSA, 50% glycerol, pH 7.4 at 25 °C
16. Proteinase K
17. 6X DNA loading dye
18. Midori green advance
19. 10% (w/v) glycerol
20. Cas9 base editor (e.g., Sth1Cas9 cytidine base editor) plasmid, Rptr plasmid, and recording plasmid for transcriptional recording
21. *E. coli* strain TOP10 for cloning
22. *E. coli* strain for transcriptional recording (e.g., CB414 from the Beisel lab)

2.2 Equipment

1. Thermocycler
2. Spectrophotometer
3. Electroporator
4. Gel imaging system
5. Electrophoresis system

3 Methods

3.1 Design of Sensed RNA, Rptr, and Targeted Plasmid for In Vitro RNA Detection

To demonstrate RNA detection with LEOPARD, an in vitro DNA cleavage assay is performed using SpyCas9 protein, a sensed RNA, its corresponding Rptr, and linear DNA target (*see* Fig. 1). A T7 promoter is used to drive transcription for the RNA-of-interest and its corresponding Rptr (*see* **Note 1**). The T7-transcribed sensed RNA and its Rptr are annealed before being bound by SpyCas9 to form a ribonucleoprotein complex, which subsequently cut its complementary DNA sequence. The resulting products are then visualized on an agarose gel to indicate the presence of different RNAs-of-interest (*see* Figs. 2, 3, and 4; **Note 2**).

Rptrs are designed as previously described with minor modifications [17]. The anti-repeat region of the native tracrRNA is replaced with the sequences complementary to the sensed RNAs by site-directed mutagenesis, and extra nts is introduced into the anti-repeat region to form a bulge, which is necessary to maintain the activity of SpyCas9 [19]. The regions in sensed RNAs bound by Rptrs are called Rptr targets. The total length of the Rptr target is 38 nts for SpyCas9. The 20-nt sequences upstream of these Rptr target regions are treated as guides (*see* Fig. 2).

1. For the sensed RNA, take the synthetic RNA fragment from CJ8421_04975 as an example. The forward primer containing the T7 promoter (CJpr0711) and the reverse primer

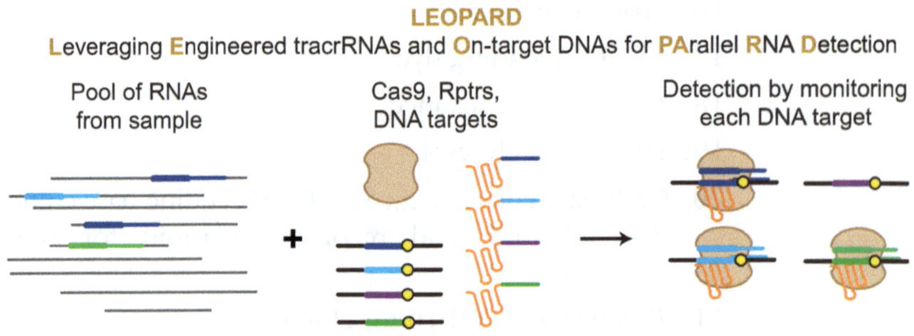

Fig. 1 Overview of the multiplexed diagnostic platform LEOPARD. Rptrs hybridize with pooled RNA samples and convert them into ncrRNAs, which guide Cas9 to their DNA targets. The binding or cleaving of matching DNA targets indicates the presence of RNAs-of-interest. The unique sequences of monitored DNA targets enable multiplexed RNA detection in one reaction. (Adapted from Ref. [17]. Copyright CC BY 4.0)

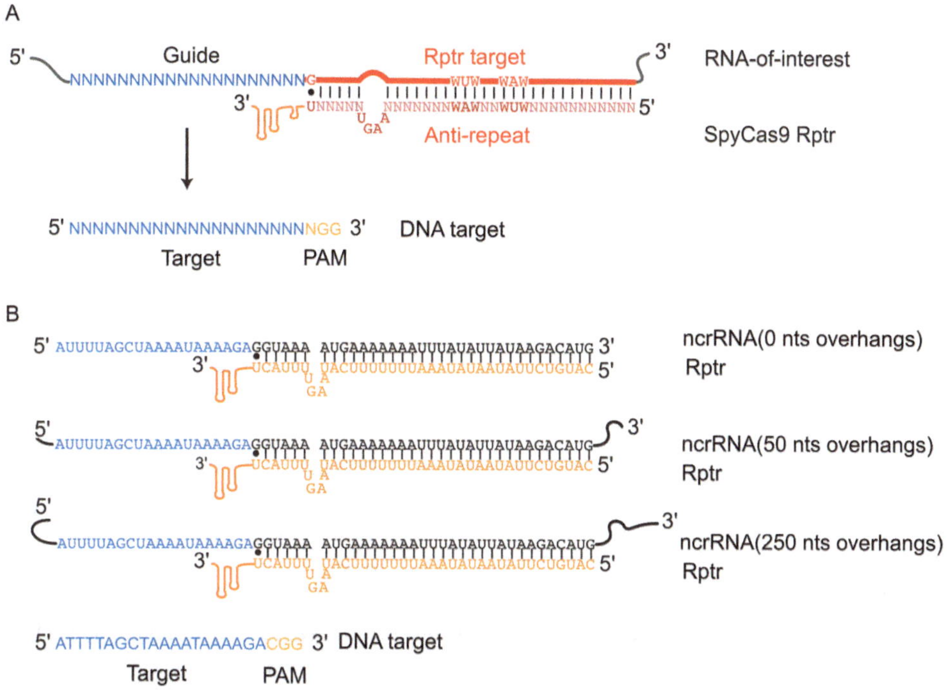

Fig. 2 ncrRNA/Rptr design for the CJ8421_04975 mRNA. (**a**) Design scheme for Rptr used by SpyCas9. W stands for A or U. When designing a Rptr, the AU-rich sequence in the sensed mRNA is selected as the Rptr target due to the preference of RNase III to cleave double-stranded RNA in AU-rich region [27]. The native anti-repeat sequence of SpyCas9 tracrRNA is replaced by the reverse complementary sequence of the Rptr target. AAGU is inserted between the sixth and seventh nucleotide counting from the 3′ end of the new anti-repeat. The 20-nt sequence in blue upstream of the Rptr target is treated as the guide sequence. The same sequence along with the SpyCas9 consensus PAM sequence (5′-NGG-3′) in yellow is used as the DNA target. (**b**) ncrRNA with 50-nt or 250-nt overhangs at both ends for CJ8421_04975 mRNA, its Rptr and DNA target

(CJpr0712) are used to amplify the DNA fragment encoding the RNA-of-interest using the *C. jejuni* CG8421 genome DNA as a template (*see* **Note 3**). See the sequence via the Benchling link in Table 1.

2. For Rptr design, open the sensed RNA file in Benchling, copy the reverse complementary strand of the 38-bp region hybridizing with the Rptr, replace the anti-repeat sequence of native tracrRNA using the copied sequence, and insert AAGU between the sixth and seventh nucleotide counting from 3′-end of the anti-repeat region (*see* Table 1, **Note 4**).
3. For the target DNA design, copy the 20-bp sequence upstream of the 38-bp Rptr target sequence in the sensed RNA, and paste the 20-bp sequence along with a SpyCas9 PAM (5′-CGG-3′) somewhere in the targeted plasmid (*see* Table 1, **Note 5**).

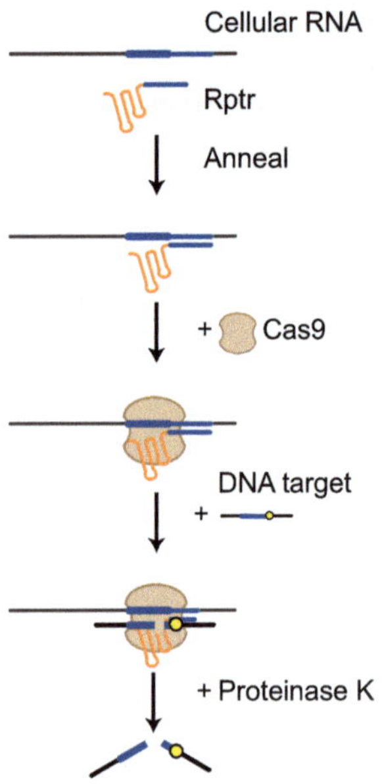

Fig. 3 Workflow for the in vitro DNA cleavage assay. Purified sensed RNA and Rptr are annealed in a thermocycler before assembling with Cas9 to form the Ribonucleoprotein (RNP) complex. The resulting complex cuts the input DNA targets. Proteinase K is used to digest the Cas9 to release cleaved products before visualization. (Adapted from Ref. [17]. Copyright CC BY 4.0)

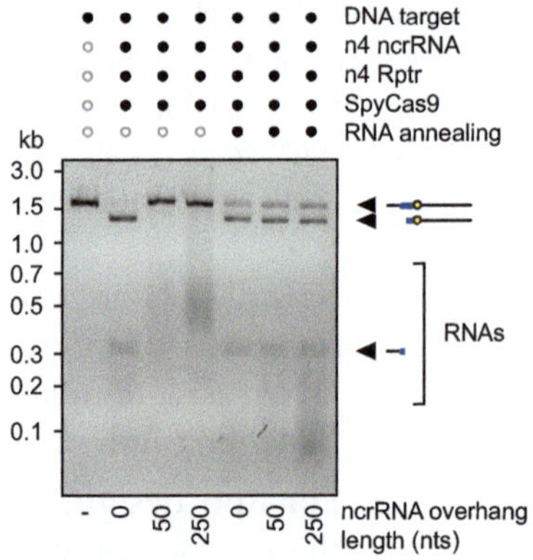

Fig. 4 Impact of the annealing step and overhangs on ncrRNA activity. Different lengths of overhangs (0, 50, and 250 nts) at both ends of n4 ncrRNA of CJ8421_04975 were used during the test. Without an annealing step, overhangs abolish ncrRNA activity, while the activity is recovered with an annealing step. (From Ref. [17]. Copyright CC BY 4.0)

4. To obtain the linear DNA target fragment, PCR-amplify with pCJ211 as the template and the primer pair of CJpr0713 and CJpr0255 (*see* Tables 1 and 2).

Table 1
List of plasmids and linear DNAs used in figures

Name	Public ID	Internal ID	Benchling link	Source	Comments
Figs. 2 and 4					
DNA encoding CJ8421_04975 n4 Rptr	NA	CJfr0302	https://benchling.com/s/seq-8gPBqcHzef5MVGTKZVVA	[17]	PCR extension of CJpr0703 and CJpr0704
DNA encoding CJ8421_04975 n4 ncrRNA (0-nt overhangs)	NA	CJfr0304	https://benchling.com/s/seq-HUwcNVZoCbiGAJDQIH5N	[17]	oligo annealing of CJpr0733 and CJpr0734
DNA encoding CJ8421_04975 n4 ncrRNA (50-nt overhangs)	NA	CJfr0348	https://benchling.com/s/seq-dLAOZDI2DwAEPqYg8C8J	[17]	PCR-amplified using template of *C. jejuni* CG8421 genome DNA and primer pair of CJpr0751 and CJpr0752
DNA encoding CJ8421_04975 n4 ncrRNA (250-nt overhangs)	NA	CJfr0306	https://benchling.com/s/seq-0SoMqw8Wdk3QdWgnPzmD	[17]	PCR-amplified using template of *C. jejuni* CG8421 genome DNA and primer pair of CJpr0711 and CJpr0712
DNA target	NA	CJfr0321	https://benchling.com/s/seq-YdHNSsSuMOxZlTqPTkDk	[17]	PCR-amplified using template of pCJ211 and primer pair of CJpr0713 and CJpr0255
n4 targeted plasmid	pCBS1182	pCJ211	https://benchling.com/s/seq-zG0BbNJ2wf3lc1NCSq7U	[17]	
Figs. 7 and 8					
tracrRNA-NT crRNA	pCBS1844	pCJ378	https://benchling.com/s/seq-IYbbjRy3zKNT1Aqdm1rW	[17]	

(continued)

Table 1
(continued)

Name	Public ID	Internal ID	Benchling link	Source	Comments
tracrRNA(scr)-mRNA	pCBS1832	pCJ375	https://benchling.com/s/seq-6S51WH1r9aYEcdhyQM50	[17]	
CJ8421_04975 n2 recording plasmid	pCBS1252	pCJ240	https://benchling.com/s/seq-X76TLbnldcyhqePOw7Be	[17]	
CJ8421_04975 n2 Rptr-mRNA	pCBS1247	pCJ235	https://benchling.com/s/seq-Ft3EWcfXoBnhiMrRZoYJ	[17]	
CJ8421_04975 n2 tracrRNA-crRNA	pCBS1827	pCJ370	https://benchling.com/s/seq-dnAX8hBg6SRSN6zRqCVT	[17]	
			https://benchling.com/s/seq-eUC2IxwAVeVdaJJzJYdn		
Sth1nCas9 base editor	pCBS2149	pCJ452	https://benchling.com/s/seq-WcpoY9RG2AvCfzX7wZGF		Sth1Cas9 D9A nickase-rAPO43_1 cytosine base editor, used for all Rptr-based recording experiments

Table 2
Reaction setup and thermocycling conditions for high-fidelity PCR

Components	50 μl reaction	Final concentration
Q5 High-Fidelity 2X Master Mix	25 μl	1X
10 μM forward primer	5 μl	0.5 μM
10 μM reverse primer	5 μl	0.5 μM
Template DNA	Variable	<1000 ng
Nuclease-free water	to 50 μl	
Step	**Temperature**	**Time**
Initial denaturation	98 °C	30 s
25–35 cycles	98 °C *50–72 °C 72 °C	10 s 15 s 2–3 kb/min
Final extension	72 °C	2 min
Hold	4 °C	

*It is recommended to use NEB Tm Calculator to calculate the Tm

Table 3
Reaction setup for T7 in vitro RNA transcription

Components	Volume	Final concentration
10X reaction buffer	2 μl	1X
ATP (100 mM)	2 μl	10 mM
GTP (100 mM)	2 μl	10 mM
UTP (100 mM)	2 μl	10 mM
CTP (100 mM)	2 μl	10 mM
Template DNA	2 μl	100 ng total
T7 RNA polymerase mix	2 μl	
Nuclease-free water	6 μl	
Total	20 μl	

3.2 In Vitro RNA Transcription

1. Assemble the reaction listed in Table 3.
2. Vortex the mix thoroughly, and pulse-spin in a microcentrifuge.
3. Incubate at 37 °C for 2 h in a thermocycler (*see* **Note 6**).

Table 4
Reaction setup for in vitro DNA cleavage assay

Components	Volume	Final concentration
Nuclease-free water	9 μl	
10X NEBuffer 3.1	1.5 μl	1X
Rptr (3000 nM)	0.5 μl	100 nM
ncrRNA (3000 nM)	0.5 μl	100 nM
Annealing at 95 °C for 2min and then cooling to 4 °C over 30 min		
SpyCas9 (1 μM)	1.5 μl	100 nM
Incubate at 25 °C for 25 min		
Target DNA (100 nM)	1.5 μl	10 nM
Incubate at 37 °C for 1 h		
Proteinase K (800 U/ml)	0.5 μl	26.67 units/ml
Incubate at room temperature for 15 min		
Total reaction volume	15 μl	

4. Purify transcribed RNA using a RNA Clean and Concentration Kit with residual DNA removed by on-Column DNase I treatment (*see* **Note** 7).

3.3 In Vitro DNA Cleavage Assay

1. Prepare 1 μM SpyCas9 by diluting the 20 μM stock with the 1X diluent buffer. Place the solution on ice for later use.
2. Assemble the following reactions listed in Table 4 (*see* Fig. 3).
3. Anneal Rptrs and sensed RNAs by heating up to 95 °C for 2 min and gradually cooling to 4 °C over 30 min in a thermocycler (*see* **Note 8**).
4. Add 1.5 μl of 1 μM SpyCas9 to the mixture. Mix evenly, and pulse-spin in a microcentrifuge. Incubate at 25 °C for 25 min in a thermocycler.
5. Add 1.5 μl of 100 nM target DNA to the mixture. Mix evenly, and pulse-spin in a microcentrifuge. Incubate at 37 °C for 1 h (*see* **Note 9**).
6. Add 0.5 μl of proteinase K (800 U/ml). Mix evenly, and pulse-spin in a microcentrifuge. Incubate at room temperature for 15 min.
7. Add 3 μl of 6X DNA loading dye. Mix evenly, and pulse-spin in a microcentrifuge.
8. Take 10 μl of the mixture and resolve it on a 2% agarose gel stained with Midori Green Advance for 40–50 min at 80 V. Visualize the gel on a gel imager (*see* Fig. 4 and **Note 10**).

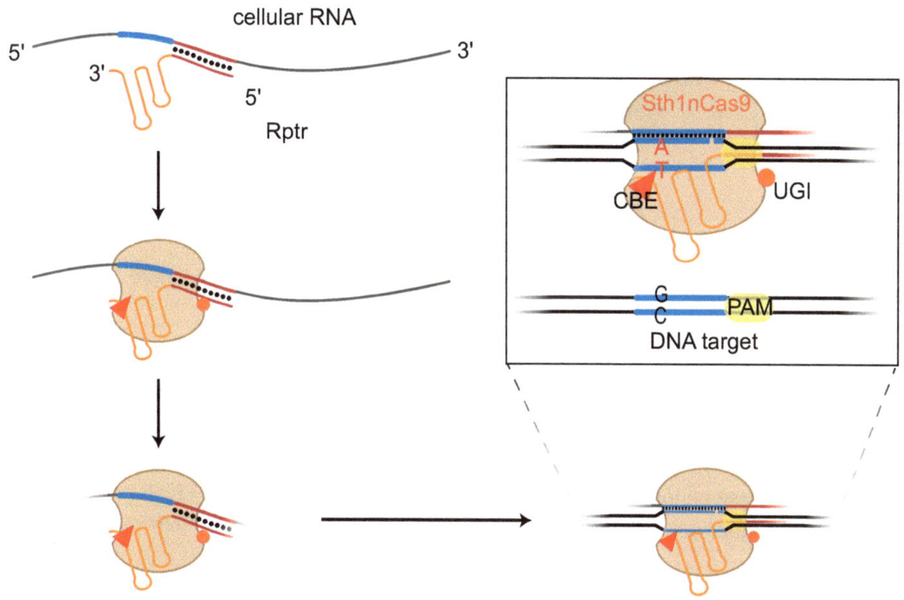

Fig. 5 Overview of the in vivo RNA detection platform. Rptr binds to a cellular RNA-of-interest and converts it into an ncrRNA, which guides Cas9 nickase to its DNA target. The cytosine base editor fused with Cas9 nickase accesses the single-stranded DNA bubble formed during target recognition by Cas9 and converts cytosine (C) to uracil (U). Another fused protein uracil glycosylase inhibitor (UGI) prevents uracil excision at the edited DNA strand, and the U subsequently turns to T after DNA replication. Cas9 nickase nicks the non-edited DNA strand, triggering mismatch repair that ensures the nicked strand rather than the edited strand is corrected

3.4 Design and Construction of Rptr and Targeted Plasmids for RNA Detection In Vivo

A hybrid inducible promoter Plac/ara-1 is used to drive expression of the base editor (rAPOBEC 1 cytosine base editor-Sth1nCas9 D9A nickase-uracil glycosylase inhibitor fusion (UGI) domain) to minimize leaky expression and subsequent cellular toxicity associated with Cas9 or the cytidine deaminase domain (*see* Figs. 5 and 6) [20, 21]. The base editor is induced with both L-arabinose and isopropyl β-D-1-thiogalactopyranoside (IPTG). The Rptr hybridizes to a cellular transcript-of-interest and converts it into an ncrRNA, which guides the base editor to its DNA target in the recording plasmid. The fused cytidine deaminase converts cytosine (C) to uracil (U) in the single-stranded DNA bubble created by the Cas9-RNA-target DNA interaction. The UGI and Cas9 nickase activity increase the editing efficiency and product purity by preventing uracil excision on the edited DNA strand and by driving repair of the nicked strand with the edited strand [22, 23]. With C-to-T conversion in the DNA target, the environmental stimuli and the cellular state in the form of the transcript are sensed and stored permanently in DNA (*see* Figs. 7 and 8).

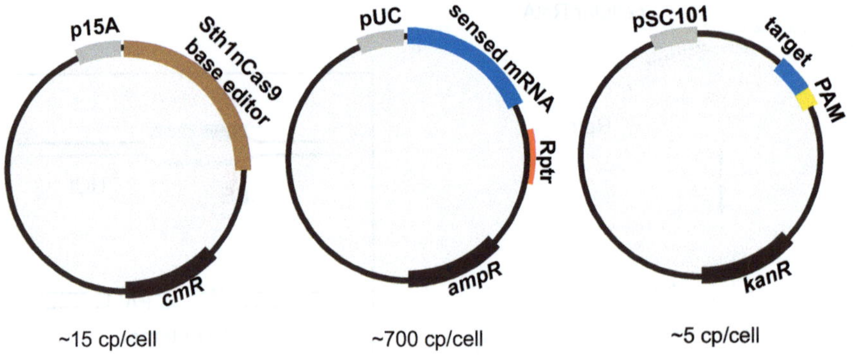

Fig. 6 Plasmids used for RNA detection in *E. coli*. The base editor is a protein fusion comprising rAPOBEC1 cytosine base editor, the Sth1nCas9 D9A nickase, and a uracil glycosylase inhibitory domain. The expression of the base editor was driven by Plac/ara-1 activated by IPTG and L-arabinose. Chloramphenicol is used to maintain the p15A-ori base editor plasmid. Sensed mRNA and its Rptr are put in a high-copy pUC-ori plasmid with an ampicillin resistance gene. A target sequence along with a PAM is inserted into a low-copy pSC101-ori recording plasmid

For in vivo RNA detection, the Cas9 from *Streptococcus thermophilus* CRISPR locus 1 (Sth1Cas9) with a long and stringent PAM (5′-NNAGAAW-3′) was used [24, 25] (*see* **Note 5**). The stringent PAM is less likely to appear downstream of genomic DNA sequence encoding the putative guide in the transcript-derived ncrRNA, avoiding potential self-targeting in the sensed gene. The stringent PAM is also less likely to appear elsewhere in the genome, reducing the number of potential off-target sites. Rptrs were designed in a similar way as described in subheading 3.1. The key is to maintain the secondary structure of the duplex between Rptrs and sensed transcripts (*see* Fig. 7). Unlike SpyCas9, the total length of the Rptr target hybridized with the Rptr anti-repeat is 36 nts for Sth1Cas9. The 20-nt sequences upstream of these Rptr target regions are treated as the guides of the resulting ncrRNA. The ncrRNA guides should contain at least one C positioned between fourth and tenth nt counting from the Rptr target-distal end; this region corresponds to the optimal editing window in the DNA target for Sth1Cas9 base editor [26] (*see* Fig. 7). In cases where a C is not present in the optimal editing window of the equivalent target DNA, a C can be introduced into the editing window of the DNA target in the recording plasmid. It is also recommended to include a T upstream the editable C for better editing performance [22]. By default, the Rptr alone or with its sensed RNA is encoded in a high-copy plasmid (pUC ori, ~700 copies/cell) and the DNA target is in a low-copy plasmid (pSC101 ori, ~5 copies/cell) (*see* Fig. 6). In some scenarios, e.g., recording the upregulated sensed RNA responding to a certain stimulus, the background expression and induced expression of this RNA cannot be differentiated due to oversaturated editing with the low-copy recording plasmid. For

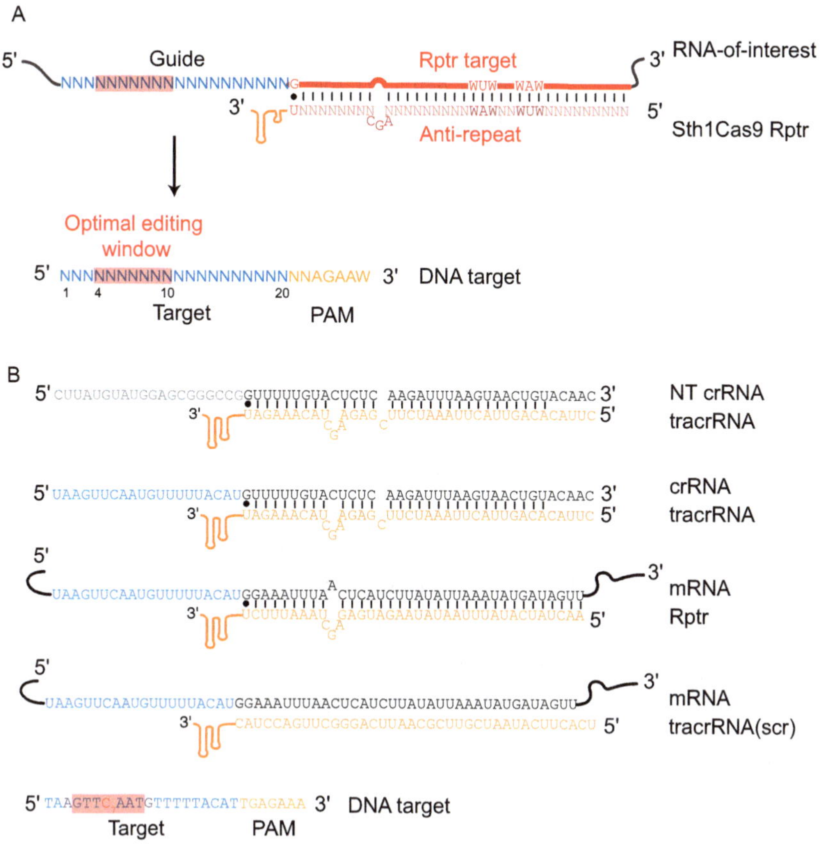

Fig. 7 Rptr and target design for in vivo RNA detection. (**a**) Design scheme for Rptr and its DNA target used by SpyCas9. W stands for A or U. When designing a Rptr, the AU-rich sequence in the sensed mRNA is selected as the Rptr target due to the preference of RNase III to cleave double-stranded RNA in AU-rich region [27]. The native anti-repeat sequence of Sth1Cas9 tracrRNA is replaced by the reverse complementary sequence of the Rptr target. The tenth nucleotide from the 3′ end of the new anti-repeat is replaced by AGC. The 20-nt sequence in blue upstream of the Rptr target is treated as the guide sequence. The same sequence along with the Sth1Cas9 consensus PAM (5′-NNAGAAW-3′) in yellow is used as the DNA target. The Sth1nCas9 base editor has optimal activity in the red boxed region in the target sequence [26]. (**b**) Sequence for CJ8421_04975 ncrRNA/Rptr, its controls and DNA target. NT crRNA, nontargeting crRNA with a random guide sequence; mRNA(mut), mRNA encoding the ncrRNA with point mutations in the predicted "seed" region of the guide. tracrRNA(scr), tracrRNA with a scrambled sequence for the anti-repeat region. C_7 in red is the editing site converted to T by the base editor

better differentiation, the Rptr plasmid can be placed in a low-copy plasmid (pSC101 ori, ~5 copies/cell), and the DNA target can be placed in a medium-copy plasmid (pBR322 ori, ~20 copies/cell).

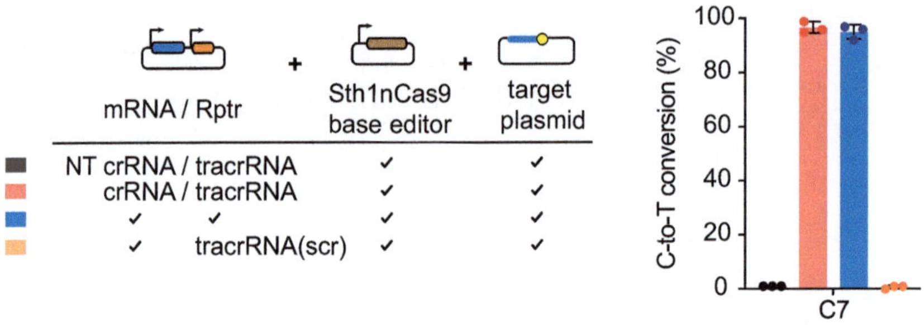

Fig. 8 Detection of heterologous transcription of the CJ8421_04975 mRNA in *E. coli*

3.5 *E. coli* Transformation

1. Inoculate single colonies of *E. coli* strain CB414 (Table 5) from an LB agar plate into LB liquid medium, and grow overnight at 37 °C and 220 rpm.
2. Back dilute the overnight cultures in a ratio of 1:50 into 25 ml fresh LB medium, and grow the cells at 37 °C and 220 rpm until the OD_{600} reaches ~0.8, which takes around 1.5 h.
3. Harvest the cells by centrifugation for 3 min at 4500 × g and 4 °C, and discard the supernatant.
4. Resuspend the cells using 15 ml of 10% ice-cold glycerol by inverting the 50 ml Falcon tube a few times, and centrifuge for 3 min at 4500 × g and 4 °C, and discard the supernatant.
5. Repeat step 4 (*see* **Notes 11** and **12**).
6. Resuspend the pellet using 1 ml of 10% ice-cold glycerol and transfer to a 1.5 ml Eppendorf tube, and spin down at 15,000 × g and 4 °C for 1 min.
7. Discard the supernatant by pipetting, resuspend the pellet using 500 μl of 10% ice-cold glycerol, and make 40 μl aliquots into new 1.5 ml Eppendorf tubes.
8. Add plasmids (50–100 ng for Sth1nCas9 base editor plasmid, Rptr plasmid, and recording plasmid) (Tables 1 and 5) into the aliquot, mix evenly by pipetting, and transfer to the precooled 0.1 cm cuvette (*see* **Note 13**).
9. Electroporate using the following parameters: capacitance = 25 μF, resistance = 200 ohm, and voltage = 1.8 kV in 0.1 cm cuvettes. Then add 500 μl of SOC medium to the cuvettes and transfer the mixture to new 1.5 ml Eppendorf tubes for 1 h recovery at 37 °C and 220 rpm (*see* **Note 14**).
10. Plate 10 μl of recovered cultures on LB+Cm+Amp+Kan plates, and incubate overnight at 37 °C (*see* **Notes 13 and 15**).

3.6 Induction and In vivo Detection

1. Inoculate three colonies per strain into LB+Amp+Kan+Cm liquid medium, Amp to maintain the Rptr-mRNA plasmid, Cm to maintain the Sth1nCas9 base editor plasmid, and Kan

Table 5
List of strains used in figures

Name	Ancestor strain	Plasmids	Resistance	Details/description	Comments
TOP10	*Escherichia coli* TOP10	-	StrR	mcrAΔ(mrr-hsdRMS-mcrBC) Φ80lacZ ΔM15 ΔlacX74 deoR recA1araD139Δ(ara-leu) 7697 galU galK rpsL endA1 nupG	Used for cloning
CB414	BW25113	–	–	BW25113 Δcas3-CRISPR1 ΔPlacI-lacZ	Used for transcriptional recording
CJ1046	CB414	pCJ452+378+240	Cm+Amp +Kan	NT crRNA peak113 mRNA	Used for transcriptional recording
CJ1047	CB414	pCJ452+370+240	Cm+Amp +Kan	crRNA2 peak113 mRNA	Used for transcriptional recording
CJ1048	CB414	pCJ452+235+240	Cm+Amp +Kan	Rptr2 peak113 mRNA	Used for transcriptional recording
CJ1049	CB414	pCJ452+355+240	Cm+Amp +Kan	mRNA(mut) peak113 mRNA	Used for transcriptional recording
CJ1050	CB414	pCJ452+375+240	Cm+Amp +Kan	tracrRNA(scr) peak113 mRNA	Used for transcriptional recording

to maintain the recording plasmid. Grow overnight at 37 °C and 220 rpm (*see* Table 5, **Notes 15** and **16**).

2. Prepare induction medium LB+Amp+Kan+Cm+IPTG+L-arabinose. Store at 4 °C for later use (*see* **Note 15**).
3. Back dilute overnight cultures in a ratio of 1:20 (250 μl of overnight cultures into 4.75 ml of inducing medium), and culture for 8 h at 37 °C and 220 rpm.
4. Collect 2 ml of culture, perform plasmid miniprep using a Plasmid Miniprep Kit, and elute with 30 μl of nuclease-free water (*see* **Note 17**).
5. Measure the concentration of extracted plasmids using a spectrophotometer.
6. Perform OneTaq PCR using primers CJpr0001 and CJpr002 and around 50 ng extracted plasmids as the template (Table 6).

Table 6
Reaction setup and thermocycling conditions for OneTaq PCR

Components	25 μl reaction	Final concentration
OneTaq 2X Master Mix with standard buffer	12.5 μl	1X
10 μM forward primer	0.5 μl	0.2 μM
10 μM reverse primer	0.5 μl	0.2 μM
Template DNA	variable	~50 ng
Nuclease-free water	to 25 μl	

Step	Temperature	Time
Initial denaturation	94 °C	30 s
32 cycles	94 °C	15 s
	*45–68 °C	30 s
	68 °C	1 kb/min
Final extension	68 °C	5 min
Hold	4 °C	

7. Clean PCR product using the DNA Cleanup and Concentration Kit, and elute with 25 ul of nuclease-free water.
8. Adjust DNA concentration based on the size, e.g. ~1.5 ng/μl per 100 bp, prior to Sanger sequencing.
9. Evaluate editing efficiency using the online tool EditR 1.0.10 (https://moriaritylab.shinyapps.io/editr_v10/). Follow the instructions provided on the website to troubleshoot the potential errors.

4 Notes

1. 19-bp minimal T7 promoter sequence (5-′-TAATACGACTCACTATAGG-3′) containing 2Gs instead of 1G or 3Gs at the 3′ end was used for efficient transcription and minimal 5′-transcript heterogeneity.
2. In addition to an agarose gel, a polyacrylamide gel electrophoresis (PAGE) gel or Bioanalyzer could also be used as a readout, especially for multiplexed detection. For a PAGE gel, fluorophore-labeled DNA targets were used to increase the throughput such that only uncut substrates and smaller cut products are visible.

3. DNA fragments encoding sensed RNA could also be ordered directly in the form of gene fragments, e.g., gBlocks from Integrated DNA Technologies.
4. Because the Rptr expressing cassette (110bp) is shorter than the minimal size requirement (125 bp) for some gene fragments (e.g., gBlocks from Integrated DNA Technologies), two strategies could be used. One is to add at least 15 bp of random sequence in front of the Rptr expression cassette to fulfill the requirement. The other is to synthesize one pair of primers with 22 bp overhang, and the whole Rptr expressing cassette could be obtained by PCR extension.
5. The consensus PAM for SpyCas9 is 5′-NGG-3′. For other nucleases, such as CjeCas9 or Sth1Cas9, NGG should be changed to their consensus PAMs, e.g., 5′-NNNNACAC-3′ for CjeCas9, and 5′-NNAGAAW-3′, W = A or T for Sth1Cas9.
6. It is recommended to extend the incubation time up to 4 h if the yield of the transcribed RNA is low.
7. It is recommended to check the size and integrity of transcribed RNAs on a gel and optionally purify the RNAs if multiple bands appear.
8. Our further study showed that the 30-min cooling step could be minimized by directly placing the heated sample on ice without compromising activity.
9. The incubation time could be adjusted based on the Rptr activity and the user's needs. For the case of SNP detection, fine-tune the incubation time to obtain appropriate cleavage of products. Excessive incubation time might lead to some cleavage of the mismatched DNA target in addition to the matched DNA target, which might lead to an incorrect interpretation.
10. For better visualization, using lower voltage for a long-running time with a fresh buffer is recommended, and adjust the gel concentration based on the user's product sizes.
11. After the second wash, the pellet on the wall of the tube is easy to dissolve into liquid, so discard the supernatant immediately.
12. The second wash step is recommended to minimize potential arcing due to residual ions from the medium.
13. For some hard-to-transform strains, plate all recovered cultures to get substantial colonies or transform the three plasmids serially.
14. A time constant of 5–6 ms is expected for higher transformation efficiency. Consider redoing the electroporation if the time constant is less than 5.
15. Antibiotics were added where appropriate at final concentrations of 100 μg/ml for ampicillin, 34 μg/ml for

chloramphenicol, and 50 μg/ml for kanamycin. Final concentrations of 20 mM L-arabinose and 1 mM IPTG were used to induce the expression of the Sth1nCas9 base editor.

16. For consistent performance between batches, it is recommended to pick freshly transformed colonies or revive strains by streaking them from a cryostock onto the plates.
17. Collect more culture, e.g., 4–5 ml for plasmid extraction if the culture is not turbid.

Acknowledgments

We thank Scott Collins and Sandra Gawlitt for providing the constructs necessary to construct the Sth1Cas9n base editor. We also thank Daphne Collias and Claas Reckstadt for helpful feedback on the manuscript. This work was supported by the European Research Council Consolidator Award (865973 to C.L.B.) and the SPP 2141 priority program of the Deutsche Forschungsgemeinschaft (BE 6703/1-1 to C.L.B.).

References

1. Adli M (2018) The CRISPR tool kit for genome editing and beyond. Nat Commun 9:1911
2. Li Y, Li S, Wang J, Liu G (2019) CRISPR/Cas systems towards next-generation biosensing. Trends Biotechnol 37:730–743
3. Nouri R, Tang Z, Dong M, Liu T, Kshirsagar A, Guan W (2021) CRISPR-based detection of SARS-CoV-2: a review from sample to result. Biosens Bioelectron 178:113012
4. East-Seletsky A, O'Connell MR, Knight SC, Burstein D, Cate JHD, Tjian R et al (2016) Two distinct RNase activities of CRISPR-C2c2 enable guide-RNA processing and RNA detection. Nature 538:270–273
5. Chen JS, Ma E, Harrington LB, Da Costa M, Tian X, Palefsky JM et al (2018) CRISPR-Cas12a target binding unleashes indiscriminate single-stranded DNase activity. Science 360:436–439
6. Gootenberg JS, Abudayyeh OO, Lee JW, Essletzbichler P, Dy AJ, Joung J et al (2017) Nucleic acid detection with CRISPR-Cas13a/C2c2. Science 356:438–442
7. Gootenberg JS, Abudayyeh OO, Kellner MJ, Joung J, Collins JJ, Zhang F (2018) Multiplexed and portable nucleic acid detection platform with Cas13, Cas12a, and Csm6. Science 360:439–444
8. Bruch R, Urban GA, Dincer C (2019) CRISPR/Cas powered multiplexed biosensing. Trends Biotechnol 37:791–792
9. Sheth RU, Wang HH (2018) DNA-based memory devices for recording cellular events. Nat Rev Genet 19:718–732
10. Ishiguro S, Mori H, Yachie N (2019) DNA event recorders send past information of cells to the time of observation. Curr Opin Chem Biol 52:54–62
11. Tang W, Liu DR (2018) Rewritable multi-event analog recording in bacterial and mammalian cells. Science 360:eaap8992
12. Farzadfard F, Gharaei N, Higashikuni Y, Jung G, Cao J, Lu TK (2019) Single-nucleotide-resolution computing and memory in living cells. Mol Cell 75:769–780
13. Perli SD, Cui CH, Lu TK (2016) Continuous genetic recording with self-targeting CRISPR-Cas in human cells. Science 353:aag0511
14. Schmidt F, Cherepkova MY, Platt RJ (2018) Transcriptional recording by CRISPR spacer acquisition from RNA. Nature 562:380–385
15. Shipman SL, Nivala J, Macklis JD, Church GM (2016) Molecular recordings by directed CRISPR spacer acquisition. Science 353:aaf1175

16. Silas S, Mohr G, Sidote DJ, Markham LM (2016) Direct CRISPR spacer acquisition from RNA by a natural reverse transcriptase–Cas1 fusion protein. Science 351:aad4234
17. Jiao C, Sharma S, Dugar G, Peeck NL, Bischler T, Wimmer F et al (2021) Noncanonical crRNAs derived from host transcripts enable multiplexable RNA detection by Cas9. Science 372:941–948
18. Liao C, Beisel CL (2021) The tracrRNA in CRISPR biology and technologies. Annu Rev Genet 55:161–181
19. Briner AE, Donohoue PD, Gomaa AA, Selle K, Slorach EM, Nye CH et al (2014) Guide RNA functional modules direct Cas9 activity and orthogonality. Mol Cell 56:333–339
20. Lutz R, Bujard H (1997) Independent and tight regulation of transcriptional units in *Escherichia coli* via the LacR/O, the TetR/O and AraC/I1-I2 regulatory elements. Nucleic Acids Res 25:1203–1210
21. Chen Y, Ho JML, Shis DL, Gupta C, Long J, Wagner DS et al (2018) Tuning the dynamic range of bacterial promoters regulated by ligand-inducible transcription factors. Nat Commun 9:1–8
22. Komor AC, Kim YB, Packer MS, Zuris JA, Liu DR (2016) Programmable editing of a target base in genomic DNA without double-stranded DNA cleavage. Nature 533:420–424
23. Wang L, Xue W, Yan L, Li X, Wei J, Chen M et al (2017) Enhanced base editing by co-expression of free uracil DNA glycosylase inhibitor. Cell Res 27:1289–1292
24. Horvath P, Romero DA, Coûté-Monvoisin A-C, Richards M, Deveau H, Moineau S et al (2008) Diversity, activity, and evolution of CRISPR loci in *Streptococcus thermophilus*. J Bacteriol 190:1401–1412
25. Leenay RT, Maksimchuk KR, Slotkowski RA, Agrawal RN, Gomaa AA, Briner AE et al (2016) Identifying and visualizing functional PAM diversity across CRISPR-Cas systems. Mol Cell 62:137–147
26. Agudelo D, Carter S, Velimirovic M, Duringer A, Rivest J-F, Levesque S et al (2020) Versatile and robust genome editing with *Streptococcus thermophilus* CRISPR1-Cas9. Genome Res 30:107–117
27. Altuvia Y, Bar A, Reiss N, Karavani E, Argaman L, Margalit H (2018) *In vivo* cleavage rules and target repertoire of RNase III in *Escherichia coli*. Nucleic Acids Res 46:10530–10531

Chapter 14

Harnessing CRISPR-Cas9 for Epigenetic Engineering

Rosa S. Guerra-Resendez and Isaac B. Hilton

Abstract

Epigenome editing has become more precise and effective by coupling epigenetic effectors to the dCas9 protein and targeting regulatory regions such as promoters and enhancers. Here, we describe a basic methodology for performing an epigenome editing experiment, starting from gRNA design and cloning to transiently transfecting the gRNA plasmid and the CRISPR/dCas9-based epigenetic effector and finalizing with chromatin immunoprecipitation (ChIP) to validate changes in epigenetic state at a targeted genomic region.

Key words Synthetic biology, Epigenome editing, Guide RNA, CRISPR, Deactivated Cas9

1 Introduction

Targeting specific loci in the human genome has become more precise and easier with the optimization of the CRISPR-Cas9 system in recent years. While this system has been mainly used for genome editing, which involves permanently changing the nucleotide sequence, other forms of gene regulation by depositing chemical modifications on the DNA as well as recruiting of specific transcription factors have been possible with the CRISPR-Cas9 technology in what is now commonly called epigenome editing [1]. By tethering epigenetic effectors to a catalytically inactive version of the Cas9 protein or deactivated Cas9 (dCas9), epigenetic marks have been delivered to promoters [2], enhancers [3], and other regulatory sites such as insulators [4] in mammalian cells.

1.1 CRISPR-Cas9-Based Epigenome Editing Tools

Transcriptional activation and repression have been achieved using different CRISPR activation (CRISPRa) and CRISPR interference (CRISPRi) methods. For instance, robust activation has been accomplished by using tripartite fusions of transactivating domains such as dCas9-VPR (VP64/p65/Rta) [5] or the recently described CRISPR-DREAM platform [6], while transcriptional silencing has

James Chappell and Melissa K. Takahashi (eds.), *Riboregulator Design and Analysis*,
Methods in Molecular Biology, vol. 2518, https://doi.org/10.1007/978-1-0716-2421-0_14,

been obtained by targeting dCas9 fused to the KRAB domain to the transcriptional start sites of genes [7].

Additional ways to regulate transcription have been attained by depositing histone acetylation marks with dCas9-p300 [3], histone methylation marks with dCas9-EZH2 [8], DNA methylation and demethylation with dCas9-DNMT3A/3L [9] and dCas9-TET [2], respectively, and most recently histone phosphorylation with dCas9-dMSK1 [10]. Manipulation of the three-dimensional chromatin architecture has been used to insulate specific genomic regions and to force promoter-enhancer interactions, thus regulating the expression of corresponding genes. This has been possible by employing either the light-activated dynamic looping (LADL) system [11] or the chemically induced forced chromatin looping with the CLOuD9 system [12]. The most appropriate epigenome editing tool will depend on the overall objective of each experiment. Thus, exhaustive research must be done on which tool is the best fit for your project, which has been addressed in several reviews [13–15].

1.2 Strategies for Optimized Epigenome Editing in Mammalian Cells

A typical experiment aimed at evaluating the effect of an epigenome editing tool in a certain locus generally needs to take three aspects into consideration: the design and location of guide RNA (gRNA), delivery method of the CRISPR-Cas9-based tool, and the downstream experiments that will comprise the output data of the experiment. Here, we describe a general protocol for designing an epigenome editing experiment, recommendations of how to optimize each of the considered aspects of the experiment, and suggested assays to generate the most reliable output data to present changes in the epigenome.

Designing gRNAs for an epigenome editing experiment can be a straightforward task with current bioinformatic tools such as CRISPOR [16], which will enable gRNA design for a wide variety of genomes and protospacer adjacent motif (PAM) sequences depending on the type or species of Cas protein being used. This tool will rank gRNAs from an input sequence based on scores considering potential off-target events in the genome of interest and on-target activity. An optimal gRNA would be one that has the lowest score for off-target events while retaining on-target activity, even for nuclease-deficient Cas9. Another aspect to consider while designing gRNAs is their location relative to the target site and to endogenous epigenetic marks. It is recommended to have evenly spaced gRNAs with 100 bp between them when trying to enrich for several dCas9-based epigenetic effectors in the same locus. For most epigenome editing experiments, targeting the −400 bp to −50 bp region surrounding the transcriptional start site with evenly spaced gRNAs will be enough to exhibit an increase in gene expression with CRISPRa [17]. Likewise, targeting from −50 bp to

+100 bp has resulted in optimal transcriptional silencing with CRISPRi [18]. Multiplexing gRNAs can also be a useful strategy to target several genes in a signaling pathway or different regions in a specific locus. The enrichment of epigenetic marks and chromatin state should also be assessed for gRNA design. For instance, DNaseI hypersensitivity sites, which indicate open chromatin, have been shown to frequently be ideal binding regions for CRISPR/dCas9-based epigenetic effectors [19]. Further, CRISPRi has been shown to have increased effectiveness in silencing regions when targeted to sites enriched with H3K27ac, H3K9ac, and H3K4me3, among other active transcription epigenetic marks [20]. Despite the bioinformatic tools and recommendations on gRNA design, identifying the most effective set of gRNAs is still an empirical process which will require to tailor the method to each unique experiment.

Delivering the CRISPR/dCas9-based epigenetic effector into cells will depend on several considerations such as the size of the construct, the target cell type, and stability of the effector. Nonviral delivery methods include liposomes and non-liposome reagents, polymers, and nanoparticles. While most of them do not display high toxicity to cell as other physical or viral vectors, delivery efficiency is low compared to other methods [21]. Viral vectors such as lentivirus, gammaretrovirus, and adeno-associated virus have different delivery efficiencies depending on the cell type (e.g., dividing vs. nondividing) as well as the packaging limit. A comparison between these viral vectors has been comprehensively analyzed elsewhere [22–24]. Finally, physical methods such as electroporation provide high delivery efficiency in most cell lines albeit with decreased cell viability [25].

Besides generating an effect in transcriptional output levels, another important consideration in an epigenome editing experiment is the validation that a particular epigenetic mark has been deposited or depleted at the desired locus. This is especially relevant for experiments using CRISPR/dCas9-based epigenetic writers and erasers of histone acetylation, methylation, phosphorylation, and DNA methylation. Several assays have been developed to assess the epigenetic state of a specific region or the whole genome. For instance, chromatin immunoprecipitation followed by sequencing (ChIP-seq) has become a key strategy for research groups involved in epigenomic studies since it can provide base-pair resolution of enrichment of epigenetic marks as well as location of DNA-binding proteins such as CRISPR/dCas9-based tools [26]. We recommend performing this assay for most epigenome editing experiments in addition to strategies evaluating transcriptional levels and protein expression, such as RT-qPCR and Western blotting, respectively. Additionally, several approaches have originated to assess DNA methylation locally, such as methylation-sensitive restriction enzyme qPCR (MSRE-qPCR) and bisulfite-PCR (BS-PCR) [27], as well as globally with whole-genome bisulfite sequencing (WGBS) and reduced representation bisulfite sequencing (RRBS) [28]. The Assays for Transposase-Accessible Chromatin and DNase-I hypersensitive sites followed by sequencing, also called ATAC-seq and DNase-seq, are two procedures for genome-wide identification of accessible chromatin [29]. Higher order chromatin organization has been visualized by chromosome conformation capture (3C) and its derived techniques such as Hi-C [30] and Capture-C [31], enabling characterization and analysis of long-distance DNA interactions.

Here, we outline a basic methodology for performing an epigenome editing experiment (*see* Fig. 1), presenting the steps from gRNA cloning to transiently transfecting the vectors to performing ChIP to validate changes in epigenetic state of a genomic region.

2 Materials

2.1 HEK293T Cell Culture

1. Complete DMEM (cDMEM) media: DMEM supplemented with 10% FBS and 1% penicillin/streptomycin. Store at 4 °C.
2. 1x PBS. Store at room temperature.
3. 0.25% trypsin-EDTA. Store at 4 °C.
4. 0.4% trypan blue solution. Store at room temperature.

2.2 Guide Cloning

1. Two complementary oligonucleotides with the target sequence to be used as gRNA and with overhangs to the destination vector (*see* **Note 1**). Store at −20 °C.
2. gRNA cloning destination vector pSPgRNA (Addgene #47108) (*see* **Note 2**). Store at −20 °C.
3. BbsI Restriction Enzyme. Store at −20 °C.
4. 10x restriction enzyme reaction buffer. Store at −20 °C.
5. Shrimp alkaline phosphatase (rSAP). Store at −20 °C.
6. Gel extraction kit. Store at room temperature.
7. T4 polynucleotide kinase. Store at −20 °C.
8. T4 ligase. Store at −20 °C.
9. 10x T4 ligase buffer. Store at −20 °C.
10. Stbl3 chemically competent cells. Store at −80 °C.
11. SOC medium: 20 g/L tryptone, 5 g/L yeast extract, 0.5 g/L NaCl.
12. LB agar petri dishes with antibiotic used for selection. Store at 4 °C.
13. Agarose. Store at room temperature.

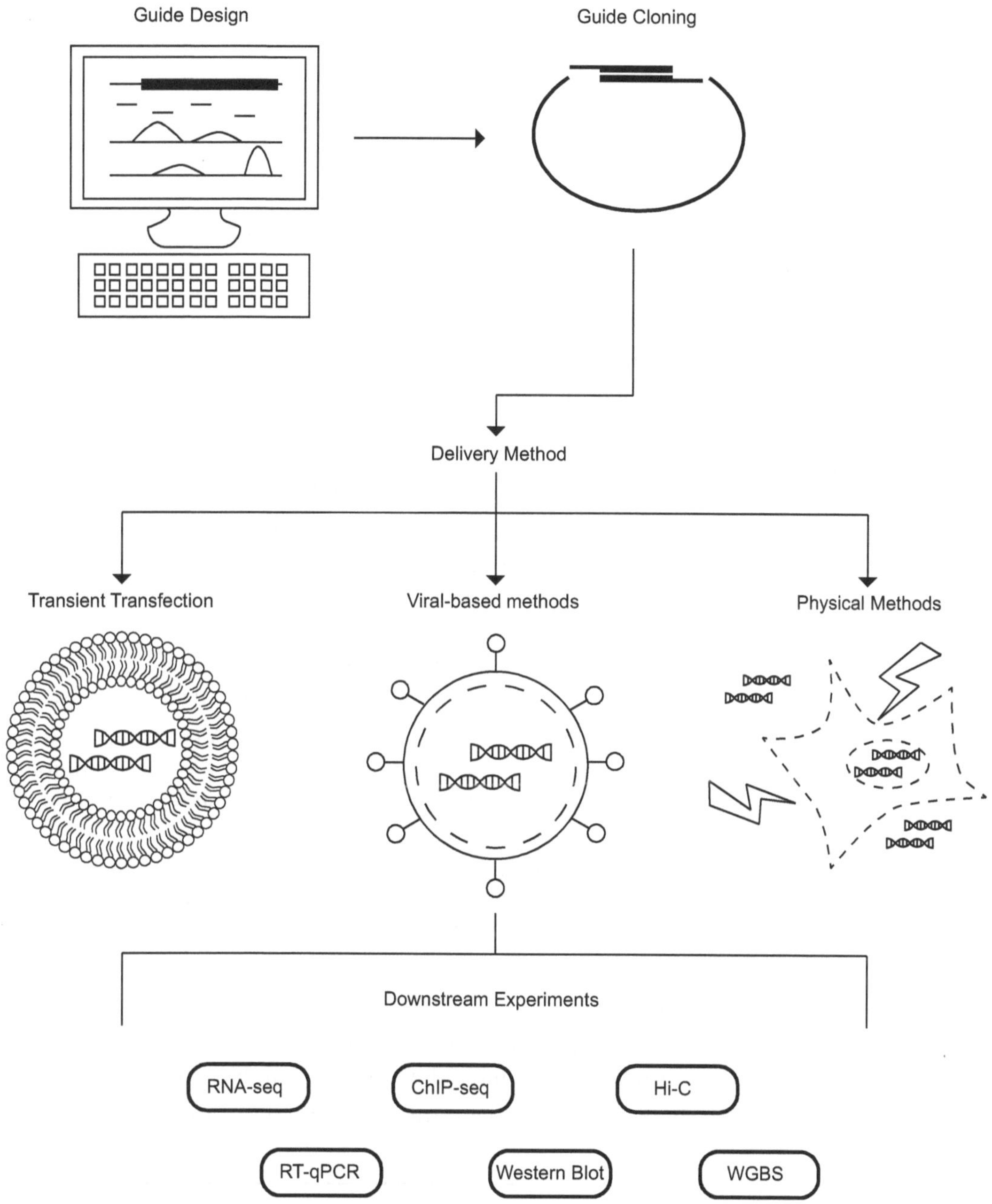

Fig. 1 General experimental design for evaluating the effect of a CRISPR-Cas9-based epigenetic effector at a certain locus. Special attention must be paid when designing the gRNAs that will be used for the specific experiment, keeping in mind adjacent epigenetic marks, transcription start site, as well as enhancers and other regulatory sites. Delivery method of the CRISPR tool must be assessed depending on the cell type to be used. Transient transfection/nucleofection would generate accurate data for most cells. For primary cells, other viral-based delivery methods such as lentivirus or gammaretrovirus must be used. For most epigenome editing experiments, the output data would be in form of relative expression levels of the targeted gene(s), confirming the expression of the synthetic epigenetic effectors and assessing changes of specific epigenetic marks in the targeted locus

2.3 Transient Transfection

1. Lipofectamine 3000 transfection reagent (*see* **Note 3**). Store at 4 °C.
2. OptiMEM media. Store at 4 °C.
3. Sterile, tissue culture-treated 24-well plates.
4. Sterile 1.7 mL centrifuge tubes.

2.4 Chromatin Immunoprecipitation (ChIP)

All reagents are prepared fresh before use unless otherwise stated.

1. 20% formaldehyde solution.
2. 2.5 M glycine.
3. 1X PBS at 4 °C.
4. Farnham lysis buffer at 4 °C: 5 mM PIPES, pH 8.0, 85 mM KCl, 0.5% NP-40. Filter the buffer with a 0.45 μm filter and add a protease inhibitor cocktail tablet just before use. Store buffer at 4 °C for no more than 4–6 weeks.
5. RIPA buffer at 4 °C: 1X PBS, 1% NP-40, 0.5% sodium deoxycholate, 0.1% SDS. Filter the buffer with a 0.45 μm filter and add a protease inhibitor cocktail tablet just before use. Store buffer at 4 °C for no more than 4–6 weeks.
6. Magnetic beads specific for antibodies of interest (e.g., anti-mouse IgG magnetic beads for antibodies made in mouse).
7. PBS/BSA at 4 °C: 1X PBS, 5 mg/ml BSA (fraction V). Filter the buffer with a 0.45 μm filter, and add a protease inhibitor cocktail tablet just before use. Store buffer at 4 °C for no more than 4–6 weeks.
8. LiCl immunoprecipitation wash buffer at 4 °C: 100 mM Tris-HCl, pH 7.5, 500 mM LiCl, 1% NP-40, 1% sodium deoxycholate, and filtered with a 0.45 μm filter prior to use.
9. 1X TE buffer at 4 °C: 10 mM Tris-HCl, pH 7.5, 0.1 mM Na_2EDTA.
10. Immunoprecipitation elution buffer: 1% SDS, 0.1 M $NaHCO_3$, and filtered with a 0.45 μm filter prior to use.
11. PCR purification kit. Store at room temperature.

3 Methods

3.1 Guide Cloning in Destination Vector

All reactions must be set up on ice unless otherwise stated. This guide cloning protocol has been adapted from the following references [32, 33].

1. Digest and dephosphorylate the gRNA cloning destination vector pSPgRNA by setting up the following reaction (*see* **Note 4**):

gRNA cloning destination vector pSPgRNA	1 μg
10x restriction enzyme buffer	2 μL
BbsI Restriction Enzyme	1 μL
rSAP	1 μL
Water (molecular biology grade)	Up to 20 μL

2. Incubate the reaction at 37 °C for 1 h.
3. Run the reaction in a 1% agarose gel and cut the largest band corresponding to the digested and phosphorylated plasmid backbone. Store in a sterile 1.7 mL centrifuge tube.
4. Gel purify the digested and dephosphorylated plasmid backbone using a commercial gel purification kit.
5. Phosphorylate and anneal the complementary oligonucleotide guide pair by setting up the following reaction (*see* **Note 5**):

Oligonucleotide #1 – sense (100 μM)	1 μL
Oligonucleotide #2 – antisense (100 μM)	1 μL
10x T4 ligation buffer	1 μL
T4 PNK	0.5 μL
Water (molecular biology grade)	6.5 μL

6. Incubate the reaction in a thermocycler with the following program:
   - 37 °C for 30 min.
   - 95 °C for 5 min.
   - Let temperature drop to 25 °C at a ramp rate of 0.1 °C/s.
7. Dilute the annealing reaction at a 1:200 dilution with molecular biology-grade water or TE buffer.
8. Ligate the diluted annealed oligos with the plasmid backbone by setting up the following reaction:

BbsI-digested plasmid backbone	50 ng
1:200 diluted annealed oligos or water as negative control (*see* **Note 6**)	1 μL
10x T4 ligation buffer	1 μL
T4 DNA ligase	1 μL
Water (molecular biology grade)	Up to 11 μL

9. Incubate the reaction at room temperature for 30 min to 1 h (*see* **Note 7**).
10. Transform ~4 μL of the ligation reaction into ~50 μL of Stbl3 chemically competent *E. coli* cells:
    - Incubate Stbl3 cells with ligation reaction on ice for 30 min on ice.
    - Heat shock for 45 s at 42 °C.
    - Incubate on ice for another 2 min.
    - Add 200 μL of SOC medium.
    - Incubate for 1 h at 37 °C, 220 rpm.
11. Plate transformation reaction and negative control in an LB agar plate with appropriate antibiotic.
12. Check for colonies ~16 h post-transformation.

3.2 Transient Transfection of Expression Vector for CRISPR/dCas9-Based Epigenetic Effector and gRNA Plasmid with Lipofectamine 3000 in HEK293T Cells

The following steps must be performed in a tissue culture hood with prewarmed media at 37 °C unless otherwise stated. Use HEK293T cells below passage 20 for best transfection efficiency results.

Day 0: plating HEK293T in 24-well plate for transfection

1. Aspirate media from tissue culture dish with HEK293T cells.
2. Add 2 mL of 1x PBS and rock the plate gently to wash the adherent cells.
3. Aspirate the PBS and add 2 mL of prewarmed 0.25% trypsin-EDTA.
4. Incubate at 37 °C for 2 min.
5. Rock the plate gently to detach the cells and add 8 mL of cDMEM, pipetting up and down to ensure detaching and removing clumps from cells.
6. Transfer the cells to a 15 mL centrifuge tube and centrifuge for 5 min at 250 rpm.
7. Aspirate the media and resuspend the pellet with 10 mL of cDMEM.
8. Count the cells and use a 0.4% trypan blue solution to assess viability. If the viability is less than 90%, do not use the cells for transfection.
9. Plate ~80,000 cells per well in a 24-well plate and incubate at 37 °C, 5% CO_2.

Day 1: transient transfection using Lipofectamine 3000

10. Change media of plated cells several hours before transfection.
11. In one 1.7 mL centrifuge tube, dilute 375 ng of the vector containing the CRISPR/dCas9-based epigenetic effector and 125 ng in 25 μL of prewarmed OptiMEM media.

12. Add 1 μL of p3000 reagent to the diluted DNA plus Opti-MEM (*see* **Note 8**).
13. In a separate tube, add 1.5 μL of Lipofectamine reagent to 25 μL of prewarmed OptiMEM.
14. Mix the contents of both tubes and incubate at room temperature for 15 min.
15. Add 50 μL of the mix to the seeded HEK293T cells in the 24-well plate well in a dropwise manner.
16. Incubate at 37 °C, 5% CO_2 for 48–72 h before harvesting the cells.

3.3 Chromatin Immunoprecipitation (ChIP)

Chromatin Cross-Linking

1. Remove culture plates from the incubator and place at room temperature on the bench.
2. Add formaldehyde to a final concentration of 1%, swirl gently, and incubate at room temperature for 10 min.
3. Stop the cross-linking reaction by adding glycine to a final concentration of 0.125 M and swirl gently to mix.
4. Scrape cells into 50 mL centrifuge tube and centrifuge at 2000 rpm for 5 min at 4 °C.
5. Aspirate media and resuspend cells in 10 mL of cold 1X PBS.
6. Centrifuge cells at 2000 rpm for 5 min at 4 °C.
7. Aspirate the PBS and resuspend cells in 5 mL cold Farnham lysis buffer.
8. Centrifuge cells at 2000 rpm for 5 min at 4 °C.
9. Carefully remove supernatant, aliquot 2×10^7 cells in tubes (*see* **Note 9**), and either proceed to sonication step or flash-freeze in liquid nitrogen and store at −80°C.

Chromatin Fragmentation by Sonication

Samples should be kept as close to 4 °C as possible during sonication with an ice water bath in a cold room or a circulator chiller water bath (*see* **Note 10**).

10. Fully resuspend each fresh or frozen pellet on ice in 1 mL Farnham lysis buffer and mix gently by flicking the tube.
11. Collect the crude nuclear lysate by centrifuging at 2000 rpm at 4 °C for 5 min.
12. Remove the supernatant and resuspend the pellet in 300 μL RIPA buffer.
13. Process samples in Bioruptor with circulating water bath according to the manufacturer's instructions for temperature

and speed of circulation at high setting. Perform three rounds of 15 min each, allowing 15 min rest between every three rounds (*see* **Note 10**).

14. Centrifuge the sonicated mixture at 14,000 rpm for 15 min at 4 °C and collect the supernatant (*see* **Note 11**).
15. Snap-freeze the samples in liquid nitrogen and store at −80 °C, or do not freeze and continue with the immunoprecipitation steps.

Immunoprecipitation

Perform all steps in an ice bucket or in the cold room at 4 °C.

To couple the primary antibody for each transcription factor or chromatin protein to magnetic beads:

16. Add 200 μL of resuspended magnetic bead slurry to a 1.5 mL microfuge tube on ice containing 1 mL cold PBS/BSA. Vortex briefly to mix well.
17. Place the microfuge tubes on the magnet rack and remove supernatants.
18. Resuspend the beads in 1 mL cold PBS/BSA.
19. Repeat **steps** (**2**) and (**3**) three times.
20. Add 200 μL PBS/BSA to beads.
21. Add 5 μg primary antibody. Do not vortex beads after adding the antibody.
22. Gently mix on a rotator platform for at least 2 h at 4 °C.
23. Repeat **steps** (**2**) and (**3**) three times, resuspending the beads by inverting the tubes during each wash.
24. Resuspend in 100 μL PBS/BSA.

To form the bead-antibody complex with fragmented, cross-linked chromatin:

25. Add 100 μL of antibody-coupled beads (from **step 9**) to each 300 μL chromatin preparation (from **step** 3.3, **15**), and incubate on a rotator for 1 h at room temperature, followed by 1 h at 4 °C.
26. Collect beads containing immuno-bound chromatin by placing the microfuge tube on a magnet rack.
27. Remove and discard supernatant.
28. Wash beads five times with 1 mL LiCl wash buffer, mixing 3 min for each wash on a rotator.
29. Add 1 mL TE buffer to beads and mix for 1 min on rotator.
30. Place tubes on magnet rack to collect beads and discard supernatant.

31. Resuspend the bead pellet in 200 μL IP elution buffer at room temperature. Vortex to mix.

To reverse cross-link and recover ChIP DNA and prepare a control DNA sample:

32. Incubate bead pellet (from **step 16**) in a 65 °C water bath for 1 h, and shake or vortex every 15 min to elute the immuno-bound chromatin from the beads.
33. Spin at 14,000 rpm in a microfuge at room temperature for 3 min.
34. Collect the supernatant, which contains the ChIP DNA by placing the tubes on the magnet to facilitate supernatant recovery.
35. Incubate the supernatant containing the ChIP DNA in a 65 °C water bath overnight to complete the reversal of the formaldehyde cross-links.
36. To prepare a control reverse cross-link chromatin library, incubate a sample of sheared chromatin representing at least 1×10^6 cells in a 65 °C water bath overnight to reverse the cross-links. Bring the volume up to 200 μL with 1X PBS, add 20 ul of proteinase K, and incubate at room temperature for 2 min. Add 4 μL of RNase A, then add five volumes of binding buffer (from PCR purification kit), and incubate at room temperature for 20 min. Skip **step** (**22**) and proceed with **step** (**23**).
37. Add five volumes of binding buffer (from PCR purification kit) to one volume of ChIP DNA (*see* **Note 12**).
38. Add half (~600 μL) of the solution to a PCR purification column, centrifuge per manufacturer's instructions, and then repeat with rest of the solution to bind the whole sample to the column.
39. Wash the column with 750 μL of wash buffer (from PCR purification kit), centrifuge per manufacturer's instructions, discard flow-through, and centrifuge the column containing the bound DNA at maximum speed for 2 min to allow it to dry completely.
40. Elute the DNA from the column with two 30 μL aliquots of warmed (~55 °C) elution buffer (from PCR purification kit), following the manufacturer's instructions (*see* **Note 13**).
41. Measure DNA concentration by using a Qubit fluorometer or BioAnalyzer (*see* **Note 14**).

Recommended: Checking the Size of Sonicated Fragments

To determine the correct sonication conditions for each cell line, we carry out a shearing time course in the Bioruptor. We typically process the chromatin for two to five 10-min cycles of 30 s ON, 30 s OFF, and then assess the size of the fragments after reversing the cross-links, isolating the DNA, and visualizing it on an agarose gel and a BioAnalyzer. The shearing time must be determined empirically for each cell line before proceeding with immunoprecipitation.

42. Load 1 μg of DNA from each sonication time point from samples prepared as described in **steps** 3.3, **36** to 3.3, **41** on a 1% agarose gel. Load and run a 100 bp ladder size marker along with the samples.
43. After electrophoresis, examine the gel and compare the midpoint of the sample's size distribution to 100 bp ladder size markers. The ideal condition should yield an enrichment of fragments in the 100–500 bp range, but larger fragments should still be present. We typically choose the shortest time point that shows noticeable enrichment in the 100–500 bp range (*see* **Note 15**).
44. Samples can also be visualized on a BioAnalyzer to confirm fragment size. Use a high sensitivity DNA chip per the manufacturer's instructions. Due to discrepancies and possible size biases on the BioAnalyzer, we recommend that gel electrophoresis be used as the primary way to visualize fragmented DNA.

4 Notes

1. The oligonucleotides must have 5′ and 3′ overhangs to the destination gRNA vector. An excellent explanation of how to design the overhangs can be found in the Zhang lab protocols for lentiviral CRISPR-Cas9 and single guide RNA protocols [32, 33].
2. Any gRNA destination vector can be used; the one listed is used as an example. If using a gRNA multiplex system, please refer to the specific cloning method and destination vector to use.
3. Other transfection reagents (liposome or non-liposome based) can be used for fast and effective (>80% effectivity) transient transfection. Please refer to other specific methods if using any other transfection reagents as the DNA quantity and incubation times might differ from the one stated.
4. Some restriction enzyme buffers are compatible with the alkaline phosphatase; thus, the digestion and dephosphorylation

reactions can be done in one step. For instances where there is buffer incompatibility between the enzymes, the alternative protocol can be followed: set up the same digestion reaction (without the rSAP) and incubate at 37 °C for 30 min, and then add the rSAP and appropriate buffer to the digestion reaction and incubate at 37 °C for 1 h.

5. Successful phosphorylation of the reaction relies on the T4 PNK activity; thus, it is recommended that the reaction buffer contains ATP, such as the 10x T4 ligase buffer. If using a supplied T4 PNK buffer which does not contain ATP, you can supplement it to 1 mM ATP.
6. It is strongly recommended to perform a negative control of the ligation reaction using water instead of the annealed oligos. This can be used as an estimation of colonies that contain religated plasmids without the desired guide sequence.
7. Alternatively, the ligation reaction can be incubated at 16 °C overnight. However, high ligation efficiency has been observed with only 30 min incubation at room temperature.
8. We recommend a ratio of 3:1 CRISPR/dCas9-based epigenetic effector vector to gRNA plasmid. If several gRNA plasmids are being used as a pooled gRNA mix, add the same mass for each of them to sum 125 ng total of gRNA plasmid. It is also recommended to setup mastermix reactions for the p3000 reagent mix and the Lipofectamine reagent mix especially when using several CRISPR/dCas9-based epigenetic effector vectors under the same gRNA.
9. Since chromatin fragmentation and ChIP are performed in aliquots of 2×10^7 cells, cells should be stored in 2×10^7 cell aliquots after the final collection. Frozen cross-linked cells appear to be stable indefinitely. Therefore, this is a convenient step in the protocol to collect several samples for testing multiple antibodies or biological replicates and controls.
10. This protocol describes sonication with a Bioruptor. A titration experiment with cross-linked chromatin should be performed to determine the appropriate conditions (time and intensity) for fragmenting chromatin before attempting a ChIP experiment.
11. When processing more than one tube of the same type of chromatin (i.e., same cell type and biological replicate), we recommend mixing all the sonicated chromatin at this step, removing 50 μL to prepare a reverse cross-link library (**step** 3.3, **36**), and then re-aliquoting into equal aliquots of ~2×10^7 cell equivalents. For example, process 12 tubes of 2×10^7 cell aliquots on the Bioruptor. After the centrifugation step, mix the contents of the 12 tubes, save 50 μl, and then divide the remaining chromatin back out into 12 equal

aliquots. We implement this procedure to reduce sample to sample variability in sonication. The reverse cross-link chromatin control library made from the 50 μl aliquot is used as the control for all the ChIPs performed from the mixed batch of chromatin.

12. Most PCR purification kits contain a binding buffer with a pH indicator. If the sample is not at the correct pH indicated by the manufacturer's instructions, the pH should be adjusted with 3 M sodium acetate.
13. We recommend adding the buffer, allow to sit on column for 2 min, centrifuge, and repeat the elution with the second volume of prewarmed buffer.
14. DNA recovery varies by antibody used for ChIP and ranges from 5 to more than 50 ng from a chromatin prep of 2×10^7 cells. Total chromatin for the control sample recovers about 20 μg. Store at −80 °C. DNA is ready for Illumina library construction.
15. We have found that overfragmentation of chromatin causes ChIP-seq to fail, possibly due to damage caused to the protein epitope targeted by the antibody.

References

1. Goell JH, Hilton IB (2021) CRISPR/Cas-based epigenome editing: advances, applications, and clinical utility. Trends Biotechnol 39:678–691. https://doi.org/10.1016/j.tibtech.2020.10.012
2. Choudhury SR, Cui Y, Lubecka K et al (2016) CRISPR-dCas9 mediated TET1 targeting for selective DNA demethylation at BRCA1 promoter. Oncotarget 7:46545–46556. https://doi.org/10.18632/oncotarget.10234
3. Hilton IB, D'Ippolito AM, Vockley CM et al (2015) Epigenome editing by a CRISPR/Cas9-based acetyltransferase activates genes from promoters and enhancers. Nat Biotechnol 33:510–517. https://doi.org/10.1038/nbt.3199
4. Tarjan DR, Flavahan WA, Bernstein BE Epigenome editing strategies for the functional annotation of CTCF insulators. Nat Commun:1–8. https://doi.org/10.1038/s41467-019-12166-w
5. Chavez A, Scheiman J, Vora S et al (2015) Highly-efficient Cas9-mediated transcriptional programming. Nat Methods 12:326–328. https://doi.org/10.1038/nmeth.3312
6. Mahata B, Li J, Cabrera A et al. Compact engineered human transactivation modules enable potent and versatile synthetic transcriptional control. bioRxiv. https://doi.org/10.1101/2022.03.21.485228
7. Gilbert LA, Larson MH, Morsut L et al (2013) CRISPR-mediated modular RNA-guided regulation of transcription in eukaryotes. Cell 154:442–451. https://doi.org/10.1016/j.cell.2013.06.044
8. O'Geen H, Ren C, Nicolet CM et al (2017) dCas9-based epigenome editing suggests acquisition of histone methylation is not sufficient for target gene repression. Nucleic Acids Res 45:9901–9916. https://doi.org/10.1093/nar/gkx578
9. Stepper P, Kungulovski G, Jurkowska RZ et al (2017) Efficient targeted DNA methylation with chimeric dCas9–Dnmt3a–Dnmt3L methyltransferase. Nucleic Acids Res 45: 1703–1713. https://doi.org/10.1093/nar/gkw1112
10. Li J, Mahata B, Escobar M et al (2021) Programmable human histone phosphorylation and gene activation using a CRISPR/Cas9-based chromatin kinase. Nat Commun 12: 896. https://doi.org/10.1038/s41467-021-21188-2
11. Kim JH, Rege M, Valeri J et al (2019) LADL: light-activated dynamic looping for endogenous gene expression control. Nat Methods

16:633–639. https://doi.org/10.1038/s41592-019-0436-5

12. Morgan SL, Mariano NC, Bermudez A et al (2017) Manipulation of nuclear architecture through CRISPR-mediated chromosomal looping. Nat Commun 8:15993. https://doi.org/10.1038/ncomms15993
13. Pulecio J, Verma N, Mejía-Ramírez E et al (2017) CRISPR/Cas9-based engineering of the epigenome. Cell Stem Cell 21:431–447. https://doi.org/10.1016/j.stem.2017.09.006
14. Thakore PI, Black JB, Hilton IB, Gersbach CA (2016) Editing the epigenome: technologies for programmable transcription and epigenetic modulation. Nat Methods 13:127–137. https://doi.org/10.1038/nmeth.3733
15. Sgro A, Blancafort P (2020) Epigenome engineering: new technologies for precision medicine. Nucleic Acids Res 48:12453–12482. https://doi.org/10.1093/nar/gkaa1000
16. Concordet J-P, Haeussler M (2018) CRISPOR: intuitive guide selection for CRISPR/Cas9 genome editing experiments and screens. Nucleic Acids Res 46:W242–W245. https://doi.org/10.1093/nar/gky354
17. Chakraborty S, Ji H, Kabadi AM et al (2014) A CRISPR/Cas9-based system for reprogramming cell lineage specification. Stem Cell Rep 3:940–947. https://doi.org/10.1016/j.stemcr.2014.09.013
18. Gilbert LA, Horlbeck MA, Adamson B et al (2014) Genome-scale CRISPR-mediated control of gene repression and activation. Cell 159:647–661. https://doi.org/10.1016/j.cell.2014.09.029
19. Kuscu C, Arslan S, Singh R et al (2014) Genome-wide analysis reveals characteristics of off-target sites bound by the Cas9 endonuclease. Nat Biotechnol 32:677–683. https://doi.org/10.1038/nbt.2916
20. Radzisheuskaya A, Shlyueva D, Müller I, Helin K (2016) Optimizing sgRNA position markedly improves the efficiency of CRISPR/dCas9-mediated transcriptional repression. Nucleic Acids Res 44:e141. https://doi.org/10.1093/nar/gkw583
21. Li L, He Z-Y, Wei X-W et al (2015) Challenges in CRISPR/CAS9 delivery: potential roles of nonviral vectors. Hum Gene Ther 26:452–462. https://doi.org/10.1089/hum.2015.069
22. Cooray S, Howe SJ, Thrasher AJ (2012) Chapter three – retrovirus and lentivirus vector design and methods of cell conditioning. In: Friedmann T (ed) Methods in enzymology. Academic Press, pp 29–57
23. Xu CL, Ruan MZC, Mahajan VB, Tsang SH (2019) Viral delivery systems for CRISPR. Viruses 11:28. https://doi.org/10.3390/v11010028
24. Rittiner JE, Moncalvo M, Chiba-Falek O, Kantor B (2020) Gene-editing technologies paired with viral vectors for translational research into neurodegenerative diseases. Front Mol Neurosci 13. https://doi.org/10.3389/fnmol.2020.00148
25. Li Y, Wu M, Zhao D et al (2015) Electroporation on microchips: the harmful effects of pH changes and scaling down. Sci Rep 5:17817. https://doi.org/10.1038/srep17817
26. Park PJ (2009) ChIP–seq: advantages and challenges of a maturing technology. Nat Rev Genet 10:669–680. https://doi.org/10.1038/nrg2641
27. Abhimanyu, Ontiveros CO, Guerra-Resendez RS et al (2021) Reversing post-infectious epigenetic-mediated immune suppression. Front Immunol 12:2093. https://doi.org/10.3389/fimmu.2021.688132
28. Rauluseviciute I, Drabløs F, Rye MB (2019) DNA methylation data by sequencing: experimental approaches and recommendations for tools and pipelines for data analysis. Clin Epigenetics 11:193. https://doi.org/10.1186/s13148-019-0795-x
29. Li Z, Schulz MH, Look T et al (2019) Identification of transcription factor binding sites using ATAC-seq. Genome Biol 20:45. https://doi.org/10.1186/s13059-019-1642-2
30. Lieberman-Aiden E, van Berkum NL, Williams L et al (2009) Comprehensive mapping of long range interactions reveals folding principles of the human genome. Science 326:289–293. https://doi.org/10.1126/science.1181369
31. Hughes JR, Roberts N, McGowan S et al (2014) Analysis of hundreds of cis-regulatory landscapes at high resolution in a single, high-throughput experiment. Nat Genet 46:205–212. https://doi.org/10.1038/ng.2871
32. Sanjana NE, Shalem O, Zhang F (2014) Improved vectors and genome-wide libraries for CRISPR screening. Nat Methods 11:783–784. https://doi.org/10.1038/nmeth.3047
33. Shalem O, Sanjana NE, Hartenian E et al (2014) Genome-scale CRISPR-Cas9 knockout screening in human cells. Science 343:84–87. https://doi.org/10.1126/science.1247005

Chapter 15

RNA Structure Prediction, Analysis, and Design: An Introduction to Web-Based Tools

Raphael Angelo I. Zambrano, Carmen Hernandez-Perez, and Melissa K. Takahashi

Abstract

Understanding RNA structure has become critical in the study of RNA in their roles as mediators of biological processes. To aid in these studies, computational algorithms that utilize thermodynamics have been developed to predict RNA secondary structure. Due to the importance of intermolecular interactions, the algorithms have been expanded to determine and predict RNA-RNA hybridization. This chapter discusses popular webservers with the tools for RNA secondary structure prediction, RNA-RNA hybridization, and design. We address key features that distinguish common-functioning programs and their purposes for the interests of the user. Ultimately, we hope this review elucidates web-based tools researchers may take advantage of in their investigations of RNA structure and function.

Key words RNA secondary structure prediction, Noncoding RNA, RNA design tools

1 Introduction

Noncoding RNA (ncRNA) molecules, or RNAs that are not translated, have emerged as key components of both prokaryotic and eukaryotic cells that play active roles in many processes including RNA processing [1] and various aspects of gene expression [2–4]. How ncRNAs carry out these functions is directly related to the structures that the ncRNAs form. Thus, the ability to predict RNA structure can play a key role in interpreting or designing experiments to elucidate the mechanism or function of a ncRNA. RNA structure is defined at three levels: primary or the sequence of the RNA, secondary or the set of canonical base pairs, and tertiary, which is the three-dimensional structure of the RNA. Computational tools have been developed where the goal is to predict an

Authors Raphael Angelo I. Zambrano and Carmen Hernandez-Perez have equally contributed to this chapter.

James Chappell and Melissa K. Takahashi (eds.), *Riboregulator Design and Analysis*,
Methods in Molecular Biology, vol. 2518, https://doi.org/10.1007/978-1-0716-2421-0_15,

RNA's secondary or tertiary structure from its sequence to aid in the elucidation of the structure-function relationship of RNAs.

In this chapter, we introduce computational tools that predict the secondary structure of RNAs using thermodynamic principles. This includes information regarding the influence of experimental data and alignment of homologous RNA sequences on structure prediction. We then describe programs that build upon the secondary structure prediction algorithms to aid in specific areas of RNA research, namely, predicting natural ncRNA targets and designing synthetic RNAs. Due to the rising interest in ncRNAs, many tools have been developed; thus, we have limited the scope of this review to those available as webservers and those most cited in literature.

2 RNA Secondary Structure Prediction

2.1 Structure Prediction of Single Sequences via Free Energy Minimization

RNA secondary structure prediction utilizes dynamic programming algorithms and thermodynamics to determine possible secondary structures for a given RNA sequence. The algorithms determine the intramolecular base-pairing that provides the lowest free energy change between the unfolded and folded state of the RNA. The structure with the lowest Gibbs free energy change is assumed to be the most prevalent at equilibrium. The free energy of a sequence can be estimated using a nearest neighbor model whereby the free energy change for forming a base pair depends on the sequence identities of that pair and the base pairs immediately neighboring it. Nearest neighbor parameters have been measured experimentally by several groups using optical melting studies [5–10]. Separate rules exist for predicting stabilities of different structures such as helices, hairpin loops, bulge loops, etc.

The algorithms search the entire set of possible secondary structures to find the lowest free energy structure; however, the assumption that an RNA is at equilibrium or of the suitability of the nearest neighbor parameters may not hold true for all RNAs. A partition function, the sum of equilibrium constants for all possible secondary structures for a given sequence, is calculated to help determine the quality of the structure prediction. The probability of a specific structure is then the equilibrium constant for that structure divided by the partition function. A similar calculation can be done to determine the probability of a specific base pair occurring. This allows for the minimum free energy structure to be annotated with base pair probabilities as done by programs such as **Fold** [11, 12], **RNAfold** [13, 14], and **NUPACK Analysis** [15] (*see* Table 1). An alternative approach is to assemble a structure composed of the highest probable pairs as calculated from the partition function. Base pairs that exceed a given probability threshold are represented in the structure. **MaxExpect** is a program that calculates these structures which are also known as maximum

Table 1
RNA secondary structure prediction programs

Program	Websuite	Type of predicted structure	Options	URL
Fold	RNAstructure	Minimum free energy, interval of energy	Pairing restraints, modifications, SHAPE data, temperature	https://rna.urmc.rochester.edu/RNAstructureWeb/Servers/Fold/Fold.html
RNAfold	Vienna	Minimum free energy, base pair probabilities	Structure constraints, SHAPE data, temperature, energy parameters	http://rna.tbi.univie.ac.at/cgi-bin/RNAWebSuite/RNAfold.cgi
RNA Folding Form	UNAFold/mFold	Minimum free energy	Folding constraints, temperature (V2.3)	http://www.unafold.org/mfold/applications/rna-folding-form.php
Analysis	NUPACK	Minimum free energy, base pair probabilities	Temperature, energy parameters, pseudoknots, specify ordered complexes	http://nupack.org/partition/new
MaxExpect	RNAstructure	Highest probable pairs		https://rna.urmc.rochester.edu/RNAstructureWeb/Servers/MaxExpect/MaxExpect.html
Stochastic	RNAstructure	Ensemble of structures from stochastic sampling		https://rna.urmc.rochester.edu/RNAstructureWeb/Servers/stochastic/stochastic.html

expected accuracy structures [16]. A third method accounts for suboptimal, low free energy structures using stochastic sampling. This is particularly useful for RNAs that are known to populate more than one structure such as riboswitches. Structures are selected from the ensemble according to their probability of occurring [17]. One program that implements this method is **Stochastic** from RNAstructure [18].

2.2 Prediction of Pseudoknots

An important aspect of RNA secondary structure is the formation of stem-loops that play a key role in the interactions of an RNA with other molecules. One such interaction creates a pseudoknot where the apical loop in the stem-loop structure base pairs with another region of the same or different RNA strand. These pseudoknots play important roles in RNA processing and gene expression regulation mechanisms [19]. The prediction of pseudoknots is considered computationally complex (NP-hard); thus, to reduce

Table 2
Programs that predict pseudoknots

Program	Notes	URL
NUPACK (Analysis)	Enable pseudoknots in advanced options	http://nupack.org/partition/new
HotKnots	Webserver limits RNA length to 100 nucleotides	http://www.cs.ubc.ca/labs/beta/Software/HotKnots/
ProbKnot	Multiple iterations may improve prediction	https://rna.urmc.rochester.edu/RNAstructureWeb/Servers/ProbKnot/ProbKnot.html
IPknot	Allows multiple alignment input	http://rtips.dna.bio.keio.ac.jp/ipknot/

computation, the programs listed in Table 1 ignore pseudoknot formation by default to simplify RNA secondary structure prediction. However, given the importance of pseudoknots in ncRNA function, several approaches have been developed to predict RNA structures with pseudoknots (*see* Table 2) [20–26].

NUPACK Analysis has an advanced option that allows for prediction of pseudoknots with limited topology in noninteracting strands [20, 21]. **HotKnots** employs a heuristic algorithm that adds substructures one at a time to partially formed structures. Several different substructures are considered at once, and the structures with lower free energies are provided [24]. Version 2.0 of HotKnots provides updated energy parameters that improve prediction accuracy; however, a higher miss rate was also observed [27]. **ProbKnot** first predicts base pair probabilities using a partition function that excludes pseudoknotted structures. ProbKnot then evaluates the individual helices predicted in the unpseudoknotted structures and assembles a type of maximum expected accuracy structure that may include multiple helices [25]. **IPknot** uses an integer programming algorithm to determine the maximum expected accuracy structure. IPknot also allows for the prediction of common secondary structures with pseudoknots when a multiple alignment of RNA sequences is provided [26] (*see* Subheading 2.3 for multiple alignment). In comparisons of prediction accuracy on sets of known RNA structures, HotKnots outperformed NUPACK [24] while ProbKnot outperformed HotKnots [25]. IPKnot generally had improved accuracy over ProbKnot except in the case of 5S rRNA sequences [28, 29].

2.3 Incorporating Experimental Data

One method to improve the accuracy of RNA secondary structure prediction is to incorporate experimental data. Experimental methods for analyzing RNA structure utilize enzymatic and chemical probes. Enzymatic probes are nucleases that cleave RNAs at specific structures or base patterns [30]. Chemical probes react with RNAs, and a variety of probes have been developed to provide structural information under specific conditions [31]. Results from enzymatic digest experiments can be incorporated as constraints or restraints where specific nucleotides can be set to be either single-stranded or base-paired (see Table 1).

A popular chemical probing method that RNA structure prediction programs have incorporated is SHAPE (Selective 2-'-Hydroxyl Acylation analyzed by Primer Extension). SHAPE utilizes chemical probes that react with the RNA in a structure-dependent manner [32, 33]. The location of the modifications are identified using capillary gel electrophoresis [32] or high-throughput sequencing [34]. Chemical probes that are compatible with SHAPE have been reviewed by Strobel et al. [31], some of which can be used in vivo. The output of a SHAPE experiment is a set of reactivities that indicate the likelihood of a nucleotide being paired or unpaired. Three methods for incorporating SHAPE data involve (1) the addition of pseudo-free energy terms to restrain the algorithm [35], (2) the use of pseudo-energy terms for each nucleotide position rather than only for nucleotides in base stacking positions [36], and (3) the use of an iterative algorithm that perturbs the energy model to better match the SHAPE data [37]. **Fold** from RNAstructure uses the Deigan et al. [35] method, while all three methods can be selected from using Vienna's **RNAfold** or **RNAprobing** [38]. While all methods accounting for SHAPE data improve the accuracy of structure prediction compared to no experimental data, the optimal method varies by RNA.

2.4 Conserved Secondary Structure Prediction

Accuracy improves by predicting a secondary structure common to multiple or homologous sequences. The two most common approaches to execute this type of prediction is to either simultaneously fold and align two or more sequences or to first align the sequences and then fold them. Although not reviewed here, RNA-shapes [39] implements a third approach that first folds the sequences and then aligns the structures. A summary of the programs discussed can be found in Table 3.

In the fold and align strategy, programs simultaneously find the lowest free energy structure of each sequence and the optimum alignment of sequences. **Dynalign** is one such program that rules out base pairs that can only exist in structures with high free energies or alignments that are unlikely [40]. Dynalign, however, can only be used to align and fold two RNA sequences. For three or more sequences, **Multilign** builds upon Dynalign by using Dynalign to do progressive calculations for additional sequences [41].

Another approach to the simultaneous fold and align method is implemented by **LocARNA** which maximizes base pair probabilities and performs a local pairwise structural alignment [42, 43]. **PARTS** (Probabilistic Alignment for RNA joinT Secondary structure prediction) uses a matched helical region motif to

Table 3
Programs that predict conserved secondary structures

Program	Notes	URL
Dynalign	Align and fold. Limited to two sequences. 500 nucleotide limit	https://rna.urmc.rochester.edu/RNAstructureWeb/Servers/dynalign/dynalign.html
Multilign	Align and fold. Maximum of 10 sequences. 600 nucleotide limit	https://rna.urmc.rochester.edu/RNAstructureWeb/Servers/multilign/multilign.html
LocARNA	Align and fold. 2000 nucleotide limit	http://rna.informatik.uni-freiburg.de/LocARNA/Input.jsp
PARTS	Align and fold. Limited to 2 sequences. 500 nucleotide limit	https://rna.urmc.rochester.edu/RNAstructureWeb/Servers/PARTS/PARTS.html
RNAalifold	Align then fold. Requires alignment input	http://rna.tbi.univie.ac.at/cgi-bin/RNAWebSuite/RNAalifold.cgi
TurboFold	Align then fold. Maximum of 12 sequences. 2500 nucleotide limit	https://rna.urmc.rochester.edu/RNAstructureWeb/Servers/TurboFold/TurboFold.html

align conserved structures, provides base pair probabilities, and, as an option, can be set to do stochastic sampling of structures [18, 44].

In the align and then fold strategy, a multiple sequence alignment is performed prior to the prediction of the lowest free energy structure from the consensus sequence. The advantage of this approach is speed; however, the quality of the predicted structure depends on the quality of the alignment. The **RNAalifold** webserver is one such program that requires a pre-calculated alignment as an input [45]. Alternatively, **TurboFold** starts with a set of homologous RNA sequences and outputs base-pairing probabilities for each of the sequences [46]. The algorithm iteratively performs a pairwise alignment and computes the base-pairing probabilities using a nearest neighbor thermodynamic model. The contribution of each of the other sequences is obtained by mapping the estimated base-pairing probabilities for the sequences from the previous iteration. The final output utilizes a maximum expected accuracy algorithm that allows for variable folding domains in some sequences [46].

2.5 Comparing Folding Tools

When determining which structure prediction tool to use, it is important to remember that none of the programs described here can perfectly predict all bases-pairs in all types of RNA structures. As discussed above, accuracies of predictions improve when experimental data can be provided, or homologous sequences can be aligned. When investigators develop new algorithms or compare existing tools, the methods are evaluated by comparing predicted structures to solved crystal structures using two metrics: sensitivity and positive predictive value (PPV). Sensitivity is the fraction of base pairs in the known structure that were predicted, and PPV is the fraction of predicted base pairs that are correct. Sensitivity and PPV are calculated as follows:

$$\text{Sensitivity} = \text{true positive pairs}/(\text{true positive pairs} + \text{false negative pairs}) \quad (1)$$

$$\text{PPV} = \text{true positive pairs}/(\text{true positive pairs} + \text{false positive pairs}) \quad (2)$$

As an example of how predictions are improved when aligning sequences, Dynalign improves accuracy of prediction over Fold for the small subunit rRNA from 47.4% to 73.3% in sensitivity and from 47.5% to 73.1% in PPV [47]. One can look into comparison studies, which are often presented when a new algorithm is developed, or independent studies such as the one presented by Puton et al. [48]. Ultimately, it is worth using multiple programs or approaches and comparing results.

3 RNA-RNA Interactions

3.1 Predicting RNA-RNA Interactions

The approaches for predicting the secondary structure of a single RNA discussed in Subheading 2.1 have also been applied to interacting RNA molecules. Instead of providing the details behind the algorithms here, we describe the approach taken to include RNA-RNA interactions and discuss what some programs allow researchers to investigate. A summary of the programs can be found in Table 4.

One type of program takes two RNA sequences and folds them into their lowest hybrid free energy structure; however, it does not calculate intramolecular base pairs in either of the sequences. This is implemented in **DuplexFold** [12] and **Two State Melting Hybridization** [49]. Other programs that do allow intramolecular base-pairing include **bifold** [12], **RNAcofold** [50], **RNAup** (ViennaRNA) [51], and **NUPACK Analysis** [15].

In addition to predicting hybridized heterodimer structures, NUPACK Analysis and RNAcofold will predict if homodimer structures will form. Both programs also take into consideration user-defined concentrations of the input RNA sequences and will calculate equilibrium concentrations of each molecular species [14, 15, 50]. Furthermore, NUPACK allows input of up to ten strands where the user can also set the maximum complex size [15].

Table 4
Programs for RNA-RNA interaction prediction

Program	Intramolecular pairing	URL
DuplexFold		https://rna.urmc.rochester.edu/RNAstructureWeb/Servers/DuplexFold/DuplexFold.html
bifold	√	https://rna.urmc.rochester.edu/RNAstructureWeb/Servers/bifold/bifold.html
Two State Melting Hybridization		http://www.unafold.org/Dinamelt/applications/two-state-melting-hybridization.php
RNAcofold	√	http://rna.tbi.univie.ac.at/cgi-bin/RNAWebSuite/RNAcofold.cgi
RNAup	√	http://rna.tbi.univie.ac.at//cgi-bin/RNAWebSuite/RNAup.cgi
NUPACK (Analysis)	√	http://nupack.org/partition/new

The programs listed above do not account for pseudoknots, and therefore some RNA-RNA interaction motifs, such as kissing hairpins, cannot be predicted. RNAup accounts for interactions with unpaired sequences within hairpin, interior, or multibranch loop structures. This is done by calculating the energy required to open or unfold intramolecular structures for stretches of bases within the sequence [14, 51].

The choice of program will depend on the question being asked and the available information. Evaluating the output from multiple programs is often useful. In one comparison study, Umu and Gardner compare performance of all the programs discussed here except Two State Melting Hybridization plus others against experimentally verified datasets that include microRNA (miRNA), small interfering RNA (siRNA), and small RNA (sRNA) with their corresponding mRNA targets [52]. In an overall comparison, RNAup performed the best with regard to sensitivity and PPV. The authors also report results for specific subsets of data (e.g., eukaryotic vs. bacterial sRNA) and algorithm run time [52].

3.2 Applications of RNA-RNA Interaction Prediction

A common use for RNA-RNA binding prediction programs is to investigate ncRNA interactions with their mRNA targets. The interest in ncRNA regulation in both eukaryotes (siRNA, miRNA) and prokaryotes (sRNA) has led to the development of programs tailored for this application. These programs often directly use, modify, or combine some of the secondary structure and RNA-RNA interaction prediction programs described above (*see* Table 5).

Table 5
Programs for ncRNA targets

Program	Purpose	URL
RNAxs	siRNA design	http://rna.tbi.univie.ac.at/cgi-bin/RNAxs/RNAxs.cgi
OligoWalk	siRNA design	http://rna.urmc.rochester.edu/cgi-bin/server_exe/oligowalk/oligowalk_form.cgi
RNApredator	sRNA target prediction	http://rna.tbi.univie.ac.at/cgi-bin/RNApredator/target_search.cgi
IntaRNA	sRNA target prediction, RNA-RNA interaction	http://rna.informatik.uni-freiburg.de/IntaRNA/Input.jsp
CopraRNA	sRNA target prediction	http://rna.informatik.uni-freiburg.de/CopraRNA/Input.jsp
TargetRNA2	sRNA target prediction	http://cs.wellesley.edu/~btjaden/TargetRNA2/

RNA interference (RNAi) is a commonly used method to investigate the effects of knocking down expression of specific genes to study their cellular function [53]. **RNAxs** [54] and **OligoWalk** [55, 56] are two programs that evaluate the secondary structure of the provided mRNA sequence for accessibility and provide a set of potential siRNA sequences along with their hybridization energies. In addition, RNAxs provides an accessibility plot for the region of the mRNA that the siRNA targets.

Small RNAs are known to regulate gene expression through imperfect base-pairing with an mRNA target [57, 58]. The imperfect base-pairing allows sRNAs to regulate multiple targets; thus, researchers are often interested in predicting mRNA targets of known sRNA sequences. To identify potential mRNA targets, programs search through genomes, and therefore the computational time required can be significantly longer than the simpler secondary structure prediction described in Subheading 2.

RNApredator [59] utilizes RNAplex from ViennaRNA, which is another program that predicts RNA-RNA interactions; however, RNAplex is not available as a webserver and thus was not discussed above. RNAplex neglects intramolecular base pairs but does have an algorithm to evaluate accessibility of the mRNA [60]. RNApredator allows the user to select from a set of bacterial genomes, or an NCBI accession number or taxonomy ID can be entered. Results can be filtered by the position it interacts with the mRNA relative to the translation start site. Selected results can be directed for further processing by RNAup [59].

IntaRNA [61, 62] predicts interactions between two RNAs and can be used in the same manner as the programs described in Subheading 3.1. However, it was designed to predict mRNA target sites for sRNAs. IntaRNA uses a similar algorithm to RNAup where a combined energy is calculated that sums the free energy of hybridization and the free energy required for making the interaction sites accessible. In addition, IntaRNA adds a seed region, which is a short (two to eight nucleotides) stretch of RNA that is thought to be key in initiating the interaction between the ncRNA and mRNA [61]. The user can provide a target sequence or an NCBI reference genome. IntaRNA provides the top 100 interactions with their predicted base-pairings. **CopraRNA** utilizes the IntaRNA algorithm to evaluate homologous sequences [63, 64]. CopraRNA requires a minimum of three homologous sRNA sequences from distinct organisms. The IntaRNA results from each organism are clustered, and gene targets that are present in less than 50% of the investigated genomes are discarded. The algorithm also considers the relatedness of the species based on 16S rDNA phylogenetic trees [63]. While CopraRNA requires a minimum of three species, it provides the top 100 predicted mRNA targets for a single species that the user selects.

TargetRNA2 [65], updated from the original TargetRNA [66, 67], also considers conserved sequences in its algorithm to predict mRNA targets. TargetRNA2 compares the input sRNA sequences to all replicons available in RefSeq [68] and considers conserved regions more likely to be involved in the RNA-RNA interaction. The structures of both the sRNA and mRNA are evaluated for accessibility. The structure of the sRNA is evaluated using RNAfold [13, 14], while the mRNA is evaluated using RNAplfold [69]. Finally, RNAduplex [14] is used to calculate the hybridization energy between the accessible regions of the sRNA and mRNA. Like the other programs discussed here, TargetRNA2 provides ranked results with predicted interactions.

4 Designing RNA Structures

Recently there has been growing interest in designing synthetic RNA structures that range in function including riboregulators that control both transcription [70] and translation [71–73], riboswitches [74, 75], sRNAs [76, 77], and biosensors [78–81]. Computational design tools play an important role in the development of new synthetic RNA riboregulators. The design of RNA sequences that fold into a specific structure was initially posed as the inverse RNA folding problem, and many approaches to solve the problem build on the principles and programs used to predict RNA secondary structures [82–84]. Below we discuss a subset of programs that design RNA structures (*see* Table 6). A more comprehensive list of RNA design tools can be found in Churkin et al. [85].

Table 6
RNA design programs

Program	Notes	URL
RNAinverse	Single sequence design	http://rna.tbi.univie.ac.at//cgi-bin/RNAWebSuite/RNAinverse.cgi
INFO-RNA	Single sequence design	http://rna.informatik.uni-freiburg.de/INFORNA/Input.jsp
RNAiFold	Single sequence and hybridized structure design	http://bioinformatics.bc.edu/clotelab/RNAiFold/
AntaRNA	Single sequence design; allows pseudoknots; define overall GC-content	http://rna.informatik.uni-freiburg.de/AntaRNA/Input.jsp
NUPACK (Design)	Multistate sequence design	http://nupack.org/design/new

RNAinverse uses an adaptive random walk algorithm to determine the RNA sequence for a given secondary structure [14, 82]. The algorithm starts by optimizing small structures and elongates from there instead of folding the entire sequence at once. The user provides a target structure and is allowed to define specific nucleotides within that structure. A minimum free energy result and thermodynamic ensemble prediction is provided. **INFO-RNA** built upon RNAinverse by developing an improved method for selecting the initial sequences to start the search [84]. In addition to providing constraints, INFO-RNA allows the user to define the number of constraint violations that is acceptable.

RNAiFold uses a constraint programming algorithm to determine minimum free energy structures using either RNAfold or RNAcofold [83, 86]. In addition to single RNA structures, RNAiFold can design hybridized RNA structures. There are two design modes: RNA-CP design and RNA synthetic design. RNA-CP design provides RNA sequences that fold into the provided target structure and sequence constraints. RNA synthetic design will determine RNA sequences according to a consensus structure determined from the RNA families database (RFAM) [87]. The output structures for either mode can be set to report the minimum free energy structure or minimize with respect to free energy of the target structure or ensemble defect. In addition to sequence constraints, the user can provide a target amino acid sequence [86].

AntaRNA uses an ant colony optimization heuristic, which is a self-adjusting local search strategy, to design structures [88, 89]. Unlike the other programs discussed here, AntaRNA allows for design of two specific subclasses of pseudoknot structures (hairpin and kissing hairpin) by using hardcoded energy parameters. In addition to nucleotide constraints, AntaRNA allows the user to target a specific GC-content, which is known to play a role in the stability of the overall structure.

NUPACK Design optimizes results based on ensemble defect which means the calculated sequences are compared to the target structure and evaluated over the ensemble of unpseudoknotted secondary structures [90]. The resulting sequences can be directed to Analysis to determine the minimum free energy structure and base pair probabilities. In addition to nucleotide constraints, the user can forbid sequence motifs in IUPAC format. For example, strings of four or more of a single nucleotide can be prevented. The user can also set an external sequence constraint where a domain within a target structure is defined by a source sequence [15]. This capability is useful for designing structures that interact with biological sequences. Furthermore, NUPACK allows multistate sequence design, which is useful for programmable nucleic acid systems (DNA or RNA) that execute self-assembly and disassembly [91]. The user defines a reaction pathway of nucleic acid structures.

5 Conclusion

Here we have introduced computational tools that predict RNA secondary structure and ncRNA targets as well as tools that allow the user to design a synthetic RNA based on a defined secondary structure. We focused our review on programs that use thermodynamics to predict structures and on those that are widely accessible by webserver. The field of RNA structure prediction has grown to include tertiary structure prediction [92–94] and methods to predict structures based on machine learning [28, 29, 95]. A more inclusive list of RNA-based tools can be found on Wikipedia: https://en.wikipedia.org/wiki/List_of_RNA_structure_prediction_software.

Acknowledgments

R.A.I.Z. and C.H.P. were supported by the National Institute of General Medical Sciences of the National Institutes of Health under Award Number TL4GM118977. The content is solely the responsibility of the authors and does not necessarily represent the official views of the National Institutes of Health.

References

1. Tanner NK (1999) Ribozymes: the characteristics and properties of catalytic RNAs. FEMS Microbiol Rev 23:257–275. https://doi.org/10.1111/j.1574-6976.1999.tb00399.x
2. Waters LS, Storz G (2009) Regulatory RNAs in Bacteria. Cell 136:615–628. https://doi.org/10.1016/j.cell.2009.01.043

3. Chappell J, Takahashi MK, Meyer S et al (2013) The centrality of RNA for engineering gene expression. Biotechnol J 8:1379–1395. https://doi.org/10.1002/biot.201300018
4. O'Brien J, Hayder H, Zayed Y, Peng C (2018) Overview of microRNA biogenesis, mechanisms of actions, and circulation. Front Endocrinol (Lausanne) 9:402. https://doi.org/10.3389/fendo.2018.00402
5. Serra MJ, Turner DH (1995) Predicting thermodynamic properties of RNA. Methods Enzymol 259:242–261. https://doi.org/10.1016/0076-6879(95)59047-1
6. Xia T, SantaLucia J, Burkard ME et al (1998) Thermodynamic parameters for an expanded nearest-neighbor model for formation of RNA duplexes with Watson – Crick base pairs. Biochemistry 37:14719–14735. https://doi.org/10.1021/bi9809425
7. Mathews DH, Sabina J, Zuker M, Turner DH (1999) Expanded sequence dependence of thermodynamic parameters improves prediction of RNA secondary structure. J Mol Biol 288:911–940. https://doi.org/10.1006/jmbi.1999.2700
8. Mathews DH, Turner DH (2002) Experimentally derived nearest-neighbor parameters for the stability of RNA three- and four-way multibranch loops. Biochemistry 41:869–880. https://doi.org/10.1021/bi011441d
9. Mathews DH, Disney MD, Childs JL et al (2004) Incorporating chemical modification constraints into a dynamic programming algorithm for prediction of RNA secondary structure. Proc Natl Acad Sci U S A 101: 7287–7292. https://doi.org/10.1073/pnas.0401799101
10. Turner DH, Mathews DH (2009) NNDB: the nearest neighbor parameter database for predicting stability of nucleic acid secondary structure. Nucleic Acids Res 38. https://doi.org/10.1093/nar/gkp892
11. Reuter JS, Mathews DH (2010) RNAstructure: software for RNA secondary structure prediction and analysis. BMC Bioinf 11:129. https://doi.org/10.1186/1471-2105-11-129
12. Bellaousov S, Reuter JS, Seetin MG, Mathews DH (2013) RNAstructure: web servers for RNA secondary structure prediction and analysis. Nucleic Acids Res 41. https://doi.org/10.1093/nar/gkt290
13. Zuker M (2003) Mfold web server for nucleic acid folding and hybridization prediction. Nucleic Acids Res 31:3406–3415. https://doi.org/10.1093/nar/gkg595
14. Gruber AR, Lorenz R, Bernhart SH et al (2008) The Vienna RNA websuite. Nucleic Acids Res 36:W70–W74. https://doi.org/10.1093/nar/gkn188
15. Zadeh JN, Steenberg CD, Bois JS et al (2011) NUPACK: analysis and design of nucleic acid systems. J Comput Chem 32:170–173. https://doi.org/10.1002/jcc.21596
16. Lu ZJ, Gloor JW, Mathews DH (2009) Improved RNA secondary structure prediction by maximizing expected pair accuracy. RNA 15:1805–1813. https://doi.org/10.1261/rna.1643609
17. Ding Y, Lawrence CE (2003) A statistical sampling algorithm for RNA secondary structure prediction. Nucleic Acids Res 31:7280–7301. https://doi.org/10.1093/nar/gkg938
18. Harmanci AO, Sharma G, Mathews DH (2009) Stochastic sampling of the RNA structural alignment space. Nucleic Acids Res 37: 4063–4075. https://doi.org/10.1093/nar/gkp276
19. Peselis A, Serganov A (2014) Structure and function of pseudoknots involved in gene expression control. Wiley Interdiscip Rev RNA 5:803–822. https://doi.org/10.1002/wrna.1247
20. Dirks RM, Pierce NA (2003) A partition function algorithm for nucleic acid secondary structure including pseudoknots. J Comput Chem 24:1664–1677. https://doi.org/10.1002/jcc.10296
21. Dirks RM, Pierce NA (2004) An algorithm for computing nucleic acid base-pairing probabilities including pseudoknots. J Comput Chem 25:1295–1304. https://doi.org/10.1002/jcc.20057
22. Ruan J, Stormo GD, Zhang W (2004) An Iterated loop matching approach to the prediction of RNA secondary structures with pseudoknots. Bioinformatics 20:58–66. https://doi.org/10.1093/bioinformatics/btg373
23. Reeder J, Giegerich R (2004) Design, implementation and evaluation of a practical pseudoknot folding algorithm based on thermodynamics. BMC Bioinf 5:1–12. https://doi.org/10.1186/1471-2105-5-104
24. Ren J, Rastegari B, Condon A, Hoos HH (2005) HotKnots: heuristic prediction of RNA secondary structures including pseudoknots. RNA 11:1494–1504. https://doi.org/10.1261/rna.7284905
25. Bellaousov S, Mathews DH (2010) ProbKnot: fast prediction of RNA secondary structure including pseudoknots. RNA 16:1870–1880. https://doi.org/10.1261/rna.2125310

26. Sato K, Kato Y, Hamada M et al (2011) IPknot: fast and accurate prediction of RNA secondary structures with pseudoknots using integer programming. Bioinformatics 27: 85–93. https://doi.org/10.1093/bioinformatics/btr215
27. Andronescu MS, Pop C, Condon AE (2010) Improved free energy parameters for RNA pseudoknotted secondary structure prediction. RNA 16:26–42. https://doi.org/10.1261/rna.1689910
28. Wang L, Liu Y, Zhong X et al (2019) DMFold: a novel method to predict RNA secondary structure with pseudoknots based on deep learning and improved base pair maximization principle. Front Genet 10:143. https://doi.org/10.3389/fgene.2019.00143
29. Mao K, Wang J, Xiao Y (2020) Prediction of RNA secondary structure with pseudoknots using coupled deep neural networks. Biophys Rep 6:146–154. https://doi.org/10.1007/s41048-020-00114-x
30. Knapp G (1989) Enzymatic approaches to probing of RNA secondary and tertiary structure. Methods Enzymol 180:192–212. https://doi.org/10.1016/0076-6879(89)80102-8
31. Strobel EJ, Yu AM, Lucks JB (2018) High-throughput determination of RNA structures. Nat Rev Genet 19:615–634. https://doi.org/10.1038/s41576-018-0034-x
32. Merino EJ, Wilkinson KA, Coughlan JL, Weeks KM (2005) RNA structure analysis at single nucleotide resolution by Selective 2-′-Hydroxyl Acylation and Primer Extension (SHAPE). J Am Chem Soc 127:4223–4231. https://doi.org/10.1021/ja043822v
33. Mortimer SA, Weeks KM (2007) A fast-acting reagent for accurate analysis of RNA secondary and tertiary structure by SHAPE chemistry. J Am Chem Soc 129:4144–4145. https://doi.org/10.1021/ja0704028
34. Lucks JB, Mortimer SA, Trapnell C et al (2011) Multiplexed RNA structure characterization with selective 2'-hydroxyl acylation analyzed by primer extension sequencing (SHAPE-Seq). Proc Natl Acad Sci U S A 108: 11063–11068. https://doi.org/10.1073/pnas.1106501108
35. Deigan KE, Li TW, Mathews DH, Weeks KM (2009) Accurate SHAPE-directed RNA structure determination. Proc Natl Acad Sci 106: 97–102. https://doi.org/10.1073/pnas.0806929106
36. Zarringhalam K, Meyer MM, Dotu I et al (2012) Integrating chemical footprinting data into RNA secondary structure prediction. PLoS One 7:45160. https://doi.org/10.1371/journal.pone.0045160
37. Washietl S, Hofacker IL, Stadler PF, Kellis M (2012) RNA folding with soft constraints: reconciliation of probing data and thermodynamic secondary structure prediction. Nucleic Acids Res 40:4261–4272. https://doi.org/10.1093/nar/gks009
38. Lorenz R, Luntzer D, Hofacker IL et al (2016) SHAPE directed RNA folding. Bioinformatics 32:145–147. https://doi.org/10.1093/bioinformatics/btv523
39. Steffen P, Voß B, Rehmsmeier M et al (2006) RNAshapes: an integrated RNA analysis package based on abstract shapes. Bioinformatics 22:500–503. https://doi.org/10.1093/bioinformatics/btk010
40. Mathews DH, Turner DH (2002) Dynalign: an algorithm for finding the secondary structure common to two RNA sequences. J Mol Biol 317:191–203. https://doi.org/10.1006/jmbi.2001.5351
41. Xu Z, Mathews DH (2011) Multilign: an algorithm to predict secondary structures conserved in multiple RNA sequences. Bioinformatics 27:626–632. https://doi.org/10.1093/bioinformatics/btq726
42. Will S, Reiche K, Hofacker IL et al (2007) Inferring noncoding RNA families and classes by means of genome-scale structure-based clustering. PLoS Comput Biol 3:680–691. https://doi.org/10.1371/journal.pcbi.0030065
43. Smith C, Heyne S, Richter AS et al (2010) Freiburg RNA tools: a web server integrating IntaRNA, ExpaRNA and LocARNA. Nucleic Acids Res 38:W373–W377. https://doi.org/10.1093/nar/gkq316
44. Harmanci AO, Sharma G, Mathews DH (2008) PARTS: probabilistic alignment for RNA joinT secondary structure prediction. Nucleic Acids Res 36:2406–2417. https://doi.org/10.1093/nar/gkn043
45. Bernhart SH, Hofacker IL, Will S et al (2008) RNAalifold: improved consensus structure prediction for RNA alignments. BMC Bioinf 9: 1–13. https://doi.org/10.1186/1471-2105-9-474
46. Harmanci AO, Sharma G, Mathews DH (2011) TurboFold: iterative probabilistic estimation of secondary structures for multiple RNA sequences. BMC Bioinf 12:1–22. https://doi.org/10.1186/1471-2105-12-108
47. Seetin MG, Mathews DH (2012) RNA structure prediction: an overview of methods.

Methods Mol Biol 905:99–122. https://doi.org/10.1007/978-1-61779-949-5_8

48. Puton T, Kozlowski LP, Rother KM, Bujnicki JM (2013) CompaRNA: a server for continuous benchmarking of automated methods for RNA secondary structure prediction. Nucleic Acids Res 41:4307–4323. https://doi.org/10.1093/nar/gkt101
49. Markham NR, Zuker M (2008) UNAFold: software for nucleic acid folding and hybridization. Methods Mol Biol 453:3–31. https://doi.org/10.1007/978-1-60327-429-6_1
50. Bernhart SH, Tafer H, Mückstein U et al (2006) Partition function and base pairing probabilities of RNA heterodimers. Algorithms Mol Biol 1:1–10. https://doi.org/10.1186/1748-7188-1-3
51. Mückstein U, Tafer H, Hackermüller J et al (2006) Thermodynamics of RNA-RNA binding. Bioinformatics 22:1177–1182. https://doi.org/10.1093/bioinformatics/btl024
52. Umu SU, Gardner PP (2017) A comprehensive benchmark of RNA-RNA interaction prediction tools for all domains of life. Bioinformatics 33:988–996. https://doi.org/10.1093/bioinformatics/btw728
53. Han H (2018) RNA interference to knock down gene expression. Methods Mol Biol 1706:293–302. https://doi.org/10.1007/978-1-4939-7471-9_16
54. Tafer H, Ameres SL, Obernosterer G et al (2008) The impact of target site accessibility on the design of effective siRNAs. Nat Biotechnol 26:578–583. https://doi.org/10.1038/nbt1404
55. Mathews DH, Burkard ME, Freier SM et al (1999) Predicting oligonucleotide affinity to nucleic acid targets. RNA 5:1458–1469. https://doi.org/10.1017/S1355838299991148
56. Lu ZJ, Mathews DH (2008) Efficient siRNA selection using hybridization thermodynamics. Nucleic Acids Res 36:640–647. https://doi.org/10.1093/nar/gkm920
57. Gottesman S, Storz G (2011) Bacterial small RNA regulators: versatile roles and rapidly evolving variations. Cold Spring Harb Perspect Biol 3. https://doi.org/10.1101/cshperspect.a003798
58. Storz G, Vogel J, Wassarman KM (2011) Regulation by small RNAs in bacteria: expanding frontiers. Mol Cell 43:880–891. https://doi.org/10.1016/j.molcel.2011.08.022
59. Eggenhofer F, Tafer H, Stadler PF, Hofacker IL (2011) RNApredator: fast accessibility-based prediction of sRNA targets. Nucleic Acids Res 39:W149–W154. https://doi.org/10.1093/nar/gkr467
60. Tafer H, Hofacker IL (2008) RNAplex: a fast tool for RNA-RNA interaction search. Bioinformatics 24:2657–2663. https://doi.org/10.1093/bioinformatics/btn193
61. Busch A, Richter AS, Backofen R (2008) IntaRNA: efficient prediction of bacterial sRNA targets incorporating target site accessibility and seed regions. Bioinformatics 24:2849–2856. https://doi.org/10.1093/bioinformatics/btn544
62. Mann M, Wright PR, Backofen R (2017) IntaRNA 2.0: enhanced and customizable prediction of RNA-RNA interactions. Nucleic Acids Res 45:W435–W439. https://doi.org/10.1093/nar/gkx279
63. Wright PR, Richter AS, Papenfort K et al (2013) Comparative genomics boosts target prediction for bacterial small RNAs. Proc Natl Acad Sci U S A 110:E3487–E3496. https://doi.org/10.1073/pnas.1303248110
64. Wright PR, Georg J, Mann M et al (2014) CopraRNA and IntaRNA: predicting small RNA targets, networks and interaction domains. Nucleic Acids Res 42:119–123. https://doi.org/10.1093/nar/gku359
65. Kery MB, Feldman M, Livny J, Tjaden B (2014) TargetRNA2: identifying targets of small regulatory RNAs in bacteria. Nucleic Acids Res 42:W124–W129. https://doi.org/10.1093/nar/gku317
66. Tjaden B, Goodwin SS, Opdyke JA et al (2006) Target prediction for small, noncoding RNAs in bacteria. Nucleic Acids Res 34:2791–2802. https://doi.org/10.1093/nar/gkl356
67. Tjaden B (2008) TargetRNA: a tool for predicting targets of small RNA action in bacteria. Nucleic Acids Res 36:109–113. https://doi.org/10.1093/nar/gkn264
68. Pruitt KD, Tatusova T, Brown GR, Maglott DR (2012) NCBI Reference Sequences (RefSeq): current status, new features and genome annotation policy. Nucleic Acids Res 40:D130–D135. https://doi.org/10.1093/nar/gkr1079
69. Bernhart SH, Hofacker IL, Stadler PF (2006) Local RNA base pairing probabilities in large sequences. Bioinformatics 22:614–615. https://doi.org/10.1093/bioinformatics/btk014
70. Chappell J, Westbrook A, Verosloff M, Lucks JB (2017) Computational design of small transcription activating RNAs for versatile and dynamic gene regulation. Nat Commun 8: 1–12. https://doi.org/10.1038/s41467-017-01082-6

71. Mutalik VK, Qi L, Guimaraes JC et al (2012) Rationally designed families of orthogonal RNA regulators of translation. Nat Chem Biol 8:447–454. https://doi.org/10.1038/nchembio.919
72. Green AA, Silver PA, Collins JJ, Yin P (2014) Toehold switches: de-novo-designed regulators of gene expression. Cell 159:925–939. https://doi.org/10.1016/j.cell.2014.10.002
73. Rodrigo G, Landrain TE, Jaramillo A (2012) De novo automated design of small RNA circuits for engineering synthetic riboregulation in living cells. Proc Natl Acad Sci U S A 109: 15271–15276. https://doi.org/10.1073/pnas.1203831109
74. Wachsmuth M, Findeiß S, Weissheimer N et al (2013) De novo design of a synthetic riboswitch that regulates transcription termination. Nucleic Acids Res 41:2541–2551. https://doi.org/10.1093/nar/gks1330
75. Wu MJ, Andreasson JOL, Kladwang W et al (2019) Automated design of diverse standalone riboswitches. ACS Synth Biol 8: 1838–1846. https://doi.org/10.1021/acssynbio.9b00142
76. Na D, Yoo SM, Chung H et al (2013) Metabolic engineering of Escherichia coli using synthetic small regulatory RNAs. Nat Biotechnol 31:170–174. https://doi.org/10.1038/nbt.2461
77. Hoynes-O'Connor A, Moon TS (2016) Development of design rules for reliable antisense RNA behavior in E. coli. ACS Synth Biol 5: 1441–1454. https://doi.org/10.1021/acssynbio.6b00036
78. Pardee K, Green AA, Ferrante T et al (2014) Paper-based synthetic gene networks. Cell 159: 940–954. https://doi.org/10.1016/j.cell.2014.10.004
79. Ma D, Shen L, Wu K et al (2018) Low-cost detection of norovirus using paper-based cell-free systems and synbody-based viral enrichment. Synth Biol 3. https://doi.org/10.1093/SYNBIO/YSY018
80. Takahashi MK, Tan X, Dy AJ et al (2018) A low-cost paper-based synthetic biology platform for analyzing gut microbiota and host biomarkers. Nat Commun 9:3347. https://doi.org/10.1038/s41467-018-05864-4
81. Hong F, Ma D, Wu K et al (2020) Precise and programmable detection of mutations using ultraspecific riboregulators. Cell 180: 1018–1032.e16. https://doi.org/10.1016/j.cell.2020.02.011
82. Hofacker IL, Fontana W, Stadler PF et al (1994) Fast folding and comparison of RNA secondary structures. Monatshefte für Chemie Chem Mon 125:167–188. https://doi.org/10.1007/BF00818163
83. Garcia-Martin JA, Clote P, Dotu I (2013) RNAiFOLD: a constraint programming algorithm for rna inverse folding and molecular design. J Bioinforma Comput Biol 11. https://doi.org/10.1142/S0219720013500017
84. Busch A, Backofen R (2006) INFO-RNA-a fast approach to inverse RNA folding. Bioinformatics 22:1823–1831. https://doi.org/10.1093/bioinformatics/btl194
85. Churkin A, Retwitzer MD, Reinharz V et al (2018) Design of RNAs: comparing programs for inverse RNA folding. Brief Bioinform 19: 350–358. https://doi.org/10.1093/bib/bbw120
86. Garcia-Martin JA, Dotu I, Clote P (2015) RNAiFold 2.0: a web server and software to design custom and Rfam-based RNA molecules. Nucleic Acids Res 43:W513–W521. https://doi.org/10.1093/nar/gkv460
87. Nawrocki EP, Burge SW, Bateman A et al (2015) Rfam 12.0: updates to the RNA families database. Nucleic Acids Res 43: D130–D137. https://doi.org/10.1093/nar/gku1063
88. Kleinkauf R, Mann M, Backofen R (2015) AntaRNA: ant colony-based RNA sequence design. Bioinformatics 31:3114–3121. https://doi.org/10.1093/bioinformatics/btv319
89. Kleinkauf R, Houwaart T, Backofen R, Mann M (2015) antaRNA – multi-objective inverse folding of pseudoknot RNA using ant-colony optimization. BMC Bioinf 16:389. https://doi.org/10.1186/s12859-015-0815-6
90. Zadeh JN, Wolfe BR, Pierce NA (2011) Nucleic acid sequence design via efficient ensemble defect optimization. J Comput Chem 32:439–452. https://doi.org/10.1002/jcc.21633
91. Wolfe BR, Porubsky NJ, Zadeh JN et al (2017) Constrained multistate sequence design for nucleic acid reaction pathway engineering. J Am Chem Soc 139:3134–3144. https://doi.org/10.1021/jacs.6b12693
92. Sharma S, Ding F, Dokholyan NV (2008) IFoldRNA: three-dimensional RNA structure prediction and folding. Bioinformatics 24: 1951–1952. https://doi.org/10.1093/bioinformatics/btn328
93. Popenda M, Szachniuk M, Antczak M et al (2012) Automated 3D structure composition for large RNAs. Nucleic Acids Res 40: e112–e112. https://doi.org/10.1093/nar/gks339

94. Parisien M, Major F (2008) The MC-Fold and MC-Sym pipeline infers RNA structure from sequence data. Nature 452:51–55. https://doi.org/10.1038/nature06684
95. Singh J, Hanson J, Paliwal K, Zhou Y (2019) RNA secondary structure prediction using an ensemble of two-dimensional deep neural networks and transfer learning. Nat Commun 10:1–13. https://doi.org/10.1038/s41467-019-13395-9

Chapter 16

Single-Molecule FRET Studies of RNA Structural Rearrangements and RNA-RNA Interactions

Ewelina M. Małecka, Boyang Hua, and Sarah A. Woodson

Abstract

RNA-guided regulation of gene expression is found in all cell types. In this mode of regulation, antisense interactions between the regulatory RNA and its target are typically facilitated by a protein partner. Single-molecule fluorescence microscopy is a powerful tool for dissecting the conformational states and intermediates that contribute to target recognition. This chapter describes protocols for studying target recognition by bacterial small RNAs and their chaperone Hfq on the single-molecule level, using a total internal reflection fluorescence microscope. The sections cover the design of suitable RNA substrates for sRNA-mRNA annealing reactions, preparation of internally labeled mRNA for detecting conformational changes in the target, and key steps of the data analysis. These protocols can be adapted to other RNA-binding proteins that chaperone RNA interactions.

Key words Hfq, RNA chaperone, Small RNA, Antisense-RNA interactions, Single-molecule fluorescence, TIRF

1 Introduction

Noncoding RNAs directly base pair with target mRNA and thus are the key element in posttranscriptional regulation of gene expression. In eukaryotes, miRNAs act in concert with Ago searcher proteins. In bacteria, the RNA chaperone protein Hfq facilitates the base-pairing between small RNAs (sRNAs) and target mRNAs [1]. Hfq has a remarkable ability to match sRNAs with their targets, despite substantial variations in sRNA size, secondary structure, and the extent of complementarity between the "seed" region of the sRNA and its target site. Moreover, the strength or direction of regulation does not always correlate with the thermodynamic stability of the sRNA-target duplex [2]. Therefore, it is challenging to disentangle the physical features that contribute to efficient sRNA-mRNA regulation. It is possible that some of the factors mentioned above increase the chances of nonproductive sRNA-mRNA

James Chappell and Melissa K. Takahashi (eds.), *Riboregulator Design and Analysis*,
Methods in Molecular Biology, vol. 2518, https://doi.org/10.1007/978-1-0716-2421-0_16,

binding events, resulting in a weaker regulatory outcome for certain sRNA-mRNA pairs.

Total internal reflection fluorescence (TIRF) microscopy of single molecules [3–7] is an extremely powerful method for identifying transient states and competing reaction pathways that contribute to RNA-guided regulation. For example, TIRF studies established that only transient miRNA-mRNA interactions are observed if they form fewer than six base pairs [8]. Moreover, a distribution of Cas9-crRNA binding events on DNA revealed by single-molecule studies helped pinpoint the target search mechanism used by Cas9 [9, 10].

The evanescent field generated by TIR typically penetrates ~100 nm from the slide surface, and only molecules within this range will be excited. In addition, molecules in solution do not yield a persistent fluorescent signal because they diffuse too quickly to remain within a pixel for an entire 30–100 ms frame. In the examples discussed in this chapter, mRNA molecules are tethered to the slide. Consequently, it is possible to observe how other fluorophore-labeled components such as sRNA or proteins assemble and dissociate with the immobilized mRNA over time.

Binding events can be detected by colocalization of the fluorophore-labeled components with the immobilized molecules [11, 12]. However, a kinetic analysis of binding lifetimes in single-molecule experiments can be further correlated with Förster resonance energy transfer (FRET), which provides additional information about the structure of the complex. Non-radiative energy transfer from a donor to an acceptor fluorophore in a FRET pair depends on the distance between them, typically varying over 2–10 nm for most FRET pairs [13].

The cyanine dyes used in this protocol are well-established FRET pairs that have been used in TIRF experiments for many years. However, FRET pairs using fluorophores such as Alexa or Atto may be superior for some applications. For example, Cyanine 3 (Cy3) fluorescence is sometimes quenched by stacking with the end of an RNA helix or enhanced by interactions with Hfq. The latter protein-induced fluorescence enhancement (PIFE) can be exploited for detecting RNA-protein interactions [14].

Single-molecule TIRF microscopy has several limitations, which have been described in more detail elsewhere [15]. First, because free donor-labeled molecules near the slide surface are also excited, they contribute to the fluorescence background of the solution. To minimize this background, the experiments are typically carried out with no more than ~20 nM labeled component in the solution. Therefore, this approach is not suitable for visualizing weakly interacting molecules with high K_d values. Another limitation of this technique is that only interactions lasting longer than a frame can be detected. Very slow processes (>5 min) are also difficult to study if the fluorophores photobleach before many events can be observed [16].

Especially when working with multicomponent complexes, it is very advantageous to confirm that the labeled components form the expected interactions using ensemble methods such as electrophoretic mobility shift assays (EMSA) or solution fluorescence spectroscopy [17]. The EMSA results provide an estimate of the affinities of the reaction components and, thus, the concentrations that can be used in a single-molecule experiment. Moreover, ensemble experiments can provide information about interactions with components that are not labeled with a fluorophore. For example, the fraction of labeled RNA in complex with an unlabeled protein can be visualized on the gel based on differences in the electrophoretic mobilities of the complexes [17], even if this interaction cannot be detected directly in the TIRF microscopy experiments.

This chapter provides comprehensive information about the design and preparation of RNA substrates for Hfq, the execution of single-molecule experiments, and the data analysis. We focus on two case studies: (1) sRNA-mRNA annealing with Hfq protein and (2) structural rearrangements within an mRNA upon sRNA binding. These protocols can be easily adapted for studying other systems.

2 Materials

In any experiments with RNA, precautions should be taken to avoid RNase contamination. Use nuclease-free reagents and RNase-free deionized water. Samples containing a fluorophore must be shielded from light.

2.1 RNA Substrate Design and Preparation

1. Sequences of complementary sRNA-mRNA pair that will serve as substrates.
2. Information about Hfq or other protein binding sites from genetics or chemical probing experiments.
3. RNA secondary structure prediction software.
4. Custom synthesized RNA or reagents for T7 transcription.
5. Activated dyes for fluorophore labeling, as desired.

2.2 mRNA-sRNA Annealing

1. 500 nM 5′ Cy3-labeled sRNA in water or TE buffer, 10 μL.
2. 200 nM unlabeled mRNA containing a 3′ extension complementary to the DNA tether, 2 μL (*see* **Note 1**).
3. 200 nM DNA tether with biotin modification at 5′ end and Cy5 at 3′ end, 1 μL (*see* **Note 1**).

4. 1 μM hexamer Hfq protein (stored in HB buffer: 50 mM Tris-HCl pH 7.5, 250 mM NH_4Cl, 1 mM EDTA, 10% glycerol, at −20 or 4 °C), 2 μL.
5. TNK buffer (10×): 100 mM Tris-HCl (pH 7.5), 500 mM NaCl, 500 mM KCl.
6. Imaging buffer: 1× TNK, 4 mM Trolox, 0.01% octaethylene glycol monododecyl ether (Nikkol), 0.8% glucose (*see* **Note 2**).
7. Oxygen scavenging system (50×): ~10 U/μL glucose oxidase, ~100 U/μL catalase (*see* **Note 3**).
8. Ribonuclease inhibitor.

2.3 Ligation of Long mRNA with Internal Fluorophores

1. 100 μM internally Cy5-labeled RNA (5′ fragment), 10 μL.
2. 100 μM 3′ RNA fragment with monophosphate at 5′ end, 20 μL.
3. 100 μM DNA splint, 12 μL.
4. 1 M KCl.
5. T4 RNA ligase 2.
6. 1× T4 Rnl2 reaction buffer: 50 mM Tris-HCl pH 7.5, 2 mM $MgCl_2$, 1 mM DTT, 400 μM ATP.
7. Centrifugal filter units 0.5 ml, 3 kDa MWCO (optional).
8. Gel running equipment (vertical electrophoresis apparatus including glass plates, spacers, well-forming combs, and clamps).
9. 8 M urea, 8% polyacrylamide gel.
10. TBE buffer (10×): 890 mM Tris base, 890 mM boric acid, 20 mM EDTA, pH 8.3.
11. RNA elution buffer: 0.3 M Na-acetate, pH 5.4, 1 mM EDTA.
12. Ethanol 200 proof.

2.4 mRNA Structural Rearrangements During sRNA-mRNA Annealing

1. 500 nM unlabeled sRNA, 10 μL.
2. 200 nM mRNA labeled internally with Cy5, 2 μL.
3. 200 nM DNA tether with biotin modification at 5′ end and Cy3 at 3′ end (Bio-SA5-Cy3), 1 μL.
4. 1 μM hexamer Hfq protein, 2 μL.
5. TNK buffer (10×).
6. Imaging buffer.
7. Oxygen scavenging system.
8. Ribonuclease inhibitor.

2.5 Total Internal Reflection Microscopy

A detailed explanation of the prism-type TIRF microscope for single-molecule experiments has been published elsewhere

[3, 18]. Besides the prism TIRF microscope [19], one can also use objective TIRF [20], micromirror [21], or zero-mode waveguide instruments [22] to observe the emission of single fluorophores. Despite these different instrument configurations, the common components of a TIRF microscope are lasers for illumination, a microscope body and sample stage, high-numerical aperture objectives for efficient signal collection, "single-photon sensitive" cameras for signal detection (emCCD or sCMOS), and various optical components for guiding and focusing the light beams. In this method, one component of the complex must be immobilized on a clean and passivated quartz slide with a glass coverslip. Methods for slide passivation with PEG or DDS have been described elsewhere [23].

2.6 Data Analysis

1. Software for extracting trajectories for individual molecules from movies (*see* **Note 4**).
2. Software for secondary data analysis (*see* **Note 5**).

3 Methods

3.1 Design and Preparation of RNAs for Single-Molecule Experiments

1. Obtain the sequences of the natural sRNA-mRNA pair of interest or an engineered pair that contains the features of interest. Bacterial sRNAs are usually less than 250 nt, and the entire sRNA can be used for smFRET experiments. Bacterial mRNAs are usually longer than 500 nt and may not fold into a defined structure. Therefore, a suitable mRNA fragment containing the sRNA binding site and known protein interaction sites should be defined based on available genetics and chemical structure probing data (*see* **Notes 5** and **6**).
2. To study intermolecular RNA-RNA interactions, the fluorophores should be attached to each RNA, so that base-pairing brings the fluorophores closer together, resulting in high FRET (*see* Fig. 1a; **Note 7**).
3. To study conformational changes in an RNA, the fluorophores should be attached to locations that directly report on the conformational change of interest (*see* Fig. 1b). For example, if the fluorophores are attached near the base of a hairpin, they will be close to each other when the hairpin is formed (high FRET) and further apart when the hairpin unfolds (low FRET).
4. Once the dye positions have been established, the next step is to prepare the fluorophore-labeled RNAs. Short RNAs (up to 60 nt) can be synthesized commercially with dyes in any position (*see* **Note 8**). Reactive groups can be also incorporated at the 5′ end of an RNA during transcription with T7 RNA

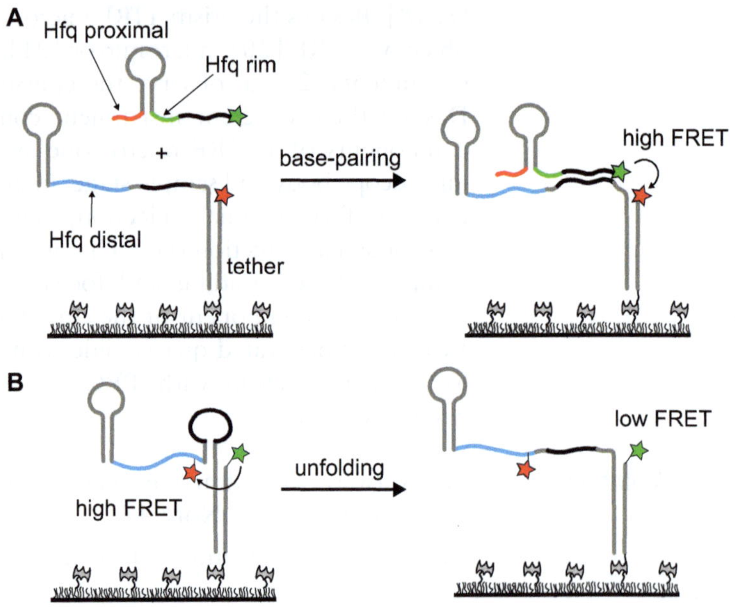

Fig. 1 Single-molecule experimental design for sRNA-mRNA interactions. mRNAs are immobilized on the passivated and neutravidin-coated slide via a biotinylated DNA tether. RNAs discussed in these protocols have a 16 bp complementary region (black) and sequence motifs recognized by Hfq (orange, proximal face; blue, distal face; green, rim). (**a**) sRNA-mRNA annealing. 5′ end of sRNA is labeled with Cy3, while Cy5 is attached at the 3′ end of DNA tether. Full base-pairing results in a high FRET state. (**b**) RNA restructuring. mRNA is internally labeled with Cy5, while Cy3 is attached at the 3′ end of the DNA tether. If the stem loop is formed, a high FRET state is observed. Opening of the stem leads to the increased distance between dyes and, thus, to a transition to a low FRET state

polymerase. Transcription is initiated with a guanosine analog, which is added to the reaction in excess over GTP. Suitable guanosine analogs for posttranscriptional labeling include GMPS (guanosine 5′-thiomonophosphate) or GMP (guanosine 5′-monophosphate). The 5′-phosphate is subsequently converted to a primary amine in a reaction with EDC (1-ethyl-3-[3-dimethylaminopropyl]carbodiimide hydrochloride) and imidazole [24].

5. Long RNAs (>40 nt) with internally located fluorophores can be prepared by ligation of two or more fragments corresponding to a 3′ OH ligation acceptor and 5′ monophosphate donor [25, 26] (*see* **Note 9**). The most efficient and frequently used enzyme for the ligation of RNA fragments is RNA ligase 2, which joins two RNAs hybridized with a DNA or RNA splint (*see* **Note 10**).

6. In addition to the fluorophores, RNAs used in single-molecule studies need to carry a unique biotin modification allowing for immobilization on a neutravidin-coated slide. The biotin can be attached directly to the RNA through custom synthesis. Alternatively, it is possible to tether the RNA to the slide via a biotinylated DNA that hybridizes with an extension of the mRNA (*see* **Note 1**). The DNA tether can be also labeled with a fluorophore. This design allows one to easily test different mRNA sequences without having to chemically modify the mRNA (*see* Fig. 1a).
7. Perform binding assays such as EMSA to confirm that the fluorophore-labeled components form the expected interactions (*see* **Note 11**).

3.2 mRNA-sRNA Annealing

This protocol is for studying the kinetics of sRNA-mRNA annealing, using model Cy3- and Cy5-labeled RNAs as illustrated in Fig. 1a. In a fully base-paired duplex, the proximity of the fluorophores results in a high FRET state.

1. Prepare a 10 μL mixture of 20 nM unlabeled mRNA and 40 nM Bio-SA5-Cy5 DNA (final concentration) in 1× TNK buffer. Denature the mRNA-DNA mixture by heating it at 75 °C for 5 min. Refold the RNA-DNA heteroduplex at 37 °C for 20 min and equilibrate at room temperature.
2. Denature 10 μL of 500 nM 5′Cy3-sRNA at 75 °C for 5 min (*see* **Note 12**). Refold at room temperature for 10–20 min.
3. Mix 4 μL refolded sRNA, 2 μL Hfq stock, 1 μL 10× TNK buffer, and water to bring the total volume to 8μl (RNA and Hfq at a final concentration of 250 nM each).
4. Dilute the refolded mRNA-DNA stock 50-fold into 150 μL imaging buffer. Add 20 U RNase inhibitor. Inject the sample into the slide channel. Incubate for 1 min.
5. Wash the channel with 150 μL imaging buffer supplemented with RNase inhibitor and 3 μl 50× oxygen scavenging system (OSS).
6. Check the density of immobilized molecules on the slide by direct excitation of Cy5 with the 640 nm laser (*see* **Note 13**).
7. Dilute the sRNA-Hfq complex 50-fold in 150 μL imaging buffer. Add 20 U RNase inhibitor and 3 μl 50× OSS. Add the solution to the chamber but do not inject it into the channel immediately (*see* **Note 14**).
8. Find a field of view on the slide with a desired density of immobilized mRNAs by brief illumination with the 640 nm laser (*see* **Note 15**). In the acquisition software, set 100 ms time resolution and the laser excitation scheme as follows: 10 frames

of 640 nm excitation, 1800 frames of 532 nm excitation, and 10 frames of 640 nm excitation at the end.

9. Start recording. When the shutter switches to 532 nm excitation, draw the solution of sRNA-Hfq complex into the slide chamber.
10. Save the movie for initial analysis as described in Subheading 3.4.

3.3 Visualizing mRNA Structural Rearrangements During sRNA-mRNA Annealing

This protocol is for monitoring the restructuring of mRNA upon annealing with sRNA, as illustrated in Fig. 1b. Cy5 is located internally in the mRNA, while Cy3 is attached to the 3′ end of the DNA tether. When the hairpin is in a closed conformation, a high FRET state is observed. When the hairpin opens, a transition into a low FRET state is observed.

First, prepare the internally labeled mRNA:

1. Prepare a mixture of 1 nmol 5′ RNA fragment, 2 nmol 3′ RNA fragment, and 1.2 nmol DNA splint in 100 μl of 20 mM KCl (*see* **Note 16**).
2. Heat the mixture at 95 °C for 40 s and slowly cool to 25 °C (−1 °C/40 s or ~50 min total).
3. Add T4 Rnl2 reaction buffer and 250 U T4 RNA ligase 2 at 500 μL final volume.
4. Incubate at 25 °C for 1 h.
5. Quench the reaction with 2 mM EDTA (final concentration).
6. Concentrate the reaction using centrifugal filter units to ~20 μl (*see* **Note 17**).
7. Add an equal volume of 8–10 M urea. Denature for 3 min at 95 °C and load on an 8 M urea 8% polyacrylamide gel (*see* **Note 18**).
8. Excise the desired band of ligated product from the gel (*see* **Note 16**), and elute the RNA in the RNA elution buffer overnight at 4 °C with rocking. Precipitate the RNA with ethanol, resuspend the dried pellet in water, and store at −20 °C.

Next, proceed to RNA immobilization, sample preparation, and imaging:

9. Prepare a 10 μL mixture of 20 nM internally labeled mRNA (obtained in 3.3, steps 1–8) and 40 nM Bio-SA5-Cy3 DNA in 1× TNK buffer. Denature the mRNA-DNA mixture by heating it at 75 °C for 5 min. Refold the RNA-DNA heteroduplex at 37 °C for 20 min, and finally equilibrate the mixture at room temperature.

10. Denature 10 μL of 500 nM unlabeled sRNA at 75 °C for 5 min. Refold at room temperature for 10–20 min.
11. Mix 4 μL refolded sRNA, 2 μL Hfq stock, and 1 μL 10× TNK buffer, and add water up to 8μL (RNA and Hfq at final concentration 250 nM each).
12. Immobilize mRNA on the slide (*see* Subheading 3.2, **steps 4–5**).
13. Check the density of Cy3 and Cy5 spots on the slide (*see* **Notes 15** and **19**).
14. Record the fluorescence intensity of the immobilized molecules to check if the hairpin has formed and if it is stable (*see* Subheading 3.2, **step 8** and **Note 20**).
15. Prepare the dilution of the sRNA-Hfq complex (*see* Subheading 3.2, **step** 7).
16. Record molecules in a new field of view upon sRNA-Hfq injection (*see* Subheading 3.2, **steps 8** and **9**).

3.4 Data Analysis: Initial Data Processing

1. Decide an initial spot selection criterion based on the design of the experiments. For example, one can choose to select spots with a Cy3, Cy5, or both labels from the beginning (or last) ten frames of a movie.
2. Create a mapping file for the optical system (*see* **Note 21**). Use the mapping file to warp images from one channel and to overlay it with the other channels. From the overlaid image, one can use an algorithm to identify single molecules for analysis, based on intensity and roundness thresholds (*see* **Note 22**) established in step 1.
3. Once spots are selected, intensities at each spot are extracted and the background signal is subtracted. If a time-lapse movie is analyzed, the intensity extraction procedure is repeated for each frame, and an intensity-time trajectory is created for each selected spot.
4. Examine each intensity trajectory and select spots for further analysis. One can impose that the fluorescence intensity lasts longer than a certain number of frames and that the Cy5-Cy3 signals are anticorrelated and photobleach in a single step (*see* **Note 23**).
5. For FRET analysis, apply corrections to the donor and acceptor intensities to correct for direct excitation of the acceptor ("cross talk") and leakage of donor signal to the acceptor channel [27]. The background for long movies is corrected to make the intensities after donor photobleaching equal to 0 (*see* Fig. 2).

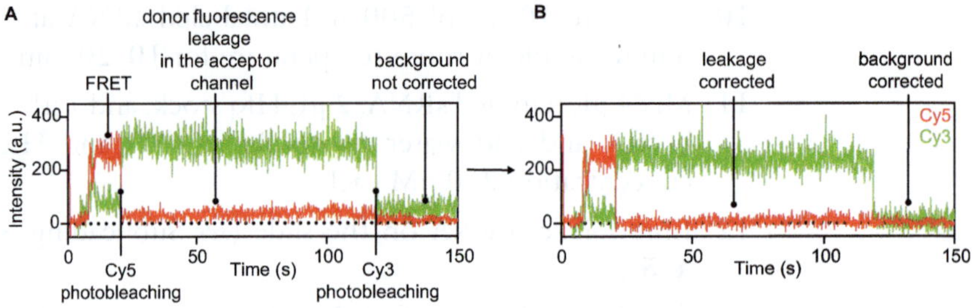

Fig. 2 Applying corrections to single-molecule FRET trajectories. Correction factors can be determined from traces with Cy5 and Cy3 photobleaching events. Example trajectories (**a**) before and (**b**) after applying corrections for background and leakage

6. After the appropriate corrections have been applied, calculate the FRET efficiency, $E = I_A/(I_A+I_D)$, in which I_A is the intensity of the acceptor and I_D is the intensity of the donor fluorophore.

3.5 Analysis of sRNA-mRNA Annealing Kinetics: Binding Lifetimes and FRET Efficiency

The changes in donor and acceptor intensities in a single-molecule experiment provide rich information about the kinetics of RNA and protein interactions. In our system, the appearance of a fluorescence signal reports on Cy3-sRNA association with the immobilized mRNA, while its disappearance reports sRNA dissociation (*see* Fig. 3a, black and gray arrows, respectively). After initial processing of the data as described in Subheading 3.4, the secondary analysis should be tailored to the system under study.

1. To obtain information about the kinetics of sRNA-mRNA interactions, first identify the peaks in fluorescence intensity for each immobilized mRNA (*see* **Note 24**). This can be done in a semiautomated way by applying an intensity threshold to create an initial selection. Selected peaks are then inspected manually.
2. This process generates a list of peaks for every trace with information about the start, end, and duration of a peak.
3. To characterize the rate of sRNA dissociation, analyze the distribution of dwell times (peak duration) using an appropriate model for the dissociation kinetics of a complex (*see* **Notes 25** and **26**). We used MLE analysis described elsewhere [28], as illustrated in Fig. 3b.

FRET efficiency provides additional information about the structures of the complexes. For example, a correlation between binding lifetimes and FRET efficiency can establish whether short-lived complexes assume a different conformation than stable complexes.

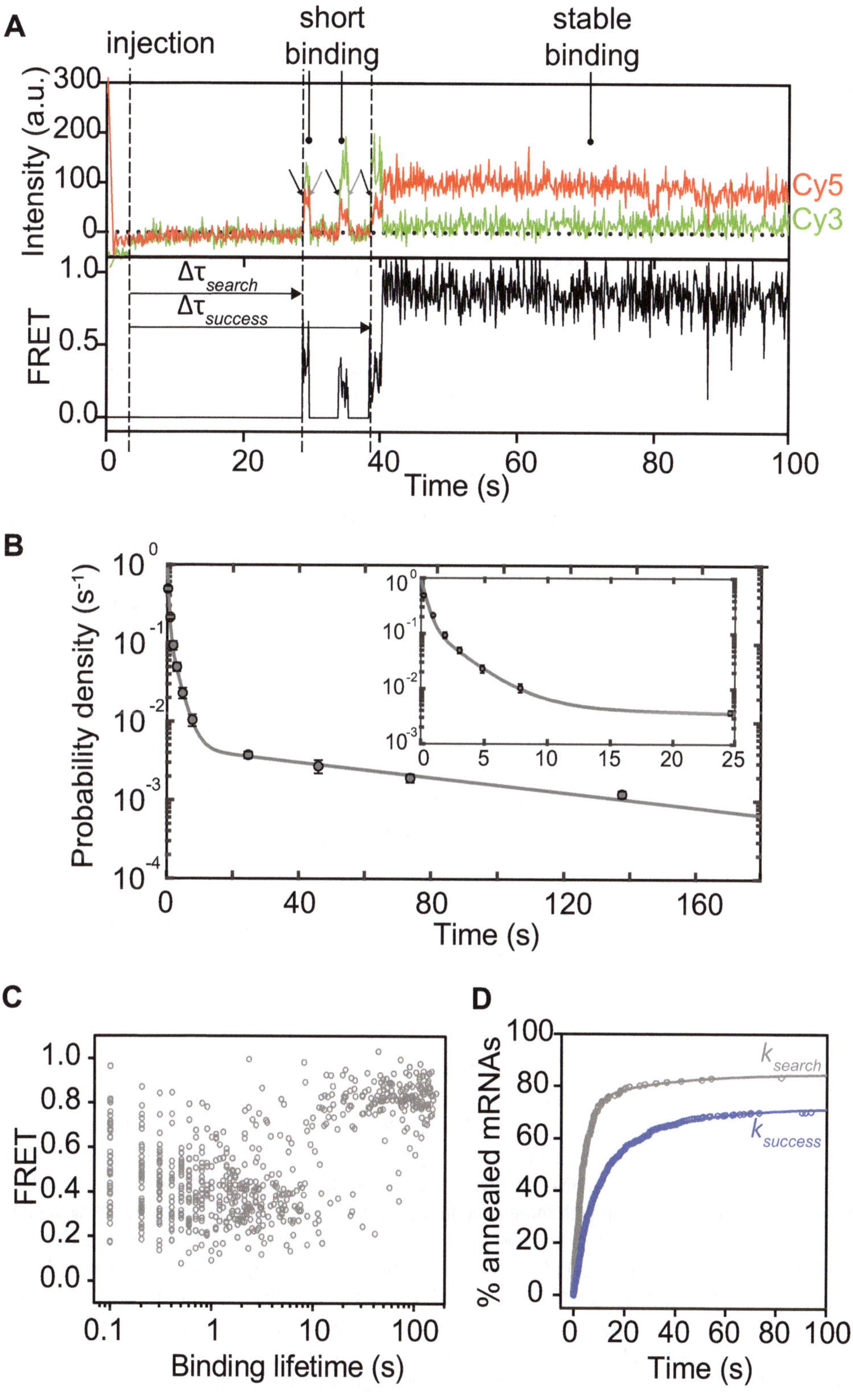

Fig. 3 Single-molecule observation of sRNA-mRNA annealing. (**a**) Representative binding sRNA-Hfq complex on mRNA (black arrow) and dissociation (gray arrow). Fluorescence intensity (top panel) and FRET

4. Create a combined list of dwell times for all of the molecules in the dataset, and calculate an average FRET efficiency for each binding event.
5. Create a scatter plot of FRET efficiency (y axis) versus binding lifetime (x axis).
6. In our experimental setup, complete sRNA-mRNA base-pairing is expected to yield a high FRET state, which was observed for stable complexes (*see* Fig. 3c). However, short-lived sRNA-mRNA complexes correlated with lower FRET efficiency (*see* Fig. 3c), indicating that these unstable complexes are not fully base paired.
7. If desired, more detailed information about the binding pathway can be obtained from transition density plots [29, 30].

The RNA-RNA association kinetics can be also measured using single-molecule data.

8. Calculate the time from injection of Cy3-sRNA to the first binding event for each immobilized mRNA (*see* Fig. 3a, $\Delta\tau_{search}$). The time of injection is marked by the increase in Cy3 background when the solution containing Cy3-labeled sRNA enters the FOV.
9. Fit the distribution of association times to a pseudo-first order rate equation to obtain the observed association rate constant.
10. Depending on the studied system, differences in lifetimes or FRET efficiencies can be exploited to distinguish the kinetics of forming different types of complexes. In this example, a further threshold was applied to select encounters that produced stable complexes (*see* Fig. 3a, $\Delta\tau_{success}$). In our case, the first binding events denote the kinetics of the target search, whereas $\tau_{success}$ reflects the frequency of successful sRNA-mRNA encounters (*see* Fig. 3d).

Fig. 3 (continued) (bottom panel) upon Cy3 excitation, with 0.1 s frames. The moment of sRNA-Hfq injection (dashed line) is marked by a slight increase in background fluorescence. The time from sRNA injection to a first binding event ($\Delta\tau_{search}$) and successful annealing ($\Delta\tau_{success}$) is indicated. (**b**) Probability density histogram of sRNA-Hfq residence lifetimes on mRNA for reversible binding events. Lines represent maximum likelihood estimate (MLE) fits to an equation with three exponential terms corresponding to populations with characteristic lifetimes: transient (0.42 ± 0.08 s), short (2.41 ± 0.44 s), and long (92 ± 12 s). Inset: expansion of the first 25 s. Error bars represent the SD of a binomial distribution. This analysis shows that some sRNA-mRNA encounters are unstable with transient (τ~0.4 s) or short (τ~3 s) lifetimes. Short-lived complexes were more prevalent when a complementary region in mRNA was embedded in a secondary structure [38]. (**c**) Scatter plot illustrating the correlation between the lifetimes of reversible sRNA-mRNA binding events and their FRET efficiency. (**d**) The distribution of first (blue) and first successful (gray) sRNA-Hfq arrival times on each mRNA fit to a double-exponential equation

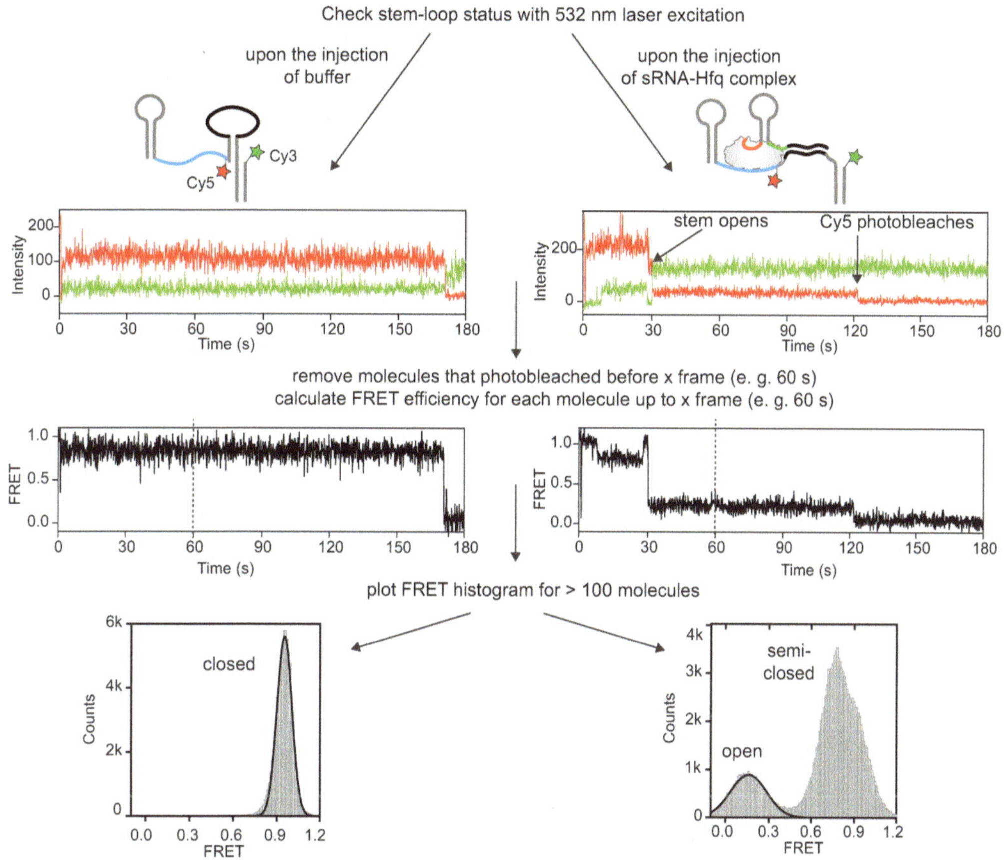

Fig. 4 Single-molecule observation of mRNA restructuring. Immobilized RNAs are imaged upon the addition of buffer (left panel) or sRNA-Hfq complex (right panel). Trajectories showing photobleaching event before the selected threshold (e.g., 60 s) are removed, and the FRET efficiency is calculated. The distribution of FRET values is then plotted on a histogram. Distinct FRET states representing the open and closed conformations are fit to Gaussian functions

3.6 Analysis of mRNA Restructuring: FRET States

Conformational changes within an RNA or RNA-protein complex are studied by analyzing the transitions between FRET states.

1. Calculate the FRET efficiency of each immobilized molecule at every frame before and after the addition of a component that is expected to induce the conformational change (*see* Fig. 4).
2. Fit the distribution of FRET values with a Gaussian function or a sum of Gaussian functions to yield the mean value for each conformational state. The width of individual peaks depends on the conformational dynamics as well as the inherent signal to noise ratio (shot noise) of the experiment (*see* **Note 27**).
3. The average FRET efficiencies can be used to estimate the distances between fluorophores, which is especially useful if the three-dimensional structure of the complex is known.

4. If dynamic transitions between conformations are observed, the exchange kinetics can be characterized by fitting the dwell times of low and high FRET states to exponential equations. Alternatively, methods such as hidden Markov modeling (HMM) are used to assign transition probabilities between FRET states in an unbiased way [29].

4 Notes

1. The 3′ end of the mRNA sequence is extended to contain a sequence complementary to the DNA tether. We use 5-′-mRNAseq- GGACGACACACUUGGACAGGACACA CAGGACACAGG-3′, which is complementary to the tether DNA Bio-SA5-Cy5: 5′-(biotin)-CCTGTGTCCTGTGTGTC CTGTCCAAAGTGTGTCGTCC-(Cy5)-3′ [31].
2. Trolox is frequently used as a triplet state quencher [32, 33]. Prepare Trolox solution by dissolving 10 mg Trolox in 10 ml water. Add 10 μl 5 M NaOH. Cover the tube in foil and incubate on tube rocker overnight at 4 °C. Octaethylene glycol monododecyl ether (Nikkol) is a nonionic detergent that minimizes nonspecific sticking of proteins to the slide. It can be prepared as 1% stock solution in water and stored at 4 °C. Imaging buffer should be filtered (0.22 μm) and prepared fresh before use.
3. To be able to image molecules for a longer time (several minutes), the photobleaching rate of the fluorophores needs to be minimized. For that, a buffer with an oxygen scavenging system (OSS) containing glucose oxidase and/or catalase is used [3, 34]. Dissolve 55 mg catalase (C40, Sigma-Aldrich) in 1 ml 1× TNK buffer. Spin down 1 min at > 16,000 × *g* at 4 °C. Transfer supernatant to a new tube. Store in aliquots in −80 °C. Dissolve 10 mg glucose oxidase in 80 μL 1× TNK buffer and add 20 ul catalase stock. Spin down for 1 min at 14000 × *g*. Transfer top solution to a new tube and store at 4 °C. The solution can be used for several days. The OSS should be added to the buffer right before sample injection to the slide.
4. Many commercial and custom software packages are available for the analysis of single-molecule data. We use IDL software (Interactive Data Language, L3Harris Geospatial) and custom scripts for analysis of movies and MATLAB (Mathworks) scripts to correct and analyze individual trajectories.
5. It is important to make sure that the RNA substrates do not fold into undesired nonnatural structures, which can be

checked using secondary structure prediction software such as RNAstructure [35].

6. Many bacterial sRNAs target the mRNA 5′ UTR. In this case, it is desirable to begin the mRNA fragment from the natural transcription start site and extend it 9–30 nt downstream of the translation initiation codon. Even 1 nt differences in length can alter the secondary structure of the mRNA fragment. Ideally, a range of mRNA fragment lengths should be tested for their ability to bind the sRNA and Hfq before proceeding with smFRET experiments.
7. For a Cy3-Cy5 donor-acceptor pair ($R_0 = 53$ Å), separation by 6–8 nt usually results in FRET efficiency ~0.8.
8. Internal dyes can be incorporated in the RNA backbone or attached to a C6 or a 2′ amino derivative of uridine during custom synthesis. Another option is to purchase RNAs with reactive groups that can be conjugated with a reactive dye by the user, which allows for flexibility in dye selection. Primary amines can be coupled with a N-hydroxysuccinimide (NHS) ester, whereas thiol groups are conjugated with maleimides [36]. Most fluorophores are available as these derivatives.
9. If the donor strand is prepared by T7 in vitro transcription, a monophosphate can be installed at the 5′ end by adding GMP to the transcription reaction. Alternatively, RNA 5′ pyrophosphohydrolase (RppH) can be used post-synthetically to remove pyrophosphate from the 5′ triphosphate.
10. The DNA splint should be fully complementary to the ends of the RNA fragments to form a nicked heteroduplex. Optimally, it forms 16–18 nt long base-pairing with each fragment which usually results in melting temperatures exceeding the temperature of the ligation reaction.
11. If the RNA sequence has been extended to hybridize with a DNA tether, be sure to hybridize the RNA with the DNA before performing EMSA experiments. This avoids complications from misfolding of the extended RNA in the absence of the complementary DNA strand.
12. A thermocycler with heated lid is ideal for programmed renaturation and for preventing evaporation of the sample. When using a regular heating block without a heated lid, the cooled liquid should be collected by brief centrifugation.
13. The density of fluorescent spots on the slide should sparse enough to view single molecules in separate pixels (~0.1–0.2 molecules per μm^2) but be dense enough for statistical analysis of the results (e.g., hundreds of molecules per FOV).
14. A simple flow system can be created using a quartz slide, glass coverslip, double-sided tape, pipette tip, and epoxy. Drill holes

near both long ends of the quartz slide (5 on each side). Clean slides and coverslips as indicated in 2.5. Attach thin strips of double-sided tape between holes and place a coverslip on top. Seal the slides using epoxy. Cut ~5 mm long pieces of the pipette tip from both sides (tip and ring). Place the tip in the hole (to create an outlet) and the ring on another side (to create an inlet or "reservoir"). Seal with epoxy. The inlet reservoir should be able to hold the solution before injection into the slide without leaking. To pull the solution, attach tubing to a 1 ml syringe and place it in the outlet tip.

15. Preparing selected field of view for the recording (e.g., focusing) should be quick. Do not excite it directly for a long time to prevent photobleaching.

16. The reaction can be easily scaled up or down. 0.5 nmol of Cy5-labeled RNA should be visible on the gel by the eye (a pink band). If a lower amount is used, UV shadowing or Typhoon scanner can be used to visualize RNA.

17. The typical concentrate volume using a 0.5 ml Amicon concentrator is 15–20 μL. The added urea will double the volume. When using this method, make sure to use a thick, wide well-forming comb when preparing the gel. Alternatively, the ligation reaction can be precipitated, and the pellet dissolved directly in 8 M urea.

18. To assess if the ligation was successful and the product is of the desired length, a control RNA should be loaded on the gel. It should be of the same size as the ligation product. If the control RNA is not fluorescently labeled, it can be visualized by UV shadowing. For that, relatively high amounts of RNA are required (~7 μM stock in 20 μL for ~100 nt long RNA), but it can be excised and eluted to recover it.

19. If the slide has visibly more Cy3 than Cy5 spots, the tether DNA and mRNA did not anneal with each other efficiently. In this case, a higher concentration of mRNA can be used in the hybridization reaction. Excess mRNA that has not hybridized with the tether DNA should wash away when the slide chamber is purged with imaging solution.

20. Because Cy3 and Cy5 dyes should be in proximity when the hairpin is formed within the mRNA, excitation with 532 nm laser should result in a high FRET state. Therefore, spots visible in the red channel should have higher intensity than corresponding spots in the green channel.

21. For each optical system, a Cy3/Cy5 channel mapping file containing a spatial transformation matrix is created before the start of experiments and is updated every time the optical system for detection (lens, mirrors) is adjusted. To generate a mapping file, one uses a slide made of fluorescent beads. The

beads are fluorescent in both the Cy3 and Cy5 channels. Three or more pairs of corresponding spatial locations are handpicked by users and a transformation matrix is calculated using the selected locations [3, 15].

22. Roundness is used to check if there is another molecule in close proximity to the molecule of interest. If so, the two spots will appear as one elongated spot instead of a round spot.
23. The number of photobleaching events provides information about the number of fluorophore-labeled molecules in a pixel. If more than the expected number of photobleaching events is observed, it is possible that the pixel of interest contains more than one molecule or complex, and such spots should not be used in the analysis.
24. Peaks should be selected based on total intensity, which should not be affected by energy transfer from the donor to acceptor. Kinetics analysis can also be based on colocalization of two or more fluorophores without FRET. For colocalization studies, control reactions without tether DNA should be done to check for specific binding to immobilized mRNA.
25. If complexes are very stable, experiments at several laser powers can be used to evaluate the contribution of fluorophore photobleaching to the measured binding lifetimes [16].
26. After equal binning of the data, the histogram can be fit to appropriate rate equations to estimate the lifetime of the complex. A method based on maximum likelihood estimation (MLE) allows for direct fitting of the unbinned data [11, 37].
27. When we probed the structure of an mRNA stem loop, the distribution of high FRET values was very narrow, indicating that the structure was stable (*see* Fig. 4). When the sRNA-Hfq complex was added, the population of the low FRET state increased in the first minute from the injection, showing that sRNA-mRNA base-pairing leads to the opening of the stem. However, the distribution of a high FRET state also became wider, which we interpret as a dynamic opening and closing of one to two bases on the base of the stem.

Acknowledgments

This work was supported by the National Institutes of Health (R35GM136351 to S.A.W.). The authors thank Subrata Panja and Andrew Santiago-Frangos for contributing to early versions of these protocols.

References

1. Wagner EGH, Romby P (2015) Small RNAs in bacteria and archaea: who they are, what they do, and how they do it. Adv Genet 90: 133–208. https://doi.org/10.1016/bs.adgen.2015.05.001
2. Fröhlich KS, Haneke K, Papenfort K, Vogel J (2016) The target spectrum of sdsr small RNA in Salmonella. Nucleic Acids Res 44: 10406–10422. https://doi.org/10.1093/nar/gkw632
3. Roy R, Hohng S, Ha T (2008) A practical guide to single-molecule FRET. Nat Methods 5:507–516. https://doi.org/10.1038/nmeth.1208
4. Jalihal AP, Lund PE, Walter NG (2019) Coming together: Rnas and proteins assemble under the single-molecule fluorescence microscope. Cold Spring Harb Perspect Biol 11. https://doi.org/10.1101/cshperspect.a032441
5. Ray S, Widom JR, Walter NG (2018) Life under the microscope: single-molecule fluorescence highlights the RNA world. Chem Rev 118:4120–4155. https://doi.org/10.1021/acs.chemrev.7b00519
6. Lerner E, Cordes T, Ingargiola A, Alhadid Y, Chung SY, Michalet X, Weiss S (2018) Toward dynamic structural biology: two decades of single-molecule förster resonance energy transfer. Science 359. https://doi.org/10.1126/science.aan1133
7. Lerner E, Barth A, Hendrix J, Ambrose B, Birkedal V, Blanchard SC, Börner R, Chung HS, Cordes T, Craggs TD, Deniz AA, Diao J, Fei J, Gonzalez RL, Gopich IV, Ha T, Hanke CA, Haran G, Hatzakis NS, Hohng S, Hong SC, Hugel T, Ingargiola A, Joo C, Kapanidis AN, Kim HD, Laurence T, Lee NK, Lee TH, Lemke EA, Margeat E, Michaelis J, Michalet X, Myong S, Nettels D, Peulen TO, Ploetz E, Razvag Y, Robb NC, Schuler B, Soleimaninejad H, Tang C, Vafabakhsh R, Lamb DC, Seidel CAM, Weiss S, Boudker O (2021) FRET-based dynamic structural biology: Challenges, perspectives and an appeal for open-science practices. eLife 10. https://doi.org/10.7554/eLife.60416
8. Chandradoss SD, Schirle NT, Szczepaniak M, Macrae IJ, Joo C (2015) A dynamic search process underlies microRNA targeting. Cell 162:96–107. https://doi.org/10.1016/j.cell.2015.06.032
9. Globyte V, Lee SH, Bae T, Kim J-S, Joo C (2019) CRISPR/Cas9 searches for a protospacer adjacent motif by lateral diffusion. EMBO J 38:e99466. https://doi.org/10.15252/EMBJ.201899466
10. Singh D, Sternberg SH, Fei J, Doudna JA, Ha T (2016) Real-time observation of DNA recognition and rejection by the RNA-guided endonuclease Cas9. Nat Commun 71(7):1–8. https://doi.org/10.1038/ncomms12778
11. Friedman LJ, Gelles J (2015) Multi-wavelength single-molecule fluorescence analysis of transcription mechanisms. Methods 86: 27–36. https://doi.org/10.1016/j.ymeth.2015.05.026
12. C van der F, AA H (2017) Methodologies for studying the spliceosome's RNA dynamics with single-molecule FRET. Methods San Diego Calif 125:45–54. https://doi.org/10.1016/J.YMETH.2017.05.011
13. Stephenson JD, Kenyon JC, Symmons MF, Lever AML (2016) Characterizing 3D RNA structure by single molecule FRET. Methods 103:57–67. https://doi.org/10.1016/J.YMETH.2016.02.004
14. Hwang H, Myong S (2014) Protein induced fluorescence enhancement (PIFE) for probing protein-nucleic acid interactions. Chem Soc Rev 43:1221–1229. https://doi.org/10.1039/c3cs60201j
15. Brinster DR, Raper SE (2004) Synchronous colon and pancreatic cancers in a patient with Peutz-Jeghers syndrome: report of a case and review of the literature
16. Joo C, McKinney SA, Nakamura M, Rasnik I, Myong S, Ha T (2006) Real-time observation of RecA filament dynamics with single monomer resolution. Cell 126:515–527. https://doi.org/10.1016/j.cell.2006.06.042
17. Panja S, Małecka EM, Santiago-Frangos A, Woodson SA (2020) Quantitative analysis of RNA chaperone activity by native gel electrophoresis and fluorescence spectroscopy. Methods Mol Biol 2106:19–39. https://doi.org/10.1007/978-1-0716-0231-7_2
18. Chauvier A, Cabello-Villegas J, Walter NG (2019) Probing RNA structure and interaction dynamics at the single molecule level. Methods 162–163:3–11. https://doi.org/10.1016/j.ymeth.2019.04.002
19. Funatsu T, Harada Y, Tokunaga M, Saito K, Yanagida T (1995) Imaging of single fluorescent molecules and individual ATP turnovers by single myosin molecules in aqueous solution. Nature 374:555–559. https://doi.org/10.1038/374555a0
20. Tokunaga M, Kitamura K, Saito K, Iwane AH, Yanagida T (1997) Single molecule imaging of

fluorophores and enzymatic reactions achieved by objective-type total internal reflection fluorescence microscopy. Biochem Biophys Res Commun 235:47–53. https://doi.org/10.1006/bbrc.1997.6732

21. Friedman LJ, Chung J, Gelles J (2006) Viewing dynamic assembly of molecular complexes by multi-wavelength single-molecule fluorescence. Biophys J 91:1023–1031. https://doi.org/10.1529/biophysj.106.084004
22. Chen J, Dalal RV, Petrov AN, Tsai A, O'Leary SE, Chapin K, Cheng J, Ewan M, Hsiung PL, Lundquist P, Turner SW, Hsu DR, Puglisi JD (2014) High-throughput platform for real-time monitoring of biological processes by multicolor single-molecule fluorescence. Proc Natl Acad Sci U S A 111:664–669. https://doi.org/10.1073/pnas.1315735111
23. Hua B, Han KY, Zhou R, Kim H, Shi X, Abeysirigunawardena SC, Jain A, Singh D, Aggarwal V, Woodson SA, Ha T (2014) An improved surface passivation method for single-molecule studies. Nat Methods 11:1233–1236. https://doi.org/10.1038/nmeth.3143
24. Rinaldi AJ, Suddala KC, Walter NG (2015) Native purification and labeling of RNA for single molecule fluorescence studies. Methods Mol Biol 1240:63–95. https://doi.org/10.1007/978-1-4939-1896-6_6
25. Stark MR, Rader SD (2014) Efficient splinted ligation of synthetic RNA using RNA ligase. Methods Mol Biol 1126:137–149. https://doi.org/10.1007/978-1-62703-980-2_10
26. Solomatin S, Herschlag D (2009) Methods of site-specific labeling of RNA with fluorescent dyes. Methods Enzymol 469:47–68. https://doi.org/10.1016/s0076-6879(09)69003-0
27. Hellenkamp B, Schmid S, Doroshenko O, Opanasyuk O, Kühnemuth R, Rezaei Adariani S, Ambrose B, Aznauryan M, Barth A, Birkedal V, Bowen ME, Chen H, Cordes T, Eilert T, Fijen C, Gebhardt C, Götz M, Gouridis G, Gratton E, Ha T, Hao P, Hanke CA, Hartmann A, Hendrix J, Hildebrandt LL, Hirschfeld V, Hohlbein J, Hua B, Hübner CG, Kallis E, Kapanidis AN, Kim J-Y, Krainer G, Lamb DC, Lee NK, Lemke EA, Levesque B, Levitus M, McCann JJ, Naredi-Rainer N, Nettels D, Ngo T, Qiu R, Robb NC, Röcker C, Sanabria H, Schlierf M, Schröder T, Schuler B, Seidel H, Streit L, Thurn J, Tinnefeld P, Tyagi S, Vandenberk N, Vera AM, Weninger KR, Wünsch B, Yanez-Orozco IS, Michaelis J, Seidel CAM, Craggs TD, Hugel T (2018) Precision and accuracy of single-molecule FRET measurements—a multi-laboratory benchmark study. Nat Methods 159(15):669–676. https://doi.org/10.1038/s41592-018-0085-0
28. Friedman LJ, Gelles J (2015) Multi-wavelength single-molecule fluorescence analysis of transcription mechanisms. https://doi.org/10.1016/j.ymeth.2015.05.026
29. McKinney SA, Joo C, Ha T (2006) Analysis of single-molecule FRET trajectories using hidden markov modeling. Biophys J 91:1941–1951. https://doi.org/10.1529/BIOPHYSJ.106.082487
30. Blanco M, Walter NG (2010) Analysis of complex single-molecule FRET time trajectories. Methods Enzymol 472:153–178. https://doi.org/10.1016/S0076-6879(10)72011-5
31. Kim H, Abeysirigunawarden SC, Chen K, Mayerle M, Ragunathan K, Luthey-Schulten Z, Ha T, Woodson SA (2014) Protein-guided RNA dynamics during early ribosome assembly. Nature 506:334–338. https://doi.org/10.1038/nature13039
32. Cordes T, Vogelsang J, Tinnefeld P (2009) On the mechanism of trolox as antiblinking and antibleaching reagent. J Am Chem Soc 131:5018–5019. https://doi.org/10.1021/JA809117Z
33. Rasnik I, McKinney SA, Ha T (2006) Non-blinking and long-lasting single-molecule fluorescence imaging. Nat Methods 3:891–893. https://doi.org/10.1038/nmeth934
34. Benesch RE, Benesch R (1953) Enzymatic removal of oxygen for polarography and related methods. Science 118:447–448. https://doi.org/10.1126/science.118.3068.447
35. Reuter JS, Mathews DH (2010) RNAstructure: software for RNA secondary structure prediction and analysis. BMC Bioinform 11. https://doi.org/10.1186/1471-2105-11-129
36. Joo C, Ha T (2012) Labeling DNA (or RNA) for single-molecule FRET. Cold Spring Harb Protoc 7:1005–1008. https://doi.org/10.1101/pdb.prot071027
37. Kaur H, Jamalidinan F, Condon SGF, Senes A, Hoskins AA (2019) Analysis of spliceosome dynamics by maximum likelihood fitting of dwell time distributions. Methods 153:13–21. https://doi.org/10.1016/J.YMETH.2018.11.014
38. Małecka EM, Woodson SA (2021) Stepwise sRNA targeting of structured bacterial mRNAs leads to abortive annealing. Mol Cell 81:1988–1999.e4. https://doi.org/10.1016/J.MOLCEL.2021.02.019

Chapter 17

Cotranscriptional RNA Chemical Probing

Courtney E. Szyjka and Eric J. Strobel

Abstract

Cotranscriptional folding is a fundamental step in RNA biogenesis and the basis for many RNA-mediated gene regulation systems. Understanding how RNA folds as it is synthesized requires experimental methods that can systematically identify intermediate RNA structures that form during transcription. Cotranscriptional RNA chemical probing experiments achieve this by applying high-throughput RNA structure probing to an in vitro transcribed array of cotranscriptionally folded intermediate transcripts. In this chapter, we present guidelines and procedures for integrating single-round in vitro transcription using *E. coli* RNA polymerase with high-throughput RNA chemical probing workflows. We provide an overview of key concepts including DNA template design, transcription roadblocking strategies, single-round in vitro transcription with *E. coli* RNA polymerase, and RNA chemical probing and describe procedures for DNA template preparation, cotranscriptional RNA chemical probing, RNA purification, and 3′ adapter ligation. The end result of these procedures is a purified RNA library that can be prepared for Illumina sequencing using established high-throughput RNA structure probing library construction strategies.

Key words Cotranscriptional RNA structure probing, In vitro transcription, RNA chemical probing

1 Introduction

RNA begins to fold as it emerges from an RNA polymerase (RNAP) during transcription [1, 2]. The 5′→3′ direction of transcription sets the order in which nascent RNA folds into secondary and tertiary structures: nucleotides that are closer to the 5′ end of a nascent transcript can participate in intra- and intermolecular interactions earlier than nucleotides that are closer to the 3′ end. The influence of transcription on RNA structure is compounded by the relative rates of RNA synthesis and folding and by the tendency of RNA to fold into kinetically trapped intermediates [3]. During unimpeded transcription, multi-subunit RNAPs catalyze nucleotide addition at 20–100 nt/s, and transcription pauses can cause RNAP to dwell on DNA without translocation for seconds [1, 2, 4, 5]. In contrast, RNA secondary structures can fold up to several orders of magnitude faster, on a microsecond timescale [3, 6]. The

James Chappell and Melissa K. Takahashi (eds.), *Riboregulator Design and Analysis*,
Methods in Molecular Biology, vol. 2518, https://doi.org/10.1007/978-1-0716-2421-0_17,

consequence of this rate difference is that local RNA structures fold efficiently as they are transcribed and can interact directly with the transcribing RNAP [7, 8]. This enables a functional interplay between transcription and RNA structure formation: the direction and rate of transcription shapes the folding landscape of nascent RNA, and nascent RNA structures can modulate RNAP activity [2].

The importance of cotranscriptional folding for RNA biogenesis and gene regulation has been established for decades [9–18]. More recently, cotranscriptional RNA folding has been used as the basis for engineering synthetic gene regulation systems [19–21] and to develop self-assembling RNA nanostructures [22–24]. Despite the biological importance and biotechnological utility of cotranscriptional RNA structure formation, the availability of experimental strategies for mapping RNA folding pathways remained limited until recently. In the past few years, several laboratories have developed biochemical and biophysical methods that can monitor cotranscriptional RNA folding [25–37], the cotranscriptional assembly of ribonucleoprotein complexes [38–42], and interactions between nascent RNA and RNAP [43, 44]. Among these methods, cotranscriptional RNA structure probing combines in vitro transcription with RNA chemical probing to facilitate the measurement of RNA folding intermediates at nucleotide resolution [25–28]. In this chapter, we discuss experimental design principles that must be considered when integrating *E. coli* RNAP in vitro transcription with RNA chemical probing and provide a detailed procedure for performing cotranscriptional RNA chemical probing reactions.

1.1 Cotranscriptional RNA Structure Probing

The goal of cotranscriptional RNA structure probing is to measure the structure of cotranscriptionally folded RNA molecules that are displayed from RNAP. This is achieved by halting transcription elongation complexes (TECs) on a DNA template so that RNA is displayed from RNAP for a structure probing reaction. Landick and coworkers performed the first cotranscriptional RNA structure probing experiment by walking RNA polymerase II to defined DNA template positions and using RNase T1 footprinting to map the rearrangement of HIV-1 nascent RNA from a pause hairpin into the transacting response hairpin [45]. Since this initial implementation, advances in transcription roadblocking and RNA chemical probing methods have facilitated the development of experiments that can comprehensively map cotranscriptional RNA folding pathways in a single in vitro transcription reaction.

Methods for mapping cotranscriptional RNA folding pathways using chemical probing require two technologies. First, *distributed transcription roadblocking* facilitates the display of RNA folding intermediates from RNAP in vitro by halting TECs at every position along a DNA template [25, 26] (*see* Fig. 1a). Second, *high-*

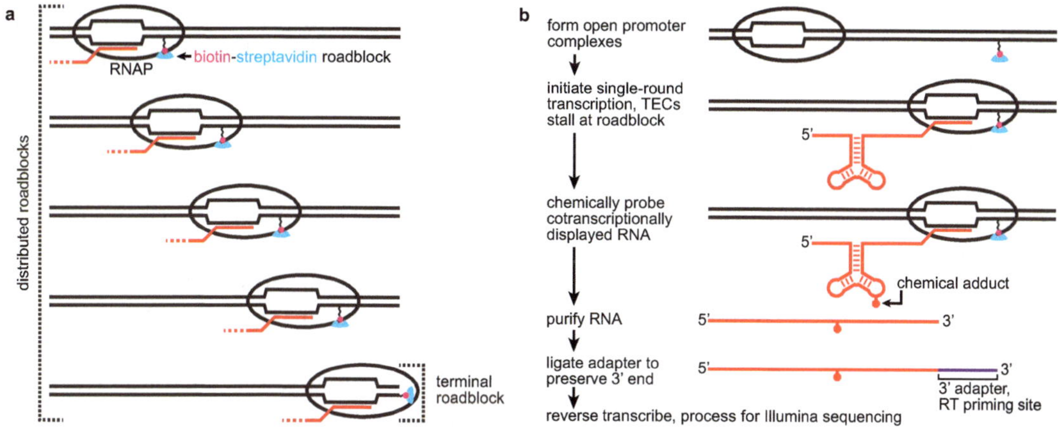

Fig. 1 Overview of cotranscriptional RNA chemical probing. (**a**) Illustration of the distributed transcription roadblocking strategy. RNAP is halted by randomly positioned biotin-streptavidin complexes in the transcribed DNA strand. Runoff transcription is prevented by a terminal biotin-streptavidin roadblock. (**b**) Cotranscriptional RNA chemical probing workflow. Open complexes are formed by incubating RNAP with DNA in the absence of Mg^{2+}. Single-round transcription is initiated by adding Mg^{2+} and rifampicin so that TECs transcribe to a biotin-streptavidin roadblock. The cotranscriptionally displayed RNA is treated with a chemical probe that covalently modifies the RNA in a structure-dependent manner. After purifying the RNA, the nascent RNA 3′ end is recorded by ligating a 3′ adapter, which also serves as a reverse transcription priming site. The RNA is then reverse transcribed and converted into an Illumina sequencing library

throughput RNA chemical probing facilitates the structural characterization of complex RNA mixtures by measuring the reactivity of RNA nucleotides with a chemical probe using high-throughput sequencing [46–48]. In a cotranscriptional RNA structure probing experiment, an array of RNA folding intermediates is generated by distributing TECs across every position of a DNA template in a single-round in vitro transcription reaction (*see* Fig. 1a). The cotranscriptionally folded RNA array is then treated with a chemical probe that covalently modifies RNA nucleotides in a structure-dependent manner (*see* Fig. 1b). The length of each transcript, which approximates the location of RNAP during the structure probing reaction, is recorded by appending a constant adapter sequence to the 3′ end of each nascent RNA. This 3′ adapter also serves as a universal reverse transcription priming site (*see* Fig. 1b). The location of chemical adducts in the RNA is then recorded in cDNA by reverse transcription: under typical RT reaction conditions, reverse transcriptases tend to stop one nucleotide before a chemical adduct to yield a truncated cDNA [49–51]. When RT reaction conditions are adjusted to promote modification bypass, some reverse transcriptases misincorporate as they traverse chemical adducts so that modified nucleotides are recorded as cDNA mutations [52, 53]. The end result of this procedure is a sequencing library that links RNA 3′ end information with RNA structural information. The library is sequenced using the Illumina platform, and the resulting reads are analyzed to determine reactivity spectra that describe the structure of each intermediate transcript of the target RNA.

1.2 Limitations of Cotranscriptional RNA Chemical Probing Experiments

The ability to systematically probe the structure of intermediate transcripts is useful for mapping RNA folding transitions that occur during transcription [25, 26, 34, 54]. However, when designing a cotranscriptional RNA structure probing experiment, it is essential to consider how methodological limitations will influence the results. First, like many other high-throughput RNA chemical probing methods, RNA structure is measured in bulk rather than at single-RNA resolution. Consequently, the chemical reactivity profile of an RNA reflects the average reactivity of each nucleotide across a population of structures. This limitation can be addressed using computational methods for multiple-structure prediction from bulk RNA structure probing data [55] and possibly by using "single-molecule" RNA structure probing methods [56–61]. Second, although RNA displayed from TECs is cotranscriptionally folded, the necessity of halting TECs at transcription roadblocks allows the nascent RNA to equilibrate to some degree. In some cases, this could cause the formation of structures that would not normally fold during unimpeded transcription. Third, the intermediate structures measured by cotranscriptional RNA structure probing are discontinuous – each structure reflects an RNA folding end point that cannot be definitively linked to upstream or downstream structures using chemical probing data. However, single-molecule force spectroscopy and single-molecule FRET have been successfully used to measure cotranscriptional and vectorial RNA folding with high temporal resolution [29–33, 35–37], and systematic functional assays can be used to test RNA folding hypotheses in some cases [54]. Cotranscriptional RNA structure probing is complementary to these approaches because it provides nucleotide-resolution secondary and tertiary structure information at the expense of temporal resolution.

1.3 Lessons from the Application of Cotranscriptional RNA Chemical Probing to Riboswitches

Cotranscriptional RNA chemical probing is a general procedure that, in principle, can be applied to any target RNA. Currently, our applications of cotranscriptional RNA chemical probing have focused primarily on riboswitches [26, 54]. Riboswitches are RNA-encoded chemical sensors that control gene expression in response to a small molecule or elemental ion ligand [62, 63]. Riboswitches are composed of a ligand-sensing *aptamer domain* and an *expression platform* that interfaces with cellular gene expression machinery; ligand recognition by the aptamer domain controls whether the expression platform permits gene expression [62, 64]. The application of cotranscriptional RNA chemical probing to riboswitches that control transcription termination has demonstrated several capabilities of the methodology. First, cotranscriptional RNA structure probing can measure out-of-equilibrium RNA structures. For both fluoride- and ZTP-sensing riboswitches, it was possible to observe antiterminated RNA structures in the presence of the full intrinsic terminator hairpin [26, 54]. In contrast, full-length RNA that was refolded before structure probing formed the terminator hairpin regardless of whether the riboswitch ligand was present. Second, cotranscriptional RNA structure probing can measure transient RNA structures. In the absence of ligand, both the fluoride and ZTP aptamers folded at intermediate transcript lengths but were disrupted by the terminator hairpin at longer transcript lengths [26, 54]. In the presence of ligand, the aptamers exhibited chemical reactivity profiles that, based on crystallographic structures, were consistent with ligand binding; the ligand-bound structures persisted even in the presence of the full terminator hairpin sequence. Furthermore, in the case of the ZTP riboswitch, it was possible to identify a non-conserved transient hairpin that later rearranged to form the ZTP aptamer [54]. Third, cotranscriptional RNA structure probing is sensitive to RNA folding perturbations. Mutations designed to disrupt and restore long-range pseudoknot base pairs changed aptamer folding in predictable ways [26, 54]. Importantly, observations made using cotranscriptional RNA chemical probing [26, 34, 54] agree with observations made using orthogonal biochemical, biophysical, and computational methods in the same or related systems [14, 32, 33, 35, 65, 66].

1.4 Scope of This Protocol

The goal of this protocol is to provide a detailed description of key considerations and procedures that are specific to cotranscriptional RNA structure probing experiments using *E. coli* RNAP. This includes the design and preparation of roadblocking DNA templates, best practices for integrating single-round in vitro transcription using *E. coli* RNAP with RNA chemical probing, purification of RNA from an in vitro transcription reaction, and appending an adapter sequence to the RNA 3′ end. The SHAPE-Seq v2.1 protocol that was used to prepare Illumina sequencing libraries in the original implementation of the cotranscriptional SHAPE-Seq method was described previously [67] but has been superseded by other methods that have been described in detail elsewhere [68, 69]. In principle, the foundational concepts and procedures for cotranscriptional RNA structure probing described here can be integrated into any high-throughput RNA chemical probing workflow.

2 Materials

All materials should be molecular biology grade and used exclusively for RNA work (*see* **Note 1**). Items that are required for multiple procedures are listed under each relevant subheading for convenience. The procedures below benefit from quantitative sample handling and should be performed using low-retention pipette tips. We have found that RNA chemical probing procedures can be sensitive to the microcentrifuge tubes used for the procedure. We currently use Posi-Click Microcentrifuge tubes (Thomas Scientific).

2.1 Equipment

1. Thermal cycler (*see* **Note 2**).
2. Refrigerated microcentrifuge.
3. Vortex.
4. Mini centrifuge.
5. Digital dry bath filled with metallic thermal beads.
6. Magnetic stand for 1.7 ml microcentrifuge tubes.
7. Qubit 4 Fluorometer.
8. Mini vertical gel electrophoresis system.
9. High voltage power supply.
10. Typhoon RGB Biomolecular Imager or equivalent laser scanner.

2.2 DNA Template Preparation

1. Nuclease-free water, store at room temp (RT).
2. Q5 High-Fidelity DNA polymerase (*see* **Note 3**), store at −20 °C.
3. 5X Q5 Buffer (provided with Q5 high-fidelity DNA polymerase), store at −20 °C.
4. Vent (exo-) DNA polymerase (*see* **Note 4**), store at −20 °C.
5. 10X ThermoPol Buffer (provided with Vent (exo-) DNA polymerase), store at −20 °C.
6. 10 mM dNTPs, store at −20 °C.
7. 0.1 mM Biotin-11-dNTP stocks; dilute 1 mM stocks using nuclease-free 10 mM Tris-HCl, pH 7.5 to make separate 0.1 mM stocks, store at −20 °C (*see* **Note 5**) .
8. Oligonucleotides PRA1_NoMod.F (5- TTATCAAAAAGAG TATTGACTCTTTTACCTCTGGCGGTGATAATGGTTG CAT -3, HPLC purified) and HP4_5bio.R (5-/5Biosg/ AATGTCTTCCAGCACACATCGCCTGACGAATCA-3) (*see* **Note 6**).
9. 3 M sodium acetate, pH 5.5, store at RT.
10. Cold 100% ethanol, store at −20 °C.
11. Cold 70% ethanol, prepared with nuclease-free water, store at −20 °C.
12. Preparative grade agarose (*see* **Note 7**).
13. 10 mg/ml ethidium bromide, store at RT.

14. QIAquick Gel Extraction Kit (*see* **Note 8**), store at RT.
15. Template DNA (prepared in Subheading 3.3, **item 1**), store at −20 °C.
16. Sera-Mag SpeedBead Carboxylate-Modified Magnetic Particles (Hydrophobic) (*see* **Note 9**), store at 4 °C.

    Caution: Sera-Mag SpeedBeads contain 0.05% sodium azide.
17. 1 M Tris-HCl, pH 8.0, store at RT.
18. 0.5 M EDTA, pH 8.0, store at RT.
19. 5 M sodium chloride, store at RT.
20. Tween 20, store at RT.
21. PEG 8000, store at RT.
22. SPRI beads (prepared in Subheading 3.3, **item 2** as described by Jolivet and Foley [70]), store at 4 °C.
23. 10 mM Tris-HCl, pH 8.0, nuclease-free, store at RT.
24. Qubit BR dsDNA assay kit, *some components* stored at 4 °C.
25. 4 μM monomeric streptavidin (*see* **Note 10**) prepared in 1X Transcription Buffer (*see* Subheading 2.4 for 10X Transcription Buffer), store working aliquot at 4 °C, other aliquots at −20 °C.
26. 6X DNA Loading Dye, +SDS, XC-only: 10 mM Tris-HCl, pH 8.0, 30% (v/v) glycerol, 0.48% SDS, 0.05% (w/v) xylene cyanol FF, store at RT.
27. 6X Native DNA loading dye: 10 mM Tris-HCl, pH 8.0, 30% (v/v) glycerol, 0.01% (w/v) bromophenol blue, store at RT.
28. 1X Tris-Borate-EDTA (TBE) Buffer: 89 mM Tris base, 89 mM boric acid, 2 mM EDTA, pH 8.3, store at RT.
29. 2X TBE Buffer: 178 mM Tris base, 178 mM boric acid, 4 mM EDTA, pH 8.3, store at RT.
30. 30% Acrylamide, store at RT.
31. 10% (w/v) ammonium persulfate, prepared using nuclease-free water, store at −20 °C.
32. TEMED (Tetramethylethylenediamine), store at RT.
33. SYBR Gold Nucleic Acid Gel Stain, store at −20 °C.

2.3 Pre-adenylation of the 3′ Adapter Oligonucleotide

1. Nuclease-free water, store at RT.
2. 100 μM 9N_VRA3 oligonucleotide (/5Phos/rNrNrNrNrNr NNNNGATCGTCGGACTGTAGAACTCTGAAC / 3AmMO/) or suitable alternative (*see* **Note 11**), dissolve in Nuclease-Free 1X TE (10 mM Tris-HCl, pH 7.5, 0.1 mM EDTA) and store as aliquots at −20 °C.
3. 5′ DNA adenylation kit (*see* **Note 12**), store at −20 °C.

4. TRIzol LS Reagent, store at 4 °C.
5. Chloroform, store at RT.
6. Cold isopropanol, ≥99.5%, store an aliquot at −20 °C.
7. GlycoBlue Coprecipitant, store at −20 °C.
8. Cold 70% ethanol, prepared with nuclease-free water, store at −20 °C.
9. 10 mM Tris-HCl, pH 8.0, nuclease-free, store at RT .
10. Qubit ssDNA assay kit, *some components* stored at 4 °C.

2.4 In Vitro Transcription and Cotranscriptional RNA Structure Probing

1. Nuclease-free water, store at RT.
2. 10X Transcription Buffer (*see* **Note 13**): 200 mM Tris-HCl, pH 8.0, 500 mM KCl, 10 mM DTT (0.2 μm filter-sterilized), 1 mM EDTA, pH 8.0, store 1 ml aliquots at −20 °C.
3. 25 mM NTPs (*see* **Note 14**), store at −20 °C.
4. 5 mg/ml molecular biology-grade BSA, store at −20 °C (*see* **Note 15**).
5. Randomly biotinylated DNA Template (prepared in Subheading 3.3, **item 3**), store at −20 °C.
6. *E. coli* RNA Polymerase, holoenzyme (*see* **Note 16**), store at −20 °C.
7. 100 mM $MgCl_2$, RNase-Free, store at RT.
8. 2 mg/ml Rifampicin in DMSO, store at −20 °C protected from light. **Caution:** handle the DMSO stock bottle in a chemical fume hood.
9. 4 μM monomeric streptavidin (*see* Subheading 2.2, *see* **Note 10**), store at 4 °C.
10. Benzoyl cyanide, store desiccated at RT.

    Caution: benzoyl cyanide is acutely toxic and must be handled in a chemical fume hood while wearing appropriate PPE, including a lab coat and thick nitrile gloves.

11. DMSO, anhydrous, ≥99.9%, store at RT.
12. TRIzol LS Reagent, store at 4 °C.
13. Household bleach, store at RT.

2.5 RNA Purification

1. Nuclease-free water, store at RT.
2. TRIzol LS Reagent, store at 4 °C.
3. Chloroform, store at RT.
4. Cold isopropanol, ≥99.5%, store an aliquot at −20 °C.
5. GlycoBlue Coprecipitant, store at −20 °C.

6. Cold 70% ethanol, prepared with nuclease-free water, store at −20 °C.
7. TURBO DNase (*see* **Note 17**), store at −20 °C.
8. 10X TURBO DNase Buffer (provided with TURBO DNase), store at −20 °C.

2.6 3′ Adapter Ligation

1. Nuclease-free water, store at RT.
2. 10% DMSO, prepared with nuclease-free water, store at RT. **Caution:** handle the DMSO stock bottle in a chemical fume hood.
3. Pre-adenylated 3′ adapter (prepared in Subheading 3.4), store at −20 °C.
4. SuperaseIN RNase Inhibitor, store at −20 °C.
5. T4 RNA Ligase 2, truncated KQ (*see* **Note 18**); store at −20 °C.
6. 10X T4 RNA Ligase Reaction Buffer (provided with T4 RNA Ligase 2, truncated KQ); store at −20 °C.
7. 50% PEG 8000 (provided with T4 RNA Ligase 2, truncated KQ), store at −20 °C.
8. SPRI beads (prepared in Subheading 3.3, **item 2** as described by Jolivet and Foley [70]), store at 4 °C.
9. Anhydrous isopropanol in a bottle with a septum, store at, RT.
10. 80% ethanol, prepared with nuclease-free water, store at RT.
11. 10 mM Tris-HCl, pH 8.0, nuclease-free, store at RT.

2.7 Denaturing Polyacrylamide Gel Electrophoresis

1. Formamide loading dye: 90% (v/v) deionized formamide, 1X Transcription Buffer (*see* Subheading 2.4), 0.01% (w/v) bromophenol blue, store at 4 °C. **Caution:** handle the formamide stock bottle in a chemical fume hood.
2. Low range ssRNA ladder (*see* **Note 19**); store at −20 °C.
3. UreaGel 19:1 Denaturing Gel System (*see* **Note 20**), store at RT.
4. 10% ammonium persulfate, prepared using nuclease-free water, store at −20 °C.
5. TEMED, store at RT.
6. 1X TBE Buffer (*see* Subheading 2.2), store at RT.
7. SYBR Gold Nucleic Acid Gel Stain, store at −20 °C.

3 Methods

3.1 In Vitro Transcription DNA Template Design

E. coli RNAP in vitro transcription reactions are typically performed using a linear dsDNA template that contains, at minimum, a σ^{70} promoter and the sequence of the RNA of interest (*see* Fig. 2). Promoter choice can profoundly impact the success of single-round in vitro transcription experiments because it is a critical determinant of transcription efficiency. We use a custom σ^{70} promoter, called P_{RA1} (5-TTATCAAAAAGAGTATTGACTCTTTTACCTCTGGC GGTGATAATGGTTGCA$_{+1}$T-3), which is primarily derived from the phage promoter $\lambda\underline{P_R}$ (+2 to -35, A-29C; from the nucleotide downstream of the transcription start site (TSS) through the -35 element) and contains the proximal UP-element subsite (−36 to −50) of the T7<u>A1</u> promoter [73] (*see* Fig. 2). Notably, P_{RA1} contains the λP_R discriminator element, which supports the formation of open promoter complexes with long lifetimes [74, 75]. In the DNA sequence context that was used to develop P_{RA1}, 52% of open promoter complexes transitioned to productive elongation in our standard single-round in vitro transcription conditions [73].

When designing a DNA template for in vitro transcription, it is critical to ensure that RNAP primarily initiates at the intended TSS. TSS selection by *E. coli* RNAP depends on the sequence of the TSS and discriminator element [76, 77]. To minimize TSS variability, the P_{RA1} promoter preserves the native λP_R TSS (A_{+1},T_{+2}) *by default*. These nucleotides can be omitted if the RNA of interest already contains an A_{+1},T_{+2} motif or if the A_{+1},T_{+2} interferes with RNA function. At a minimum, it is crucial that the sequence at or immediately downstream of the TSS is not homopolymeric, which can cause RNAP to undergo reiterative transcription initiation or "stuttering" so that the nascent RNA 5′ end contains a homopolymeric tail [78–81]. The initial transcribed sequence (ITS), defined as nucleotides +1 to ~+20, can also influence promoter escape efficiency by determining the extent of abortive RNA synthesis that occurs during transcription initiation [82, 83]. This effect is complex and depends on the presence and location of transcription pause sites in the ITS [84–86]. Under our standard in vitro transcription conditions, which use between 0.1 and 0.5 mM NTPs, abortive initiation has not been an issue.

To standardize the preparation of linear dsDNA by PCR, it is convenient to place a constant sequence that can be used as a PCR priming site downstream of the RNA of interest. For cotranscriptional RNA structure probing experiments that use the P_{RA1} σ^{70} promoter, we use a modified *E. coli his* leader pause site [87, 88], called HP4 (5- CCTGATTCGTCAGGCGATGTGTGCTGGAA GACATT-3), in which the *his* pause hairpin loop is replaced by a UUCG tetraloop (*see* Fig. 2). It is recommended, but not required, to use the P_{RA1} σ^{70} promoter and HP4 constant sequences

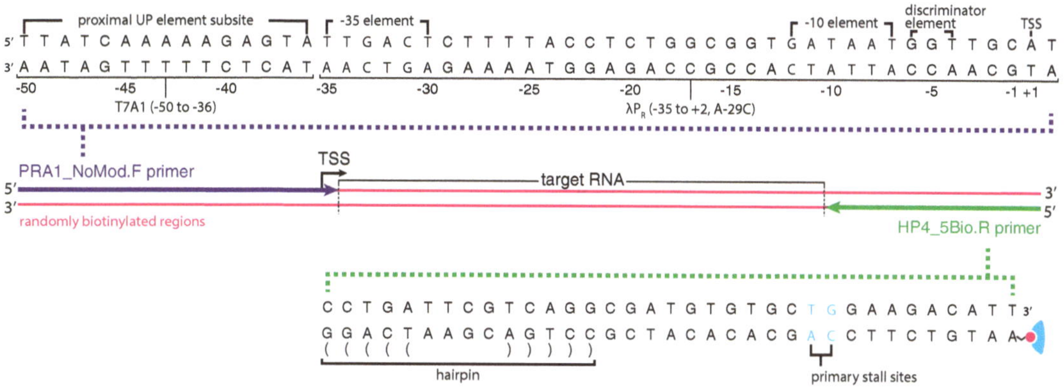

Fig. 2 Overview of in vitro transcription DNA template layout. The recommended in vitro transcription DNA template layout, including the P_{RA1} promoter, target RNA sequence, and HP4 constant region, is shown. The sequences of P_{RA1} and HP4 are annotated to indicate key elements

provided here when designing DNA templates for cotranscriptional RNA structure probing experiments. When using different sequences as PCR priming sites, it is essential to ensure that the primers do not form problematic homo- or heterodimers that could lead to the production of primer dimer during PCR.

The maximum length of the target RNA sequence is determined by several properties of the experiment. Cotranscriptional RNA chemical probing experiments require the use of a defined reverse transcription primer that anneals to the 3′ adapter so that nascent RNA 3′ end information is preserved in the sequencing library. Consequently, even when mutational profiling is used to detect chemical adducts, the maximum target RNA length is limited by the maximum sequencing read length on Illumina platforms (2 × 250 bp for a total of 500 nt on the NovaSeq platform; 2 × 150 bp for a total of 300 nt on the HiSeq and NextSeq platforms; 2 × 300 bp for a total of 600 nt on the MiSeq platform) and whether any sequences that extend the target RNA are required (e.g., molecular barcodes that are appended during sequencing library construction, constant transcribed sequences, etc.). If chemical modifications are detected as cDNA truncations, the maximum target RNA length will be reduced to 300–400 nt due to signal decay during reverse transcription.

3.2 Choosing an RNA Polymerase Roadblocking Strategy

Cotranscriptional RNA structure probing experiments begin by stalling RNAP at a transcription roadblock so that a cotranscriptionally folded nascent RNA is displayed from the arrested TEC for chemical probing [25, 26]. In order to be useful, a transcription roadblock must halt RNAP efficiently (TECs stall upon encountering the roadblock), stably (roadblocked TECs do not dissociate from DNA), and persistently (TECs do not bypass the roadblock during the experiment). Transcription roadblocking strategies can

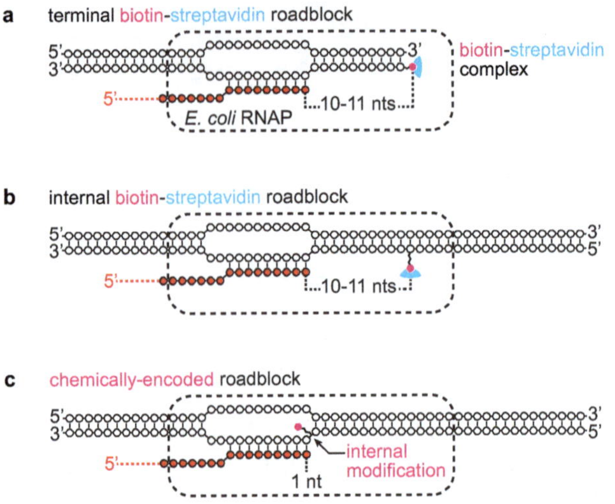

Fig. 3 Transcription roadblocking strategies. The architecture of a TEC halted at (**a**) a terminal biotin-streptavidin roadblock, (**b**) an internal biotin-streptavidin roadblock, and (**c**) a chemically encoded transcription roadblock is shown. The distance between the RNA 3′ end and the roadblock when transcription stalls is indicated

be *targeted* or *distributed*. Targeted transcription roadblocking strategies are designed to stall TECs precisely at a defined position in the DNA template and can be used in single-length cotranscriptional RNA structure probing experiments or as a terminal stall site in multi-length experiments. Distributed transcription roadblocking strategies are designed to systematically distribute TECs across every position of a DNA template in multi-length cotranscriptional RNA structure probing experiments (*see* Fig. 1a). This protocol focuses on the use of biotin-streptavidin complexes as distributed transcription roadblocks.

Biotin-streptavidin complexes that are attached to the transcribed DNA strand halt *E. coli* RNAP transcription efficiently both when placed at the downstream 5′ terminus of the DNA template and when placed internally [25, 29]. TECs that collide with a biotin-streptavidin roadblock tend to stop 7 to 13 nucleotides upstream of the biotin modification, with most complexes stalled 10 nucleotides upstream [25, 89] (*see* Fig. 3a, b). We previously showed that internal biotin-streptavidin roadblocks halt RNAP with 80–90% efficiency, that >98% of stalled TECs remain bound to DNA, and that stalled RNAP cannot bypass the biotin-streptavidin complex [25]. Biotin-streptavidin roadblocking is our preferred method for halting *E. coli* RNAP transcription in both single-length and multi-length cotranscriptional RNA structure probing experiments for two reasons (*see* **Note 21**). First, biotin-streptavidin roadblocks are sequence-independent and, consequently, compatible with all RNA targets. Second, the preparation of DNA templates for biotin-streptavidin roadblocking is inexpensive and straightforward. Terminally biotinylated DNA templates are prepared by PCR using a primer with a 5′ biotin modification. Internally biotinylated DNA templates are prepared by random incorporation of biotin-modified dNTPs during PCR. The primary disadvantage of this strategy is that RNAP tends to reverse translocate, or "backtrack" [91], 3–4 nts upon colliding with a biotin-streptavidin roadblock so that the RNA 3′ end is displaced from the RNAP active center and the footprint of RNAP is shifted [25]. When analyzing cotranscriptional structure probing data, the location of RNAP is inferred from the RNA 3′ end. Biotin-streptavidin roadblocking therefore blurs the link between the RNA 3′ end and the RNAP footprint. This effect is complex, likely sequence-dependent, and cannot be easily accounted for during data analysis. However, in our experience this does not meaningfully impact data interpretation, and the simplicity and low cost of DNA template preparation outweigh this disadvantage. Furthermore, in cases where it is necessary to resolve the structure of a specific transcript length with high precision, chemically encoded transcription roadblocks [89] can be used (*see* Fig. 1c) (*see* **Note 22**).

3.3 Preparation of Randomly Biotinylated Template DNA

Distributed biotin-streptavidin transcription roadblocking requires the preparation of linear dsDNA that contains randomly positioned biotin modifications. Randomly biotinylated DNA is prepared using a straightforward procedure in which biotin-modified dNTPs are incorporated into linear dsDNA during PCR. The PCR product is then purified and quantified.

In general, we store in vitro transcription DNA template sequences as sequence-verified plasmid DNA. If plasmid DNA is used as the template for PCR, it must be removed from the linear dsDNA preparation by UV-free agarose gel extraction. This procedure requires a large-scale (typically 500 μl) PCR to overcome the yield loss that occurs during gel extraction and is therefore not ideal for PCRs that require costly reagents, like biotin-modified dNTPs. To circumvent the need for gel extraction when preparing randomly biotinylated DNA, we perform PCR using a linear dsDNA template that was previously amplified from plasmid DNA using high-fidelity PCR and purified by UV-free gel extraction. The randomly biotinylated DNA can then be purified using a high-efficiency solid phase reversible immobilization (SPRI) bead-based protocol, **assuming that no primer dimer forms during PCR** (*see* **Note 23**). Our recommended primers do not form any detectable primer dimer when used in the PCR below.

Pre-amplification and Agarose Gel Purification of Linear dsDNA

Linear dsDNA, which will be used as a template for PCR in Subheading 3.3, **item 3**, is amplified from plasmid DNA by PCR with Q5 high-fidelity DNA polymerase and then purified by UV-free agarose gel extraction to remove the plasmid DNA and excess primers.

1. Begin preparing 520 μl of Q5 PCR Master Mix by adding 369 μl of nuclease-free water to a 1.7 ml microcentrifuge tube on ice.
2. Add 104 μl of 5X Q5 Buffer to the Master Mix.
3. Vortex the Master Mix and spin down liquid using a mini centrifuge.
4. Add 10.4 μl of 10 mM dNTP solution to the Master Mix.
5. Add 13 μl of 10 μM PRA1_NoMod.F primer to the Master Mix.
6. Add 13 μl of 10 μM HP4_5bio.R primer to the Master Mix.
7. Add 5.2 μl of 30–60 ng/μl plasmid DNA to the Master Mix.
8. Vortex the Master Mix and spin down liquid using a mini centrifuge.
9. Add 5.2 μl of Q5 high-fidelity DNA polymerase and mix the Master Mix by pipetting.
10. Split the Master Mix into 100 μl aliquots in thin-walled PCR tubes (*see* **Note 2**).
11. Perform PCR using the following thermal cycler protocol with a heated lid set to 105 °C (*see* **Note 24**): 98 °C for 30 s [98 °C for 10 s, 65 °C for 20 s, 72 °C for 30 s/kb] × 30 cycles, 72 °C for 5 min; hold at 10 °C.
12. Pool the PCRs into a 1.7 ml microcentrifuge tube, and ethanol precipitate the DNA by adding 50 μl of 3M sodium acetate (pH 5.5) and 1 ml of cold 100% ethanol and chilling at −80 °C for 30 min.
13. Pellet the precipitated DNA by centrifugation at 18,500 × g and 4 °C for 30 min.
14. Remove the supernatant without disturbing the pellet.
15. Wash the pellet by adding 1 ml of cold 70% ethanol, inverting the tube several times, and centrifuging at 18,500 × g and 4 °C for 2 min.
16. Remove the supernatant without disturbing the pellet.
17. Spin the sample in a mini centrifuge to pull down any residual liquid, and remove the residual liquid by pipetting.

18. Air-dry the sample for several minutes – do not allow the sample to dry completely.
19. Resuspend the pelleted DNA in 30 μl of 10 mM Tris-HCl (pH 8.0).
20. Mix the sample with 6 μl of 6X DNA loading dye, +SDS, XC-only.
21. Load the sample on a 1X TBE agarose gel (*see* **Note 7**) that contains 0.625 μg/ml ethidium bromide and run the gel at 100 V for 1–2 h.
22. During gel electrophoresis, a red band will become visible because ethidium bromide co-migrates with the large quantity of DNA in the sample. Without exposing the gel to UV light, excise this band with a clean razor blade and trim excess agarose. Once the band has been excised, check the gel on a UV transilluminator to confirm that no significant off-target or primer dimer products are formed during PCR.
23. Purify the DNA using the Qiagen QIAquick Gel Extraction Kit according to the manufacturer's instructions, and elute the DNA in 30 μl of 10 mM Tris-HCl (pH 8.0).
24. Quantify the purified DNA with the Qubit dsDNA BR Assay Kit using a Qubit Fluorometer.
25. Store the purified DNA at −20 °C. As needed, dilute the DNA to 1 nM using 10 mM Tris-HCl (pH 8.0) for use below in Subheading 3.3, **item 3**.

Preparation of SPRI Beads for Nucleic Acid Purification

The procedures below make extensive use of SPRI beads for nucleic acid purification. We prepare homemade SPRI beads according to the procedure of Jolivet and Foley [70], which is described below for convenience.

26. Prepare a 10% Tween-20 stock. Tare a 50 ml conical on a balance and slowly add 5.475 g of 100% Tween-20. Add 45 ml of nuclease-free water, cap the conical, and mix with rotation to dissolve the Tween-20. This typically takes about 1 h.
27. Prepare a 50% PEG 8000 stock. Tare a 50 ml conical on a balance and add 12.5 g of PEG 8000 powder. Add 14 ml of nuclease-free water, cap the conical, and shake vigorously until no clumps of solid PEG 8000 are stuck to the conical walls. Allow the solution to sit at room temperature for at least 1 h so that the PEG 8000 dissolves and air bubbles rise to the surface. Add 25 ml of water to a separate 50 ml conical, and mark the height of the liquid on the side of the tube. Using the marked 50 mL conical as a guide, gently fill the 50 ml conical that

contains the PEG 8000 solution with nuclease-free water until the height of the PEG 8000 solution matches the marked position. Mix the 50% PEG 8000 solution by inverting.

28. Prepare bead wash buffer by combining 14.75 ml of nuclease-free water, 150 μl of 1 M Tris-HCl (pH 8.0), 30 μl of 0.5 M EDTA (pH 8.0), and 75 μl of 10% Tween 20 in a 15 ml conical tube and mixing by inversion.
29. Combine 4.15 ml of nuclease-free water, 25 ml of 5 M NaCl, 500 μl of 1 M Tris-HCl (pH 8.0), and 100 μl of 0.5 M EDTA (pH 8.0) in a conical tube and mix by inversion. This is an incomplete bead binding solution that will be used below in **step 38**.
30. Mix the stock bottle of Sera-Mag SpeedBeads Carboxylate-Modified Magnetic Particles (Hydrophobic) well to resuspend the settled beads. **Caution:** Sera-Mag SpeedBeads contain 0.05% sodium azide.
31. Transfer 1 ml of the resuspended Sera-Mag bead stock to a 1.7 ml microcentrifuge tube.
32. Pull down liquid by pulse spinning the sample in a mini centrifuge. Place the tube on a magnetic stand and allow the beads to collect on the tube wall. This takes 1–2 min.
33. Gently remove and discard the supernatant.
34. Remove the tube from the magnetic stand and add 1 ml of bead wash buffer (prepared in **step 28**). Mix by vortexing for at least 15 s.
35. Pull down liquid by pulse spinning the sample in a mini centrifuge. Place the tube on a magnetic stand and allow the beads to collect on the tube wall. This takes 1–2 min.
36. Gently remove and discard the supernatant.
37. Repeat **steps 34** to **36** two additional times for a total of three washes.
38. Remove the tube from the magnetic stand and add 1 ml of the incomplete bead binding solution (prepared in **step 29**). Mix by vortexing for at least 15 s.
39. Pull down liquid by pulse spinning the sample in a mini centrifuge, and transfer the bead solution from the microcentrifuge tube to the 50 ml conical tube that contains the remaining incomplete bead binding solution.
40. Mix by vortexing the 50 ml conical tube for at least 30 s.
41. Using a 25 ml serological pipette, add 20 ml of 50% PEG 8000 to the bead solution. Dispense the 50% PEG 8000 slowly to ensure that it is fully dispensed from the pipette.
42. Add 250 μl of 10% Tween 20 to the 50 ml conical tube.

43. Cap the conical tube and mix gently by inversion until the solution is homogenous.
44. Aliquot the beads into 1.7 ml microcentrifuge tubes and store at 4 °C. When aliquoting the beads, periodically cap the conical tube and mix the beads to minimize bead settling.
45. The beads can be validated for randomly biotinylated DNA template purification by performing the procedures described in Subheading 3.3, **item 3** *including* the native PAGE quality control gel in **step 76**.

Preparation of Randomly Biotinylated DNA by Vent (exo-) PCR

The preparation of randomly biotinylated DNA by PCR requires the use of a DNA polymerase that can incorporate biotin-11-dNTPs. We prefer Vent (exo-) DNA polymerase for this purpose because it has been shown to tolerate complete substitution of unmodified dNTPs with biotin-modified dNTPs [93]. Furthermore, the moderate strand displacement activity of Vent (exo-) DNA polymerase can be useful for amplifying GC-rich DNA templates.

46. The amount of each biotin-11-dNTP added to the PCR will vary depending on the DNA sequence. Calculate the amount of each biotin-11-dNTP to add to the PCR using the following formula:

$$\mathrm{dNTP_{bio}} = \left(\frac{\mathrm{dNTP_{comb}}}{4\mathrm{N_{count}}}\right) \times h$$

where $\mathrm{dNTP_{bio}}$ is amount of biotin-11-dNTP for base N to add to the reaction in nmol, N_{count} is the number times base N occurs in the target RNA sequence in the transcribed region of both DNA strands (primer sequences are excluded), and $\mathrm{dNTP_{comb}}$ is the total amount of dNTP for base N (biotinylated and non-biotinylated) in the PCR in nmol; in a 100 μl reaction $\mathrm{dNTP_{comb}}$ is 20 nmol. h is a multiplier to set the desired number of biotin modifications per DNA template molecule; we set h to 2 by default because non-transcribed strand biotin-streptavidin complexes do not efficiently halt RNAP [25]. Multiply $\mathrm{dNTP_{bio}}$ by 10 to determine the amount of 0.1 mM biotin-11-dNTP stock solution to add to the reaction.

47. Begin preparing 220 μl of Vent (exo-) PCR Master Mix by adding nuclease-free water to a 1.7 ml microcentrifuge tube on ice so that the final reaction volume will be 220 μl (*see* **Note 25**). The volume of biotin-11-dNTPs to be added is

determined in **step 46**, and the volume of the other reagents added in the steps below is 44 μl.

48. Add 22 μl of 10X ThermoPol Buffer to the Master Mix.
49. Vortex the Master Mix and spin down liquid using a mini centrifuge.
50. Add 4.4 μl of 10 mM dNTP solution to the Master Mix.
51. Add 5.5 μl of 10 μM PRA1_NoMod.F primer to the Master Mix.
52. Add 5.5 μl of 10 μM HP4_5bio.R primer to the Master Mix.
53. Add the amount of biotin-11-dNTPs determined in **step 46** to the Master Mix.
54. Add 4.4 μl of 1 nM linear dsDNA template (prepared in Subheading 3.3, **item 1**) to the Master Mix.
55. Vortex the Master Mix and spin down liquid using a mini centrifuge.
56. Add 2.2 μl of Vent (exo-) DNA polymerase and mix the Master Mix by pipetting.
57. Split the Master Mix into 100 μl aliquots in thin-walled PCR tubes (*see* **Note 2**).
58. Perform PCR using the following thermal cycler protocol with a heated lid set to 105 °C (*see* **Note 24**): 95 °C for 3 min [95 °C for 20 s, 58 °C for 30 s, 72 °C for 1 min/kb] × 30 cycles, 72 °C for 5 min; hold at 12 °C.
59. Combine the 100 μl PCRs into a single 1.7 ml microcentrifuge tube for a total of 200 μl.
60. Resuspend the SPRI bead stock by vortexing.
61. The SPRI beads will be mixed with the 200 μl PCR at a 1:1 ratio. Add 200 μl of SPRI beads to the pooled PCR and gently mix by pipetting until the solution is homogenous.
62. Incubate the samples at room temperature for 5 min.
63. Place the microcentrifuge tubes on a magnetic stand and allow the beads to collect on the tube wall for 3 min.
64. Aspirate and discard the supernatant. Be careful not to disturb the beads and do not remove the microcentrifuge tube from the magnetic stand.
65. Without removing the microcentrifuge tubes from the magnetic stand, gently add 500 μl of 70% ethanol without disturbing the beads. **Do not resuspend the beads.**
66. Incubate the samples at room temperature for 1 min.
67. Aspirate and discard the supernatant.
68. Repeat **steps 65** to **67** to wash the beads a second time.

69. Pulse spin the beads in a mini centrifuge to pull down residual liquid and return the beads to the magnetic stand. Remove any residual liquid.
70. Place the samples in a dry bath set to 37 °C with the caps open for 15 to 30 s to evaporate residual ethanol. **Watch the samples closely** to ensure that the beads do not dry completely.
71. Resuspend the beads in 50 μl of 10 mM Tris-HCl (pH 8.0) and incubate for 3 min at room temperature to elute the DNA.
72. Place the microcentrifuge tubes on a magnetic stand and allow the beads to collect on the tube wall for 1 min.
73. Carefully recover and transfer the supernatant to a clean 2.0 ml screw-cap tube (w/ an O-ring). This supernatant contains the purified DNA template.
74. Quantify the purified DNA with the Qubit dsDNA BR Assay Kit using a Qubit Fluorometer. A typical prep yields ~25 pmol of linear dsDNA.
75. Dilute the stock to a working concentration of 100 nM using 10 mM Tris-HCl (pH 8.0). Both the stock and working samples should be stored at −20 °C.
76. If desired, the randomly biotinylated DNA template can be assessed for quality using native PAGE. We previously described a detailed procedure for performing native PAGE QC on in vitro transcription DNA templates using National Diagnostics ProtoGel Acrylamide and a Bio-Rad Mini-PROTEAN Vertical Electrophoresis Cell [92]. The same procedure can be used to QC randomly biotinylated DNA. Mix 0.5 μl of 100 nM DNA template with 1 μl 4 μM streptavidin (or 10 mM Tris-HCl (pH 8.0) as a negative control) and 3.5 μl of 10 mM Tris-HCl (pH 8.0). Incubate the streptavidin binding mixture and negative control at room temperature for 10 min, add 1 μl of 6X Native DNA loading dye, and run the samples on a 5% polyacrylamide, 1X TBE gel. Stain the gel with SYBR Gold and use a laser scanner or gel documentation system to image the gel. The negative control should migrate as a single band of the expected size, and adding streptavidin should cause the DNA to shift into a smear, which may contain a ladder of discrete bands. An example QC gel is shown in Fig. 4.

3.4 Pre-adenylation of the 3′ Adapter Oligonucleotide

Multi-length cotranscriptional RNA structure probing experiments require a 3′ adapter ligation to preserve RNA 3′ end information and append a universal priming site for reverse transcription. The 3′ adapter ligation is performed using T4 RNA Ligase 2, truncated KQ, which is modified to reduce side product formation but requires the use of a pre-adenylated linker.

Our standard 3′ adapter oligo, 9N_VRA3 (*see* Subheading 2.3), contains the VRA3 adapter (the reverse complement of the

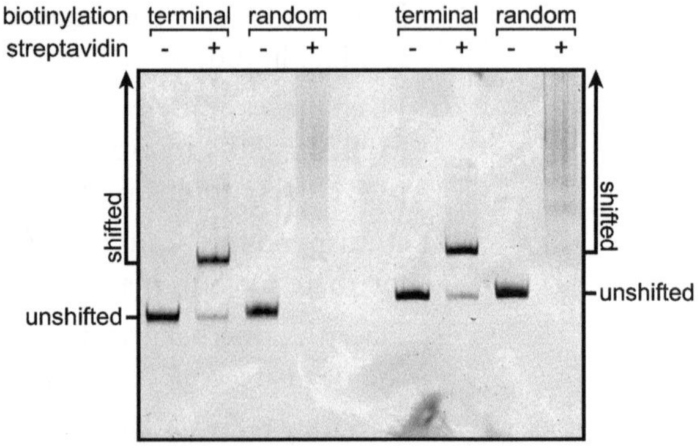

Fig. 4 Native PAGE QC gel for biotinylated DNA. Native PAGE QC for terminally and randomly biotinylated preparations of two DNA templates is shown. In the absence of streptavidin, both terminally and randomly biotinylated DNA migrate at the expected size. In the presence of streptavidin, most of the terminally biotinylated DNA template shifts as a discrete band. In this case the HP4_5bio.R primer was not HPLC purified so the unshifted DNA most likely does not contain a 5′ biotin. Streptavidin binding causes randomly biotinylated DNA to shift as a smear, although some discrete bands are visible

Illumina TruSeq Small RNA RA5 adapter [94]), a 9N unique molecular identifier, a 5′ phosphate modification to facilitate pre-adenylation, a 3′ amino modifier to block the 3′ end, and six ribonucleotides at the 5′ end to facilitate ligation by T4 RNA Ligase 2, truncated KQ. The 3′ adapter requires RNase-Free HPLC purification. The VRA3 sequence can be substituted with an alternative Illumina adapter if desired (*see* **Note 11**). 9N_VRA3 should be resuspended in nuclease-free 1X TE (10 mM Tris-HCl (pH 7.5), 0.1 mM EDTA) to 100 μM and stored at -20 °C as small aliquots in 2.0 ml screw-cap tubes with an O-ring. 9N_VRA3 is pre-adenylated using the NEB 5′ DNA adenylation kit in the following procedure:

1. Begin preparing 100 μl of 5′ adenylation Master Mix by adding 65 μl of nuclease-free water to a 1.7 ml microcentrifuge tube on ice.
2. Add 10 μl of 10X 5′ DNA adenylation buffer to the Master Mix.
3. Vortex the Master Mix and spin down liquid using a mini centrifuge.
4. Add 10 μl of 1 mM ATP to the Master Mix.
5. Add 5 μl of 100 μM 9N_VRA3 oligo to the Master Mix.
6. Vortex the Master Mix and spin down liquid using a mini centrifuge.

7. Add 10 μl of Mth RNA ligase and gently mix the completed Master Mix by pipetting.
8. Split the 100 μl DNA adenylation Master Mix into 25 μl aliquots in thin-walled 200 μl PCR tubes (*see* **Note 2**).
9. Incubate the 5′ adenylation reaction at 65 °C in a thermal cycler with a heated lid set to 105 °C for 1 h.
10. Combine the 25 μl reactions into two 50 μl pools.
11. Add 150 μl of TRIzol LS to each 50 μl pool. Vortex the sample to ensure that it is thoroughly mixed. Spin the sample down in a mini centrifuge to remove liquid from the tube cap and walls.
12. Add 40 μl of chloroform to each sample. Mix the samples by vortexing and inverting the tubes to ensure thorough mixing.
13. Centrifuge the samples at 18,500 × g and 4 °C for 5 min.
14. Transfer the aqueous (clear) phase to a new 1.7 mL tube. Be sure to avoid the organic (pink) phase.
15. Add 1 μl of GlycoBlue and 100 μl of ice-cold isopropanol. Vortex thoroughly and incubate at room temperature for 15 min to precipitate the nucleic acids.
16. Centrifuge the samples at 18,500 × g and 4 °C for 15 min.
17. Carefully aspirate and discard the supernatant while avoiding the blue pellet.
18. Add 500 μl of ice-cold 70% ethanol and gently invert the tube to wash the sample.
19. Centrifuge the samples at 18,500 × g and 4 °C for 2 min.
20. Carefully aspirate and discard the supernatant while avoiding the blue pellet.
21. Centrifuge the samples for several seconds to pull down any residual ethanol.
22. Aspirate and discard the residual ethanol and allow the pellet to dry for a few minutes.
23. Resuspend both pellets in a single 25 μl volume of 10 mM Tris-HCl (pH 8.0).
24. Quantify the concentration of the pre-adenylated oligo with the Qubit ssDNA assay kit using a Qubit Fluorometer (*see* **Note 26**).
25. Dilute the oligo to 1 μM using 10 mM Tris-HCl (pH 8.0) and store in 20 μl aliquots at −20 °C. The small volume of the aliquots is to minimize freeze-thaw cycles that can be adjusted as needed (*see* **Note 27**).

3.5 Single-Round In Vitro Transcription and Cotranscriptional RNA Structure Probing

The procedure for cotranscriptional RNA structure probing integrates roadblocked single-round in vitro transcription with RNA chemical probing. Cotranscriptional RNA structure probing experiments must be performed using *single-round* (as opposed to multi-round) in vitro transcription, which permits transcription initiation to occur a maximum of once per DNA template molecule. During the cotranscriptional RNA structure probing procedure, RNA is displayed from RNAP for chemical probing by trapping a TEC at a DNA-encoded roadblock (*see* Fig. 1b). Under single-round transcription conditions, each DNA template molecule can contain up to one roadblocked TEC. However, if transcription initiation is unrestricted, repeated initiation from the same promoter will cause multiple TECs to pile up on each DNA molecule after the first RNAP arrests at the transcription roadblock. This scenario should be avoided because it is not possible to predict how these RNAP-RNAP collisions influence RNA structure.

For cotranscriptional RNA structure probing experiments, our preferred single-round transcription format is as follows: first, open promoter complexes are formed by incubating RNAP with DNA in the absence of Mg^{2+} so that transcription initiation cannot occur. Second, a "start solution" that contains Mg^{2+} and the antibiotic rifampicin is mixed with the preformed open complexes to initiate single-round transcription. The addition of Mg^{2+} permits the preformed open promoter complexes to initiate transcription, and rifampicin binds to free RNAP so that any new open complexes cannot extend a nascent RNA chain beyond 2–3 nt [95] (*see* **Note 28**). The single-round in vitro transcription protocol below is written using this format (*see* **Note 29**).

Because the goal of cotranscriptional RNA structure probing experiments is to assess the structure of nascent RNA, it is ideal to minimize the duration of the chemical probing reaction to limit the time that the cotranscriptionally displayed RNA has to equilibrate. For this reason, our preferred reagent for cotranscriptional RNA structure probing is the SHAPE (selective 2′-hydroxyl acylation analyzed by primer extension) [98] probe benzoyl cyanide (*see* **Note 30**). SHAPE reagents react with the 2′ hydroxyl of flexible nucleotides so that structural information for all four nucleotides can be obtained from a single reaction. Benzoyl cyanide was originally used to perform time-resolved RNA structure probing experiments due to its short (250 ms) hydrolysis half-life at 37 °C [100, 101] and is ideal for cotranscriptional RNA structure probing because the chemical probing reaction self-quenches after ~1 s (*see* **Note 31**). Consequently, after a short in vitro transcription reaction, nascent RNA is rapidly probed, and the in vitro transcription reaction can be stopped immediately after probing is complete.

The transcription conditions presented here can be optimized as needed for a given RNA target but are a good starting point for most applications. Before performing the cotranscriptional RNA structure probing procedure, it is useful to confirm that the in vitro transcription reaction yields RNA product(s) of the expected length (s) and, in some cases, that the RNA functions as expected. This can be accomplished by making several minor modifications to the in vitro transcription reaction used for the cotranscriptional RNA structure probing procedure (*see* **Note 32**) and assessing the resulting RNA by denaturing PAGE (*see* Subheading 3.9).

Each cotranscriptional RNA structure probing sample will require 60 μl of in vitro transcription Master Mix which will be split into two 25 μl aliquots during the chemical probing reaction. The procedure here is written using volumes for one sample but can be scaled as needed for addition samples.

1. Begin preparing in vitro transcription Master Mix by adding 32.1 μl of nuclease-free water to a 1.7 ml microcentrifuge tube on ice (*see* **Note 33**).
2. Add 6 μl of 10X Transcription Buffer to the Master Mix.
3. Vortex the Master Mix and spin down liquid using a mini centrifuge.
4. Add 1.2 μl of 25 mM NTP solution to the Master Mix (*see* **Note 34**).
5. Add 1.2 μl of 5 mg/ml molecular biology-grade BSA to the Master Mix.
6. Add 6 μl of 100 nM randomly biotinylated DNA template (prepared in Subheading 3.3, **item 3**) to the Master Mix.
7. Vortex the Master Mix and spin down liquid using a mini centrifuge.
8. Add 1.5 μl of 1 U/μl *E. coli* RNAP holoenzyme to the Master Mix and gently mix by pipetting. At this time, the total volume of the Master Mix will be 48 μl because 10X start solution and streptavidin have not yet been added.
9. If a Master Mix for more than one sample has been prepared, aliquot one reaction volume of Master Mix (48 μl, the total volume without 10X start solution and streptavidin) into a separate 0.6 ml microcentrifuge tube for each sample.
10. Prepare 10X start solution by mixing 200 μl of 100 mM $MgCl_2$ and 10 μl of 2 mg/ml rifampicin in a 1.7 ml microcentrifuge tube. Vortex thoroughly and spin down in a mini centrifuge. Cover the 10X start solution in foil to protect from light and keep at room temperature. **10X start solution must be prepared fresh for each experiment.**
11. Prepare a 1.7 ml microcentrifuge tube containing enough TRIzol LS for each transcription reaction. Each sample requires 150 μl of TRIzol LS. TRIzol LS will be used to stop the in vitro transcription reaction after chemical probing.

12. Prepare 400 mM benzoyl cyanide as follows: tare an analytical balance with a 1.7 ml microcentrifuge tube. In a chemical fume hood, carefully transfer several milligrams of benzoyl cyanide to the 1.7 ml microcentrifuge tube and close the cap. Weigh the microcentrifuge tube that contains the benzoyl cyanide using the analytical balance. Multiply the mass in milligrams by 19.065 to determine the volume of anhydrous DMSO in microliters needed to dissolve the benzoyl cyanide to 400 mM. In a chemical fume hood, add this volume of anhydrous DMSO to the tube containing benzoyl cyanide. Close the tube tightly, wrap the cap in parafilm, and gently vortex the tube until the benzoyl cyanide has dissolved. Spin down the liquid by pulsing the tube in a mini centrifuge.

    400 mM benzoyl cyanide must be prepared fresh for each experiment.

    Caution: benzoyl cyanide is acutely toxic and must be handled in a chemical fume hood at all times while wearing appropriate PPE, including a lab coat and thick nitrile gloves. Once the experiment is complete, excess benzoyl cyanide solution can be mixed with bleach and discarded as hazardous liquid waste in the chemical fume hood.

13. For each 60 μl transcription reaction, prepare two 1.7 ml microcentrifuge tubes. Aliquot 2.78 μl of anhydrous DMSO into one tube, which will be the negative control. Aliquot 2.78 μl of 400 mM benzoyl cyanide into the second tube.

14. Perform the single-round in vitro transcription time course in a chemical fume hood. We recommend performing this reaction using a thermal cycler (that can accommodate 0.6 ml tubes) so that the reaction temperature is robust to airflow from the fume hood (*see* **Note 35**). The example described in the steps below is for one sample but can be adapted for multiple samples by staggering the steps (*see* **Note 36**).

15. Place the in vitro transcription reaction on a thermal cycler block set to 37 °C to begin forming open promoter complexes. Open promoter complex formation will proceed for 20 min.

16. After 10 min of open complex formation, add 6 μl of 4 μM streptavidin to the reaction and mix by pipetting (*see* **Note 37**).

17. After 20 min of open complex formation, start transcription by adding 6 μl of 10X start solution and mix by pipetting. Typically, transcription should proceed for at least 30 s with an upper limit of 2 min before proceeding to the next step (*see* **Note 38**).

18. Perform the structure probing reaction by transferring 25 μl of the transcription reaction to the tube containing benzoyl cyanide and transferring 25 μl of the transcription reaction to the tube containing DMSO. Then stop transcription by adding 75 μl of TRIzol LS to each reaction aliquot, vortexing briefly,

and placing the reaction at room temperature. *See* **Note 39** for additional details on the best practices for this procedure.

3.6 RNA Purification

Caution: the procedure below is performed with samples that have been treated with benzoyl cyanide. Perform this procedure inside a chemical fume hood at least until the precipitated nucleic acids have been washed once with 70% ethanol (through **step 10**).

1. At this time, the samples contain a 27.78 μl transcription reaction mixed with 75 μl of TRIzol LS.
2. Pulse spin the samples in a mini centrifuge to pull down liquid from the tube cap and walls. Add 20 μl of chloroform to each sample. Mix the samples by vortexing and inverting the tubes to ensure thorough mixing.
3. Centrifuge the samples at 18,500 × g and 4 °C for 5 min.
4. Transfer the aqueous (clear) phase to a new 1.7 mL tube. Be sure to avoid the organic (pink) phase.
5. Add 1 μl of GlycoBlue and 50 μl of ice-cold isopropanol. Vortex thoroughly and incubate at room temperature for 15 min to precipitate the nucleic acids.
6. Centrifuge the samples at 18,500 × g and 4 °C for 15 min.
7. Carefully aspirate and discard the supernatant while avoiding the blue pellet.
8. Add 500 μl of ice-cold 70% ethanol and gently invert the tube to wash the sample.
9. Centrifuge the samples at 18,500 × g and 4 °C for 2 min.
10. Carefully aspirate and discard the supernatant while avoiding the blue pellet.
11. Centrifuge the samples for several seconds to pull down any residual ethanol.
12. Aspirate and discard the residual ethanol and allow the pellet to dry for a few minutes.
13. Resuspend the pellet in 25 μl of 1X TURBO DNase buffer by pipetting.
14. Add 0.75 μl TURBO DNase to each sample and mix by pipetting.
15. Incubate the DNase reaction at 37 °C for 15 min to degrade the template DNA.
16. Add 75 μl of TRIzol LS to each sample and mix thoroughly by vortexing. Spin the samples down in a mini centrifuge to remove liquid from the tube cap and walls.
17. Repeat **steps 2–12** to TRIzol extract the sample a second time; it is not necessary to add GlycoBlue Coprecipitant again. For cotranscriptional RNA structure probing experiments, the purified RNA pellet will be dissolved in 5 μl of 10% DMSO in

Subheading 3.7. If the purified RNA will be used to assess RNA length or function by denaturing PAGE, it will be resuspended in 7 μl of formamide loading dye in Subheading 3.9.

3.7 3′ Adapter Ligation

The 3′ adapter ligation appends a constant sequence to the RNA 3′ end so that reverse transcription can be performed using a single defined primer. This procedure also preserves RNA 3′ end information so that each intermediate transcript can be differentiated during data analysis. After the ligation reaction, excess 3′ adapter is depleted using a SPRI bead-based purification.

1. Resuspend each RNA pellet from Subheading 3.6, **step 17** in 5 μl of 10% DMSO.
2. Each sample will require 15 μl of 3′ adapter ligation Master Mix. The procedure here is written using volumes for the modified and untreated channels of one sample but can be scaled as needed for additional samples. The tube containing the Master Mix can be kept at room temperature. Begin preparing the 3′ adapter ligation Master Mix by adding 11 μl of nuclease-free water to a 1.7 ml microcentrifuge tube.
3. Add 4.4 μl of 10X T4 RNA ligase buffer to the Master Mix, vortex, and spin down in a mini centrifuge.
4. Add 13.2 μl of 50% PEG 8000 to the Master Mix, vortex, and spin down in a mini centrifuge.
5. Add 2.2 μl of 1 μM pre-adenylated 9N_VRA3 oligo to the Master Mix, vortex, and spin down in a mini centrifuge.
6. Add 1.1 μl of SuperaseIn RNase Inhibitor to the Master Mix.
7. Add 1.1 μl of T4 RNA Ligase 2, truncated KQ, to the Master Mix.
8. Gently mix the completed Master Mix by pipetting.
9. Add 15 μl of 3′ adapter ligation Master Mix to each 5 μl sample, mix thoroughly by pipetting, and transfer the ligation reaction to a thin-walled 200 μl PCR tube.
10. Incubate the 3′ adapter ligation reactions for 2 h on a thermal cycler block set to 25 °C with the heated lid set to 25 °C (*see* **Note 40**).
11. Resuspend the SPRI bead stock by vortexing. For each sample, aliquot 22.5 μl of SPRI beads into a 1.7 ml microcentrifuge tube.
12. Add 17.5 μl of water and 40 μl of fresh anhydrous isopropanol to each ligation reaction (*see* **Note 41**). Mix the samples by vortexing and spin down in a mini centrifuge.
13. Add each 3′ ligation reaction to its corresponding 22.5 μl SPRI bead aliquot, and gently mix by pipetting until the solution is homogenous. The total volume will be 100 μl.

14. Incubate the samples at room temperature for 5 min.
15. Place the microcentrifuge tubes on a magnetic stand and allow the beads to collect on the tube wall for 3 min.
16. Aspirate and discard the supernatant. Be careful not to disturb the beads, and do not remove the microcentrifuge tube from the magnetic stand.
17. Without removing the microcentrifuge tubes from the magnetic stand, gently add 200 μl of 80% ethanol without disturbing the beads. **Do not resuspend the beads.**
18. Incubate the samples at room temperature for 1 min.
19. Aspirate and discard the supernatant.
20. Repeat **steps 17** to **19** to wash the beads a second time.
21. Pulse spin the beads in a mini centrifuge to pull down residual liquid and return the beads to the magnetic stand. Remove any residual liquid.
22. Place the samples in a dry bath set to 37 °C with the caps open for ~3 min to evaporate residual ethanol. **Watch the samples closely** to ensure that the beads do not dry completely.
23. Resuspend the beads in 20 μl of 10 mM Tris-HCl (pH 8.0) and incubate for 3 min at room temperature to elute the RNA (*see* **Note 42**).
24. Place the microcentrifuge tubes on a magnetic stand and allow the beads to collect on the tube wall for 1 min.
25. Carefully recover and transfer the supernatant to a clean 1.7 ml microcentrifuge tube, then perform a second SPRI bead purification, as follows. Add 11.5 μl of water, 40 μl of anhydrous isopropanol, and 6 μl of 50% PEG 8000 to each sample. Mix the samples by vortexing, spin down liquid in a mini centrifuge, and then mix each sample with 22.5 μl of SPRI beads and repeat **steps 14** to **24**. Carefully recover and transfer the supernatant to a clean 1.7 ml microcentrifuge tube. This supernatant contains the eluted RNA.

3.8 Options for Illumina Sequencing Library Construction

At this point the purified RNA contains a 3′ adapter that can be used to prime reverse transcription, which records the location of chemical adducts from the structure probing reaction in the cDNA. A second adapter is then appended to the cDNA 3′ end, and the resulting library is PCR amplified and purified for Illumina sequencing. The SHAPE-Seq v2.1 sequencing library construction protocol [67] used for the original implementation of the cotranscriptional SHAPE-Seq method has been superseded by procedures that reduce systematic biases that can occur during library preparation, which are described in detail elsewhere [52, 68, 69]. The best practices and procedures for cotranscriptional RNA structure probing described here will be compatible with other library construction protocols as long as the adapter that is appended to the RNA 3′ end above is from the same Illumina Library Preparation Kit as the adapter that is appended to the cDNA 3′ end (*see* **Note 11**). We are currently validating a sequencing library construction protocol that is designed specifically for cotranscriptional RNA structure probing applications.

3.9 Denaturing PAGE to Assess Single-Round In Vitro Transcription Reaction Products

We previously described a detailed procedure for performing denaturing PAGE using the National Diagnostics SequaGel UreaGel 19:1 Denaturing Gel System and a Bio-Rad Mini-PROTEAN Vertical Electrophoresis Cell [92]. The same procedure can be used to perform denaturing PAGE to assess the single-round in vitro transcription reaction. Resuspend each RNA pellet from Subheading 3.6, **step 17** in 7 μl of formamide loading dye, then denature the samples by heating at 95 °C, load and run the samples on a pre-warmed denaturing gel at 480 V, stain with SYBR Gold, and visualize the gel using a laser scanner or gel documentation system as described previously. When used to detect single-stranded nucleic acids with different lengths and sequences, SYBR Gold staining will not necessarily be quantitative. However, SYBR Gold is typically suitable for assessing the production of full-length transcripts or estimating RNA function in cotranscriptional RNA chemical probing conditions. An example denaturing PAGE gel that assesses ZTP riboswitch function in the conditions used for cotranscriptional RNA chemical probing is shown in Fig. 5.

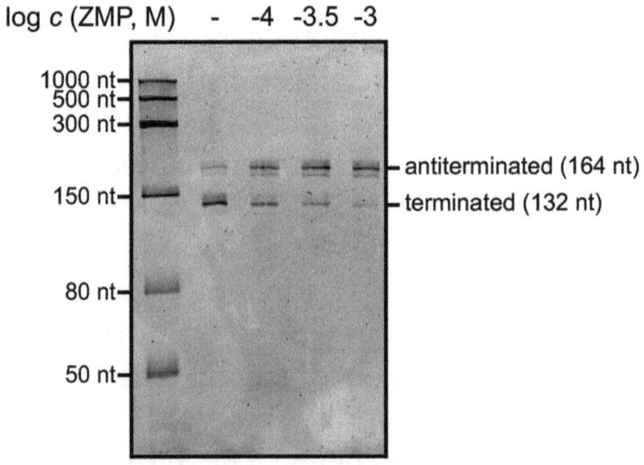

Fig. 5 Denaturing PAGE QC gel to assess in vitro transcribed RNA length. The *Clostridium beijerinckii pfl* ZTP riboswitch was in vitro transcribed under single-round conditions with variable concentrations of ZMP. The expected terminated and antiterminated bands are observed, and increasing ZMP concentration promotes antitermination. This indicates that the riboswitch is functional in the single-round in vitro transcription conditions used for RNA structure probing

4 Notes

1. For small-scale nucleic acid manipulations, we use commercially available nuclease-free reagents whenever possible to maximize the consistency of the procedures.
2. Ideally the thermal cycler used for these procedure should have a capacity of at least 100 μl. It is possible to perform these procedures using a thermal cycler with lower sample volume capacity as long as the sample volumes are kept at or below the limit of the thermal cycler.
3. Pre-amplification of a linear dsDNA template requires a high-fidelity DNA polymerase. The protocol has only been validated with Q5 DNA polymerase (New England Biolabs (NEB), catalog number M0491L), but other DNA polymerases with similar fidelity are likely suitable as well.
4. Vent (exo-) DNA polymerase (NEB, catalog number M0257L) is the preferred enzyme for incorporating biotin-11-dNTPs during PCR. The protocol has not been tested with other DNA polymerases.
5. We obtain biotin-11-dATP and biotin-11-dGTP from PerkinElmer (catalog numbers NEL540001EA and NEL541001EA) and biotin-11-dUTP and biotin-11-dCTP from Biotium (catalog numbers 40029 and 40036). Biotin is attached to biotin-11-dATP and biotin-11-dGTP by an 11 atom linker beginning with an alkyne at C7 of the 7-deaza purine base. Biotin is attached to biotin-11-dUTP and biotin-11-dCTP by an 11 atom linker beginning with an alkene at C5 of the pyrimidine base.
6. It is critical to purify full-length PRA1_NoMod.F either by purchasing the oligo with HPLC purification or performing PAGE purification in-house. HP4_5bio.R does not require purification. We purchase these oligonucleotides from Integrated DNA Technologies. /5Biosg/ is the Integrated DNA Technologies notation for a "standard" 5′ biotin modification. The PRA1_NoMod.F and HP4_5bio.R oligos should be resuspended in nuclease-free 1X TE (10 mM Tris-HCl (pH 7.5), 0.1 mM EDTA) to 100 μM; 10 μM working stocks can be prepared by diluting with 10 mM Tris-HCl (pH 8.0) and stored at −20 °C in 2.0 ml screw-cap tubes with an O-ring.
7. It is essential to use preparative-grade agarose for DNA purification. The protocol is validated for use with SeaKem GTG Agarose (Lonza, catalog number 50074).
8. The protocol is validated for use with the QIAquick Gel Extraction Kit (Qiagen, catalog number 28706).

9. The protocol is only validated for use with Sera-Mag Speed-Bead Carboxylate-Modified Magnetic Particles (Hydrophobic) (Cytiva Life Sciences, catalog number 65152105050250).
10. Streptavidin must be monomeric to avoid aggregation of randomly biotinylated DNA. The procedures described here are validated for use with streptavidin obtained from Promega (catalog number Z7041).
11. The 3′ adapter sequence provided here is the reverse complement of the RA5 adapter from the Illumina TruSeq Small RNA library kit. Using a different sequencing library adapter configuration is possible but requires changing the 3′ adapter sequence.
12. The protocol is validated for use with the NEB 5′ DNA adenylation kit (catalog number E2610L).
13. To minimize variability, we prepare 10X Transcription Buffer using commercial nuclease-free solutions.
14. It is essential to use high-purity rNTP solutions to prepare the NTP working stocks. The procedure is validated for use with high-purity rNTP solutions from Cytiva (catalog number 27202501).
15. It is essential to use molecular biology-grade BSA. The procedure is validated for use with UltraPure BSA (Invitrogen, catalog number AM2616).
16. *E. coli* RNA polymerase holoenzyme is available commercially (NEB, catalog number M0551S) or can be prepared by purifying *E. coli* RNA polymerase core [71] and σ^{70} [72] and reconstituting the holoenzyme.
17. The protocol is validated for use with TURBO DNase (Invitrogen, catalog number AM2239). In principle, any DNase I will work.
18. The protocol is validated for use with T4 RNA Ligase 2, truncated KQ (NEB, catalog number M0373L).
19. We use the NEB low range ssRNA ladder which has a size range of 50–1000 nts. Any comparable ladder will work.
20. Denaturing urea-PAGE is sensitive to the quality of the acrylamide mixture. We find that the UreaGel 19:1 Denaturing Gel System (National Diagnostics, catalog number EC-833) provides consistent results.
21. The first published implementation of cotranscriptional SHAPE-seq used a catalytically dead EcoRI variant, $\text{EcoRI}_{\text{E111Q}}$, as a transcription roadblock [26]. In contrast, to biotin-streptavidin roadblocks, collision of RNAP with $\text{EcoRI}_{\text{E111Q}}$ does not tend to induce backtracking [25]. However, $\text{EcoRI}_{\text{E111Q}}$ roadblocking requires an EcoRI site to bind DNA [90]. Consequently, performing a multi-length

cotranscriptional RNA structure probing experiment with $EcoRI_{E111Q}$ requires a DNA template preparation that uses a primer library and individual small-scale PCRs to systematically incorporate EcoRI sites that will halt RNAP at the desired transcript lengths. Furthermore, $EcoRI_{E111Q}$ roadblocking is not compatible with RNA targets that contain a natural EcoRI site. We previously reported a detailed comparison of the biotin-streptavidin and $EcoRI_{E111Q}$ roadblocking approaches in cotranscriptional RNA structure probing experiments [25].

22. Chemically encoded transcription roadblocks halt transcription with single-nucleotide precision and >99% efficiency but are not ideal for multi-length cotranscriptional RNA structure probing experiments due to cost. Chemically encoded transcription roadblocks can be used for single-length cotranscriptional RNA structure probing experiments. Detailed procedures for preparing internally modified DNA for this application have been described elsewhere[89, 92].
23. It is essential that primer dimer does not form during PCR. Not only will primer dimer reduce the PCR yield, but it may not be completely removed by the SPRI bead purification. The PRA1_NoMod.F and HP4_5bio.R primers do not form any detectable primer dimer in the PCR conditions provided in Subheading 3.3. When using different PCR primers, it is crucial to confirm that they do not form primer dimer or other undesirable products.
24. Adjust the annealing temperature if primers other than the standard sequences provided here are used. We recommend using the NEB Tm Calculator (https://tmcalculator.neb.com/#!/main) to determine annealing temperature.
25. The PCR can be scaled as needed. We find that a 200–300 μl PCR is more than sufficient for most experiments.
26. Because the 3′ adapter oligonucleotide contains a mixture of DNA and RNA nucleotides and a randomized segment, the molar concentration determined using the Qubit ssDNA assay kit is intended to standardize the amount of linker used in the 3′ adapter ligation and not to measure absolute quantity. To estimate the molar concentration of the pre-adenylated adapter, use the NEB ssDNA: Mass to Moles Convertor (https://nebiocalculator.neb.com/#!/ssdnaamt). Enter the oligo length and mass as ng in 1 μl; 1 pmol/μl is equal to 1 μM.
27. We aliquot pre-adenylated linker so that each aliquot is used within four freeze-thaw cycles.
28. Transcription can also be limited to a single round by using heparin to scavenge free RNAP. We strongly recommend using rifampicin because its mechanism of inhibition is RNAP-specific, whereas heparin will bind to many proteins that interact with nucleic acids [96].

29. As an alternative to initiating single-round transcription from preformed open promoter complexes, it is possible to "walk" RNAP to a defined location in the DNA template by omitting one or more NTPs from the reaction (e.g., omitting CTP permits RNAP to transcribe to the nucleotide that precedes the first C in the RNA). Single-round transcription can then be initiated by supplying the missing NTP(s) and rifampicin. Key considerations for performing single-round transcription using this format have been described in detail previously [96, 97].
30. Choosing an appropriate chemical probe is critical to the success of RNA chemical probing experiments. Different chemical probes measure distinct aspects of RNA structure, such as backbone flexibility, base-pairing, and solvent accessibility. Furthermore, the parameters of the chemical probing reaction, such as reaction time and whether the probing reaction needs to be quenched, depend on the reagent used for structure probing. The properties of commonly used RNA chemical probes have been described in detail previously [47, 49, 98, 99].
31. A reasonable alternative to benzoyl cyanide is the SHAPE reagent 1-methyl-7-nitroisatoic anhydride (1M7), which has a hydrolysis half-life of 14 s at 37 °C and self-quenches after ~1 min [102, 103]. Dimethyl sulfate (DMS) has also been used for cotranscriptional RNA structure probing by us and others [26, 27]. DMS is an alkylating agent that preferentially methylates N7 of guanine, N1 of adenine, and N3 of cytosine [50]. Of these modifications, N1-methyl-A and N3-methyl-C are detectable by reverse transcription because they occur on the Watson-Crick face of the base. N7-methyl-G is not detectable by reverse transcription. RNA structure probing with DMS typically requires 2–3 min of chemical modification followed by a quenching step. Although DMS can be used for cotranscriptional RNA structure probing, it is important to be aware that true single-hit kinetics are not possible in a DMS chemical probing experiment because DMS reacts more rapidly with guanosine nucleotides than adenosine or cytidine [50]. Detailed procedures for RNA structure probing using DMS have been described elsewhere [50].
32. The cotranscriptional RNA structure probing procedure described in Subheading 3.5 can be adapted to assess RNA length or function by making the following modifications: (1) use a DNA template that does not contain random internal biotin modifications; the 5′ biotin modification in the HP4_5bio.R primer is still useful for preventing RNAP from template-switching because it blocks runoff transcription. (2) Reduce the total reaction volume for each sample from 60 μl to 25 μl. (3) Because the DNA template will only contain

a 5′ biotin modification, streptavidin can be included in the in vitro transcription Master Mix. We still recommend a 20-min incubation at 37 °C to form open promoter complexes. (4) Omit all steps related to chemical probing; once the transcription reaction is complete, simply stop the reaction by adding TRIzol LS. (5) After purifying the RNA in Subheading 3.6, proceed to Subheading 3.9 to perform denaturing PAGE.

33. In some cases, other components (e.g., elongation factors, riboswitch ligands) will need to be included in the transcription reaction. Additional reagents can be added to the single-round in vitro transcription reaction in three ways: (1) include the reagent in the Master Mix so that it is present during open promoter complex formation. (2) Mix the reagent with 10X start solution so that it is absent during open promoter complex formation but present when transcription is initiated. (3) Walk RNAP to a defined position in the DNA template by omitting one or several NTPs from the reaction mixture to prepare preformed TECs [96, 97] and then include the reagent when supplying the omitted NTP(s). Which of these procedures is most appropriate will depend on the application at hand.

34. NTP concentration can be adjusted as needed; however, low NTP concentrations will reduce RNA yield. We typically use between 0.1 and 0.5 mM NTPs.

35. In vitro transcription is performed using a thermal cycler so that reaction temperature is robust to airflow from the chemical fume hood. We use an MJ Research PTC-200 with a 60 × 0.5 ml tube block, which is discontinued but can be purchased used. It is crucial that the chemical fume hood used for these experiments is kept clutter-free. If possible we recommend dedicating a fume hood for RNA chemical probing experiments; ours contains a thermal cycler, a vortex, a mini centrifuge, and containers for pipette tips and liquid waste.

36. When performing a cotranscriptional RNA structure probing experiment that requires a large number of samples, we recommend performing the transcription time course with two people: one person is responsible for transferring each sample to the thermal cycler block, adding streptavidin, and adding start solution. The other person is responsible for performing the structure probing reaction and stopping transcription. With this structure, the time points can be compressed and even overlap.

37. Streptavidin is added after 10 min of open complex formation so that RNAP has time to form open complexes on DNA that contains a biotin modification in the promoter.

38. For single-round in vitro transcription with 0.5 mM NTPs, we typically perform the structure probing reaction 30 s after initiating transcription. At lower NTP concentrations, it may be necessary to increase the transcription reaction time. We typically use an upper limit of 2 min of transcription.

39. We perform the cotranscriptional RNA chemical probing procedure in the following way: (1) preload two 200 μl pipettes with 75 μl of TRIzol LS each. (2) Add 10X start solution to the transcription reaction and briefly mix by pipetting. (3) When it is time to perform the chemical probing reaction, pipette 25 μl of the transcription reaction directly onto the 2.78 μl of benzoyl cyanide and leave the tube open. (4) Change the pipette tip, transfer 25 μl of the transcription reaction into the tube that contains 2.78 μl of DMSO, and leave the tube open. (5) Pipette 75 μl of TRIzol LS into the benzoyl cyanide-modified and DMSO control samples. (6) Close the tube caps and vortex the samples. With practice, this manipulation can be performed in 10–15 s.

40. During 3′ adapter ligation, we perform the 25 °C incubation using a thermal cycler (as opposed to leaving the reactions on the bench) to ensure that the reaction temperature is consistent across different experiments.

41. Fishman et al. described a procedure for tuning SPRI bead-based size selection using isopropanol [104]. With SPRI beads prepared as described by Jolivet and Foley [70] (using the "DNA buffer" protocol), we have found that performing two sequential SPRI bead purifications with 22.5 μl of SPRI beads and 40% isopropanol in a final volume of 100 μl (7.5% PEG 8000) efficiently depletes unligated linker oligonucleotide while retaining most of the desired transcripts. We recommend calibrating the SPRI beads by fractionating the NEB low range ssRNA ladder with a mock 3′ adapter oligo spike-in while varying isopropanol concentration (any DNA oligo ~35 nt in length will work; we calibrate our beads with a 37 nt oligo; oligos of other sizes can be included at the same concentration as needed for calibration). In a 50 μl total volume, combine nuclease-free water (to 50 μl), 1 μl of a 1:4 dilution of NEB low range ssRNA ladder (diluted in 10 mM Tris-HCl (pH 8.0)), 0.5 μl of 0.5 μM mock 3′ adapter oligonucleotide, 18.75 μl of SPRI beads (which yields a concentration of 7.5% PEG 8000), and isopropanol to the desired concentration (0% and 25–45% in 5% increments) and mix. Then perform the SPRI bead purification as described above in Subheading 3.7, **steps 14–24** except that the elution is performed using only 5 μl of 10 mM Tris-HCl (pH 8.0). Mix the 5 μl of eluate with 15 μl of formamide loading dye, and assess the samples by denaturing PAGE as described in Subheading 3.9. The mock 3′ adapter

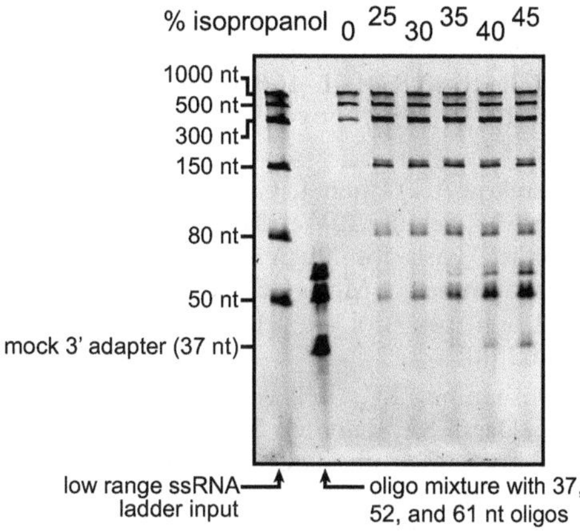

Fig. 6 SPRI bead QC gel. Example of a SPRI bead QC gel used to determine the isopropanol concentration for depleting excess 3′ adapter oligo. Low range ssRNA ladder and a DNA oligonucleotide mixture were pooled and purified using SPRI beads with variable isopropanol concentration. In this case, 40% isopropanol achieved the ideal balance between mock 3′ adapter oligo depletion and retention of longer products. Peforming this purification twice depletes the mock 3′ adapter oligo completely

oligonucleotide should be almost completely depleted, whereas RNA 50 nt or longer should be at least partially retained. A typical result of the calibration is shown in Fig. 6.

42. The buffer used to elute the purified RNA from beads can be adjusted depending on the requirements of the procedure used for sequencing library construction.

Acknowledgments

This work was supported by startup funding from the University at Buffalo (to E.J.S.).

References

1. Pan T, Sosnick T (2006) RNA folding during transcription. Annu Rev Biophys Biomol Struct 35:161–175. https://doi.org/10.1146/annurev.biophys.35.040405.102053
2. Zhang J, Landick R (2016) A two-way street: Regulatory interplay between RNA polymerase and nascent RNA structure. Trends Biochem Sci 41(4):293–310. https://doi.org/10.1016/j.tibs.2015.12.009
3. Al-Hashimi HM, Walter NG (2008) RNA dynamics: it is about time. Curr Opin Struct Biol 18(3):321–329. https://doi.org/10.1016/j.sbi.2008.04.004
4. Roberts JW, Shankar S, Filter JJ (2008) RNA polymerase elongation factors. Annu Rev Microbiol 62:211–233. https://doi.org/10.1146/annurev.micro.61.080706.093422
5. Kang JY, Mishanina TV, Landick R, Darst SA (2019) Mechanisms of transcriptional pausing in bacteria. J Mol Biol 431(20):4007–4029. https://doi.org/10.1016/j.jmb.2019.07.017

6. Mustoe AM, Brooks CL, Al-Hashimi HM (2014) Hierarchy of RNA functional dynamics. Annu Rev Biochem 83:441–466. https://doi.org/10.1146/annurev-biochem-060713-035524
7. Guo X, Myasnikov AG, Chen J, Crucifix C, Papai G, Takacs M, Schultz P, Weixlbaumer A (2018) Structural basis for NusA stabilized transcriptional pausing. Mol Cell 69(5): 816–827 e814. https://doi.org/10.1016/j.molcel.2018.02.008
8. Kang JY, Mishanina TV, Bellecourt MJ, Mooney RA, Darst SA, Landick R (2018) RNA polymerase accommodates a pause RNA hairpin by global conformational rearrangements that prolong pausing. Mol Cell 69(5):802–815 e805. https://doi.org/10.1016/j.molcel.2018.01.018
9. Kramer FR, Mills DR (1981) Secondary structure formation during RNA synthesis. Nucleic Acids Res 9(19):5109–5124. https://doi.org/10.1093/nar/9.19.5109
10. Pan T, Artsimovitch I, Fang XW, Landick R, Sosnick TR (1999) Folding of a large ribozyme during transcription and the effect of the elongation factor NusA. Proc Natl Acad Sci U S A 96(17):9545–9550. https://doi.org/10.1073/pnas.96.17.9545
11. Neugebauer KM (2002) On the importance of being co-transcriptional. J Cell Sci 115 (Pt 20):3865–3871. https://doi.org/10.1242/jcs.00073
12. Heilman-Miller SL, Woodson SA (2003) Effect of transcription on folding of the Tetrahymena ribozyme. RNA 9(6):722–733. https://doi.org/10.1261/rna.5200903
13. Meyer IM, Miklos I (2004) Co-transcriptional folding is encoded within RNA genes. BMC Mol Biol 5:10. https://doi.org/10.1186/1471-2199-5-10
14. Wong TN, Sosnick TR, Pan T (2007) Folding of noncoding RNAs during transcription facilitated by pausing-induced nonnative structures. Proc Natl Acad Sci U S A 104(46):17995–18000. https://doi.org/10.1073/pnas.0705038104
15. Perdrizet GA 2nd, Artsimovitch I, Furman R, Sosnick TR, Pan T (2012) Transcriptional pausing coordinates folding of the aptamer domain and the expression platform of a riboswitch. Proc Natl Acad Sci U S A 109(9):3323–3328. https://doi.org/10.1073/pnas.1113086109
16. Lai D, Proctor JR, Meyer IM (2013) On the importance of cotranscriptional RNA structure formation. RNA 19(11):1461–1473. https://doi.org/10.1261/rna.037390.112
17. Scull CE, Dandpat SS, Romero RA, Walter NG (2020) Transcriptional riboswitches integrate timescales for bacterial gene expression control. Front Mol Biosci 7:607158. https://doi.org/10.3389/fmolb.2020.607158
18. Scharfen L, Neugebauer KM (2021) Transcription regulation through nascent RNA folding. J Mol Biol 433(14):166975. https://doi.org/10.1016/j.jmb.2021.166975
19. Chappell J, Takahashi MK, Lucks JB (2015) Creating small transcription activating RNAs. Nat Chem Biol 11(3):214–220. https://doi.org/10.1038/nchembio.1737
20. Ameruoso A, Gambill L, Liu B, Villegas Kcam MC, Chappell J (2019) Brave new 'RNA' world—advances in RNA tools and their application for understanding and engineering biological systems. Curr Opin Syst Biol 14:32–40. https://doi.org/10.1016/j.coisb.2019.02.005
21. Carlson PD, Lucks JB (2019) Elements of RNA design. Biochemistry 58(11): 1457–1459. https://doi.org/10.1021/acs.biochem.8b01129
22. Geary C, Rothemund PW, Andersen ES (2014) A single-stranded architecture for cotranscriptional folding of RNA nanostructures. Science 345(6198):799–804. https://doi.org/10.1126/science.1253920
23. Sparvath SL, Geary CW, Andersen ES (2017) Computer-aided design of RNA origami structures. Methods Mol Biol 1500:51–80. https://doi.org/10.1007/978-1-4939-6454-3_5
24. Geary C, Grossi G, McRae EKS, Rothemund PWK, Andersen ES (2021) RNA origami design tools enable cotranscriptional folding of kilobase-sized nanoscaffolds. Nat Chem 13(6):549–558. https://doi.org/10.1038/s41557-021-00679-1
25. Strobel EJ, Watters KE, Nedialkov Y, Artsimovitch I, Lucks JB (2017) Distributed biotin-streptavidin transcription roadblocks for mapping cotranscriptional RNA folding. Nucleic Acids Res 45(12):e109. https://doi.org/10.1093/nar/gkx233
26. Watters KE, Strobel EJ, Yu AM, Lis JT, Lucks JB (2016) Cotranscriptional folding of a riboswitch at nucleotide resolution. Nat Struct Mol Biol 23(12):1124–1131. https://doi.org/10.1038/nsmb.3316
27. Incarnato D, Morandi E, Anselmi F, Simon LM, Basile G, Oliviero S (2017) In vivo probing of nascent RNA structures reveals principles of cotranscriptional folding. Nucleic

Acids Res 45(16):9716–9725. https://doi.org/10.1093/nar/gkx617

28. Chauvier A, Picard-Jean F, Berger-Dancause JC, Bastet L, Naghdi MR, Dube A, Turcotte P, Perreault J, Lafontaine DA (2017) Transcriptional pausing at the translation start site operates as a critical checkpoint for riboswitch regulation. Nat Commun 8: 13892. https://doi.org/10.1038/ncomms13892
29. Frieda KL, Block SM (2012) Direct observation of cotranscriptional folding in an adenine riboswitch. Science 338(6105):397–400. https://doi.org/10.1126/science.1225722
30. Hua B, Panja S, Wang Y, Woodson SA, Ha T (2018) Mimicking Co-transcriptional RNA folding using a superhelicase. J Am Chem Soc 140(32):10067–10070. https://doi.org/10.1021/jacs.8b03784
31. Uhm H, Kang W, Ha KS, Kang C, Hohng S (2018) Single-molecule FRET studies on the cotranscriptional folding of a thiamine pyrophosphate riboswitch. Proc Natl Acad Sci U S A 115(2):331–336. https://doi.org/10.1073/pnas.1712983115
32. Hua B, Jones CP, Mitra J, Murray PJ, Rosenthal R, Ferre-D'Amare AR, Ha T (2020) Real-time monitoring of single ZTP riboswitches reveals a complex and kinetically controlled decision landscape. Nat Commun 11(1):4531. https://doi.org/10.1038/s41467-020-18283-1
33. Fukuda S, Yan S, Komi Y, Sun M, Gabizon R, Bustamante C (2020) The biogenesis of SRP RNA is modulated by an RNA folding intermediate attained during transcription. Mol Cell 77(2):241–250 e248. https://doi.org/10.1016/j.molcel.2019.10.006
34. Yu AM, Gasper PM, Cheng L, Lai LB, Kaur S, Gopalan V, Chen AA, Lucks JB (2021) Computationally reconstructing cotranscriptional RNA folding from experimental data reveals rearrangement of non-native folding intermediates. Mol Cell 81(4):870–883 e810. https://doi.org/10.1016/j.molcel.2020.12.017
35. Yadav R, Widom JR, Chauvier A, Walter NG (2021) An anionic ligand snap-locks a long-range interaction in a magnesium-folded riboswitch. Res Square. https://doi.org/10.21203/rs.3.rs-459800/v1
36. Chauvier A, Ajmera P, Yadav R, Walter NG (2021) Dynamic competition between a ligand and transcription factor NusA governs riboswitch-mediated transcription regulation. Proc Natl Acad Sci 118(47). https://doi.org/10.1073/pnas.2109026118
37. Chauvier A, St-Pierre P, Nadon J-F, Hien EDM, Pérez-González C, Eschbach SH, Lamontagne A-M, Penedo JC, Lafontaine DA (2021) Monitoring RNA dynamics in native transcriptional complexes. Proc Natl Acad Sci 118(45). https://doi.org/10.1073/pnas.2106564118
38. Duss O, Stepanyuk GA, Grot A, O'Leary SE, Puglisi JD, Williamson JR (2018) Real-time assembly of ribonucleoprotein complexes on nascent RNA transcripts. Nat Commun 9(1): 5087. https://doi.org/10.1038/s41467-018-07423-3
39. Duss O, Stepanyuk GA, Puglisi JD, Williamson JR (2019) Transient protein-RNA interactions guide nascent ribosomal RNA folding. Cell 179(6):1357–1369 e1316. https://doi.org/10.1016/j.cell.2019.10.035
40. Rodgers ML, Woodson SA (2019) Transcription increases the cooperativity of ribonucleoprotein assembly. Cell 179(6):1370–1381 e1312. https://doi.org/10.1016/j.cell.2019.11.007
41. Rodgers ML, Woodson SA (2021) A roadmap for rRNA folding and assembly during transcription. Trends Biochem Sci. https://doi.org/10.1016/j.tibs.2021.05.009
42. Chatterjee S, Chauvier A, Dandpat SS, Artsimovitch I, Walter NG (2021) A translational riboswitch coordinates nascent transcription-translation coupling. Proc Natl Acad Sci U S A 118(16). https://doi.org/10.1073/pnas.2023426118
43. Hein PP, Kolb KE, Windgassen T, Bellecourt MJ, Darst SA, Mooney RA, Landick R (2014) RNA polymerase pausing and nascent-RNA structure formation are linked through clamp-domain movement. Nat Struct Mol Biol 21(9):794–802. https://doi.org/10.1038/nsmb.2867
44. Widom JR, Nedialkov YA, Rai V, Hayes RL, Brooks CL 3rd, Artsimovitch I, Walter NG (2018) Ligand modulates cross-coupling between riboswitch folding and transcriptional pausing. Mol Cell 72(3):541–552 e546. https://doi.org/10.1016/j.molcel.2018.08.046
45. Palangat M, Meier TI, Keene RG, Landick R (1998) Transcriptional pausing at +62 of the HIV-1 nascent RNA modulates formation of the TAR RNA structure. Mol Cell 1(7): 1033–1042. https://doi.org/10.1016/s1097-2765(00)80103-3
46. Strobel EJ, Yu AM, Lucks JB (2018) High-throughput determination of RNA structures. Nat Rev Genet 19(10):615–634. https://doi.org/10.1038/s41576-018-0034-x

47. Carlson PD, Evans ME, Yu AM, Strobel EJ, Lucks JB (2018) SnapShot: RNA structure probing technologies. Cell 175(2):600–600 e601. https://doi.org/10.1016/j.cell.2018.09.024
48. Mitchell D 3rd, Assmann SM, Bevilacqua PC (2019) Probing RNA structure in vivo. Curr Opin Struct Biol 59:151–158. https://doi.org/10.1016/j.sbi.2019.07.008
49. Merino EJ, Wilkinson KA, Coughlan JL, Weeks KM (2005) RNA structure analysis at single nucleotide resolution by selective 2-′-hydroxyl acylation and primer extension (SHAPE). J Am Chem Soc 127(12): 4223–4231. https://doi.org/10.1021/ja043822v
50. Tijerina P, Mohr S, Russell R (2007) DMS footprinting of structured RNAs and RNA-protein complexes. Nat Protoc 2(10): 2608–2623. https://doi.org/10.1038/nprot.2007.380
51. Lucks JB, Mortimer SA, Trapnell C, Luo S, Aviran S, Schroth GP, Pachter L, Doudna JA, Arkin AP (2011) Multiplexed RNA structure characterization with selective 2′-hydroxyl acylation analyzed by primer extension sequencing (SHAPE-Seq). Proc Natl Acad Sci U S A 108(27):11063–11068. https://doi.org/10.1073/pnas.1106501108
52. Siegfried NA, Busan S, Rice GM, Nelson JA, Weeks KM (2014) RNA motif discovery by SHAPE and mutational profiling (SHAPE-MaP). Nat Methods 11(9):959–965. https://doi.org/10.1038/nmeth.3029
53. Zubradt M, Gupta P, Persad S, Lambowitz AM, Weissman JS, Rouskin S (2017) DMS-MaPseq for genome-wide or targeted RNA structure probing in vivo. Nat Methods 14(1):75–82. https://doi.org/10.1038/nmeth.4057
54. Strobel EJ, Cheng L, Berman KE, Carlson PD, Lucks JB (2019) A ligand-gated strand displacement mechanism for ZTP riboswitch transcription control. Nat Chem Biol 15(11): 1067–1076. https://doi.org/10.1038/s41589-019-0382-7
55. Li H, Aviran S (2018) Statistical modeling of RNA structure profiling experiments enables parsimonious reconstruction of structure landscapes. Nat Commun 9(1):606. https://doi.org/10.1038/s41467-018-02923-8
56. Homan PJ, Favorov OV, Lavender CA, Kursun O, Ge X, Busan S, Dokholyan NV, Weeks KM (2014) Single-molecule correlated chemical probing of RNA. Proc Natl Acad Sci U S A 111(38):13858–13863. https://doi.org/10.1073/pnas.1407306111
57. Krokhotin A, Mustoe AM, Weeks KM, Dokholyan NV (2017) Direct identification of base-paired RNA nucleotides by correlated chemical probing. RNA 23(1):6–13. https://doi.org/10.1261/rna.058586.116
58. Mustoe AM, Lama NN, Irving PS, Olson SW, Weeks KM (2019) RNA base-pairing complexity in living cells visualized by correlated chemical probing. Proc Natl Acad Sci U S A 116(49):24574–24582. https://doi.org/10.1073/pnas.1905491116
59. Tomezsko PJ, Corbin VDA, Gupta P, Swaminathan H, Glasgow M, Persad S, Edwards MD, McIntosh L, Papenfuss AT, Emery A, Swanstrom R, Zang T, Lan TCT, Bieniasz P, Kuritzkes DR, Tsibris A, Rouskin S (2020) Determination of RNA structural diversity and its role in HIV-1 RNA splicing. Nature 582(7812):438–442. https://doi.org/10.1038/s41586-020-2253-5
60. Morandi E, Manfredonia I, Simon LM, Anselmi F, van Hemert MJ, Oliviero S, Incarnato D (2021) Genome-scale deconvolution of RNA structure ensembles. Nat Methods 18(3):249–252. https://doi.org/10.1038/s41592-021-01075-w
61. Samuel W, Olson Anne-Marie W, Turner JW, Arney I, Saleem CA, Weidmann DM, Margolis KM, Weeks AM, Mustoe (2022) Discovery of a large-scale cell-state-responsive allosteric switch in the 7SK RNA using DANCE-MaP. Molecular Cell. https://doi.org/10.1016/j.molcel.2022.02.009
62. McCown PJ, Corbino KA, Stav S, Sherlock ME, Breaker RR (2017) Riboswitch diversity and distribution. RNA 23(7):995–1011. https://doi.org/10.1261/rna.061234.117
63. Serganov A, Nudler E (2013) A decade of riboswitches. Cell 152(1–2):17–24. https://doi.org/10.1016/j.cell.2012.12.024
64. Garst AD, Edwards AL, Batey RT (2011) Riboswitches: structures and mechanisms. Cold Spring Harb Perspect Biol 3(6). https://doi.org/10.1101/cshperspect.a003533
65. Zhao B, Guffy SL, Williams B, Zhang Q (2017) An excited state underlies gene regulation of a transcriptional riboswitch. Nat Chem Biol 13(9):968–974. https://doi.org/10.1038/nchembio.2427
66. Binas O, Schamber T, Schwalbe H (2020) The conformational landscape of transcription intermediates involved in the regulation of the ZMP-sensing riboswitch from

Thermosinus carboxydivorans. Nucleic Acids Res 48(12):6970–6979. https://doi.org/10.1093/nar/gkaa427

67. Watters KE, Yu AM, Strobel EJ, Settle AH, Lucks JB (2016) Characterizing RNA structures in vitro and in vivo with selective 2′-hydroxyl acylation analyzed by primer extension sequencing (SHAPE-Seq). Methods 103:34–48. https://doi.org/10.1016/j.ymeth.2016.04.002
68. Smola MJ, Rice GM, Busan S, Siegfried NA, Weeks KM (2015) Selective 2′-hydroxyl acylation analyzed by primer extension and mutational profiling (SHAPE-MaP) for direct, versatile and accurate RNA structure analysis. Nat Protoc 10(11):1643–1669. https://doi.org/10.1038/nprot.2015.103
69. Ritchey LE, Su Z, Tang Y, Tack DC, Assmann SM, Bevilacqua PC (2017) Structure-seq2: sensitive and accurate genome-wide profiling of RNA structure in vivo. Nucleic Acids Res 45(14):e135. https://doi.org/10.1093/nar/gkx533
70. Jolivet P, Foley JW (2020) SPRI bead mix. protocolsio. https://doi.org/10.17504/protocols.io.bnz4mf8w
71. Svetlov V, Artsimovitch I (2015) Purification of bacterial RNA polymerase: tools and protocols. Methods Mol Biol 1276:13–29. https://doi.org/10.1007/978-1-4939-2392-2_2
72. Marr MT, Roberts JW (1997) Promoter recognition as measured by binding of polymerase to nontemplate strand oligonucleotide. Science 276(5316):1258–1260. https://doi.org/10.1126/science.276.5316.1258
73. Strobel EJ (2021) Preparation of E. coli RNA polymerase transcription elongation complexes by selective photoelution from magnetic beads. J Biol Chem 297(1):100812. https://doi.org/10.1016/j.jbc.2021.100812
74. Ruff EF, Drennan AC, Capp MW, Poulos MA, Artsimovitch I, Record MT Jr (2015) E. coli RNA polymerase determinants of open complex lifetime and structure. J Mol Biol 427(15):2435–2450. https://doi.org/10.1016/j.jmb.2015.05.024
75. Henderson KL, Felth LC, Molzahn CM, Shkel I, Wang S, Chhabra M, Ruff EF, Bieter L, Kraft JE, Record MT Jr (2017) Mechanism of transcription initiation and promoter escape by E. coli RNA polymerase. Proc Natl Acad Sci U S A 114(15):E3032–E3040. https://doi.org/10.1073/pnas.1618675114
76. Vvedenskaya IO, Zhang Y, Goldman SR, Valenti A, Visone V, Taylor DM, Ebright RH, Nickels BE (2015) massively systematic transcript end readout, “MASTER”: transcription start site selection, transcriptional slippage, and transcript yields. Mol Cell 60(6):953–965. https://doi.org/10.1016/j.molcel.2015.10.029
77. Winkelman JT, Vvedenskaya IO, Zhang Y, Zhang Y, Bird JG, Taylor DM, Gourse RL, Ebright RH, Nickels BE (2016) Multiplexed protein-DNA cross-linking: Scrunching in transcription start site selection. Science 351(6277):1090–1093. https://doi.org/10.1126/science.aad6881
78. Jacques JP, Kolakofsky D (1991) Pseudo-templated transcription in prokaryotic and eukaryotic organisms. Genes Dev 5(5):707–713. https://doi.org/10.1101/gad.5.5.707
79. Turnbough CL Jr, Switzer RL (2008) Regulation of pyrimidine biosynthetic gene expression in bacteria: repression without repressors. Microbiol Mol Biol Rev 72(2):266–300, table of contents. https://doi.org/10.1128/MMBR.00001-08
80. Murakami KS, Shin Y, Turnbough CL Jr, Molodtsov V (2017) X-ray crystal structure of a reiterative transcription complex reveals an atypical RNA extension pathway. Proc Natl Acad Sci U S A 114(31):8211–8216. https://doi.org/10.1073/pnas.1702741114
81. Liu Y, Yu L, Pukhrambam C, Winkelman JT, Firlar E, Kaelber JT, Zhang Y, Nickels BE, Ebright RH (2022) Structural and mechanistic basis of reiterative transcription initiation. Proc Natl Acad Sci 119(5). https://doi.org/10.1073/pnas.2115746119
82. Kammerer W, Deuschle U, Gentz R, Bujard H (1986) Functional dissection of Escherichia coli promoters: information in the transcribed region is involved in late steps of the overall process. EMBO J 5(11):2995–3000
83. Hsu LM, Vo NV, Chamberlin MJ (1995) Escherichia coli transcript cleavage factors GreA and GreB stimulate promoter escape and gene expression in vivo and in vitro. Proc Natl Acad Sci U S A 92(25):11588–11592. https://doi.org/10.1073/pnas.92.25.11588
84. Skancke J, Bar N, Kuiper M, Hsu LM (2015) Sequence-dependent promoter escape efficiency is strongly influenced by bias for the pretranslocated state during initial transcription. Biochemistry 54(28):4267–4275. https://doi.org/10.1021/acs.biochem.5b00272

85. Dulin D, Bauer DLV, Malinen AM, Bakermans JJW, Kaller M, Morichaud Z, Petushkov I, Depken M, Brodolin K, Kulbachinskiy A, Kapanidis AN (2018) Pausing controls branching between productive and non-productive pathways during initial transcription in bacteria. Nat Commun 9(1): 1478. https://doi.org/10.1038/s41467-018-03902-9
86. Winkelman JT, Pukhrambam C, Vvedenskaya IO, Zhang Y, Taylor DM, Shah P, Ebright RH, Nickels BE (2020) XACT-seq comprehensively defines the promoter-position and promoter-sequence determinants for initial-transcription pausing. Mol Cell 79(5): 797–811 e798. https://doi.org/10.1016/j.molcel.2020.07.006
87. Chan CL, Landick R (1989) The Salmonella typhimurium his operon leader region contains an RNA hairpin-dependent transcription pause site. Mechanistic implications of the effect on pausing of altered RNA hairpins. J Biol Chem 264(34):20796–20804
88. Chan CL, Landick R (1993) Dissection of the his leader pause site by base substitution reveals a multipartite signal that includes a pause RNA hairpin. J Mol Biol 233(1): 25–42. https://doi.org/10.1006/jmbi.1993.1482
89. Strobel EJ, Lis JT, Lucks JB (2020) Chemical roadblocking of DNA transcription for nascent RNA display. J Biol Chem 295(19): 6401–6412. https://doi.org/10.1074/jbc.RA120.012641
90. Pavco PA, Steege DA (1990) Elongation by Escherichia coli RNA polymerase is blocked in vitro by a site-specific DNA binding protein. J Biol Chem 265(17):9960–9969
91. Nudler E (2012) RNA polymerase backtracking in gene regulation and genome instability. Cell 149(7):1438–1445. https://doi.org/10.1016/j.cell.2012.06.003
92. Strobel EJ (2021) Preparation and characterization of internally modified DNA templates for chemical transcription roadblocking. Bio Protocol 11(17). https://doi.org/10.21769/BioProtoc.4141
93. Tasara T, Angerer B, Damond M, Winter H, Dorhofer S, Hubscher U, Amacker M (2003) Incorporation of reporter molecule-labeled nucleotides by DNA polymerases. II. High-density labeling of natural DNA. Nucleic Acids Res 31(10):2636–2646. https://doi.org/10.1093/nar/gkg371
94. Mahat DB, Kwak H, Booth GT, Jonkers IH, Danko CG, Patel RK, Waters CT, Munson K, Core LJ, Lis JT (2016) Base-pair-resolution genome-wide mapping of active RNA polymerases using precision nuclear run-on (PRO-seq). Nat Protoc 11(8):1455–1476. https://doi.org/10.1038/nprot.2016.086
95. Campbell EA, Korzheva N, Mustaev A, Murakami K, Nair S, Goldfarb A, Darst SA (2001) Structural mechanism for rifampicin inhibition of bacterial rna polymerase. Cell 104(6):901–912. https://doi.org/10.1016/s0092-8674(01)00286-0
96. Artsimovitch I, Henkin TM (2009) In vitro approaches to analysis of transcription termination. Methods 47(1):37–43. https://doi.org/10.1016/j.ymeth.2008.10.006
97. Landick R, Wang D, Chan CL (1996) Quantitative analysis of transcriptional pausing by Escherichia coli RNA polymerase: his leader pause site as paradigm. Methods Enzymol 274:334–353. https://doi.org/10.1016/s0076-6879(96)74029-6
98. Weeks KM, Mauger DM (2011) Exploring RNA structural codes with SHAPE chemistry. Acc Chem Res 44(12):1280–1291. https://doi.org/10.1021/ar200051h
99. Weeks KM (2010) Advances in RNA structure analysis by chemical probing. Curr Opin Struct Biol 20(3):295–304. https://doi.org/10.1016/j.sbi.2010.04.001
100. Mortimer SA, Weeks KM (2008) Time-resolved RNA SHAPE chemistry. J Am Chem Soc 130(48):16178–16180. https://doi.org/10.1021/ja8061216
101. Mortimer SA, Weeks KM (2009) Time-resolved RNA SHAPE chemistry: quantitative RNA structure analysis in one-second snapshots and at single-nucleotide resolution. Nat Protoc 4(10):1413–1421. https://doi.org/10.1038/nprot.2009.126
102. Mortimer SA, Weeks KM (2007) A fast-acting reagent for accurate analysis of RNA secondary and tertiary structure by SHAPE chemistry. J Am Chem Soc 129(14): 4144–4145. https://doi.org/10.1021/ja0704028
103. Busan S, Weidmann CA, Sengupta A, Weeks KM (2019) Guidelines for SHAPE reagent choice and detection strategy for RNA structure probing studies. Biochemistry 58(23): 2655–2664. https://doi.org/10.1021/acs.biochem.8b01218
104. Fishman A, Light D, Lamm AT (2018) QsRNA-seq: a method for high-throughput profiling and quantifying small RNAs. Genome Biol 19(1):113. https://doi.org/10.1186/s13059-018-1495-0

INDEX

James Chappell and Melissa K. Takahashi (eds.), *Riboregulator Design and Analysis*, Methods in Molecular Biology, vol. 2518, https://doi.org/10.1007/978-1-0716-2421-0,

MIX
Papier aus verantwortungsvollen Quellen
Paper from responsible sources
FSC® C105338

If you have any concerns about our products,
you can contact us on
ProductSafety@springernature.com

In case Publisher is established outside the EU,
the EU authorized representative is:
Springer Nature Customer Service Center GmbH
Europaplatz 3, 69115 Heidelberg, Germany

Printed by Libri Plureos GmbH
in Hamburg, Germany